A history of modern planetary physics

Where did we come from? Before there was life there had to be
something to live on – a planet, a solar system. During the past
200 years, astronomers and geologists have developed and tested
several different theories about the origin of the Solar System and
the nature of the Earth. Did the Earth and other planets form as a
by-product of a natural process that formed the Sun? Did the
Solar System come into being as the result of a catastrophic encounter of two stars? Together, the three volumes that make up *A
History of Modern Planetary Physics* present a survey of these theories.

The early 20th century saw the replacement of the Nebular
Hypothesis with the Chamberlin–Moulton theory that the Solar
System resulted from the encounter of the Sun with a passing star.
Fruitful Encounters follows the eventual refutation of the encounter
theory in the 1930s and the subsequent revival of a modernized
Nebular Hypothesis, which was reconstructed with the help of
nuclear physics.

The "giant-impact" theory of the Moon's origin imagines an
actual collision between the young Earth and a Mars-size planet,
with the Moon being formed from a mixture of material from the
impacting planet and the Earth's mantle. Professor Brush discusses
the role of findings from the *Apollo* space program, especially the
analysis of lunar samples, culminating in the establishment of this
theory in the 1980s.

OTHER BOOKS BY STEPHEN G. BRUSH

Boltzmann's Lectures on Gas Theory (translator)

Kinetic Theory, Volume 1, *The Nature of Gases and of Heat;* Volume 2, *Irreversible Processes;* Volume 3, *The Chapman–Enskog Solution of the Transport Equation for Moderately Dense Gases* (editor)

History in the Teaching of Physics: Proceedings of the International Working Seminar on the Role of the History of Physics in Physics Education (coeditor)

Resources for the History of Physics (editor)

Introduction to Concepts and Theories in Physical Science (coauthor, second edition)

The Kind of Motion We Call Heat: A History of the Kinetic Theory of Gases in the 19th Century. Book 1, *Physics and the Atomists;* Book 2, *Statistical Physics and Irreversible Processes*

The Temperature of History: Phases of Science and Culture in the Nineteenth Century

Maxwell on Saturn's Rings (coeditor)

Statistical Physics and the Atomic Theory of Matter, from Boyle and Newton to Landau and Onsager

Maxwell on Molecules and Gases (coeditor)

The History of Modern Physics: An International Bibliography (coauthor)

The History of Geophysics and Meteorology: An Annotated Bibliography (coauthor)

The History of Modern Science: A Guide to the Second Scientific Revolution, 1800–1950

History of Physics: Selected Reprints (editor)

The Origin of the Solar System: Soviet Research, 1925–1991 (coeditor)

Maxwell on Heat and Statistical Mechanics (coeditor)

Nebulous Earth: The Origin of the Solar System and the Core of the Earth from Laplace to Jeffreys

Transmuted Past: The Age of the Earth and the Evolution of the Elements from Lyell to Patterson

A HISTORY OF MODERN PLANETARY PHYSICS

Fruitful encounters
The origin of the Solar System and of the Moon from Chamberlin to Apollo

STEPHEN G. BRUSH
University of Maryland, College Park

CAMBRIDGE
UNIVERSITY PRESS

CAMBRIDGE UNIVERSITY PRESS
Cambridge, New York, Melbourne, Madrid, Cape Town, Singapore, São Paulo, Delhi

Cambridge University Press
The Edinburgh Building, Cambridge CB2 8RU, UK

Published in the United States of America by Cambridge University Press, New York

www.cambridge.org
Information on this title: www.cambridge.org/9780521101448

First published 1996
This digitally printed version 2009

A catalogue record for this publication is available from the British Library

Library of Congress Cataloguing in Publication data
Brush, Stephen G.
Fruitful encounters : the origin of the Solar System and of the
Moon from Chamberlin to Apollo / Stephen G. Brush.
p.   cm. – (A history of modern planetary physics ; v. 3)
Includes bibliographical references and index.
ISBN 0-521-55214-1 (hc)
1. Solar system – Origin.   2. Nebular hypothesis.   3. Moon – Origin.
4. Moon – Exploration.   5. Planetology.   I. Title.   II. Series:
Brush, Stephen G. History of modern planetary physics ; v. 3.
QB601.B89   1996 vol. 3
[QB503]
523.2′09 s – dc20
[523.2′09]   95-32973
                                                                    CIP

ISBN 978-0-521-55214-1 hardback
ISBN 978-0-521-10144-8 paperback

Contents

Illustrations

Preface

This is the third of three volumes presenting the results of my research on 19th- and 20th-century theories of the origin of the Solar System, the internal structure of the Earth, and the age of the Earth.

Nebulous Earth discusses theories of the origin of the Solar System – primarily the Nebular Hypothesis – in the 19th century, and follows ideas about the Earth's core (including the geomagnetic dynamo) up to about 1970.

Transmuted Past outlines the attempts to estimate the age of the Earth in the 19th and 20th centuries, reviews related developments in nuclear physics and stellar evolution, and also offers perspectives on the changing reputation of planetary science, as well as on the comparison of styles in scientific and humanistic research.

Fruitful Encounters surveys the development of theories of the origin of the Solar System in the 20th century, including especially the impact of the *Apollo* lunar missions on ideas about the origin of the Moon.

Readers who are not already familiar with these subjects are advised to begin by reading the first chapter in each part, to get an overall view of the subjects discussed in that part.

This book has benefited greatly from the assistance and criticism of many historians and scientists, especially Michael A'Hearn, Edward Anders, Gustaf Arrhenius, James G. Baker, Alan Binder, Louis Brown, A. G. W. Cameron, Bibhas De, W. A. Fowler, Viktor Frenkel, Michael Gardner, Lawrence Grossman, Alan Harris, William K. Hartmann, Norriss Hetherington, William Kaula, Aleksey Levin, Alan Musgrave, John A. O'Keefe, Morris Podolak, R. T. Reynolds, A. E. Ringwood, Evgenia Ruskol, Christopher Russell, Victor Safronov, Susan Schultz, David Stern, David Stevenson, S. R. Taylor, William R. Ward, Paul H. Warren, Gerald Wasserburg, John Wasson, S. J. Weidenschilling, Don Wilhelms, John Wood, and M. M. Woolfson.

Libraries and librarians are essential to any historical research project. I have been fortunate to have had access to the collections of Harvard University, the Institute for Advanced Study at Princeton, the Library of Congress, the National Oceanic and Atmospheric Administration, Princeton University, the Smithsonian Institution, the U.S. Geological Survey, the University of California at Los Angeles, the University of Maryland at College Park, and the University of Minnesota-Twin Cities; I thank the staff at all those libraries for their generous and efficient help.

My research was supported by grants from the History and Philosophy of

Science Program of the National Science Foundation. This book was completed while I was a Member of the School of Social Science at the Institute for Advanced Study, Princeton, with financial support from the General Research Board of the University of Maryland and the Andrew W. Mellon Foundation.

For permission to use previously published material, I thank the following owners of the copyrights: the American Association for the Advancement of Science, publisher of *Science* (parts of Chapter 4.3); the American Astronautical Society, publisher of *Spacelab, Space Platforms and the Future* (parts of Chapter 4.3); the American Institute of Physics, publisher of *Reviews of Modern Physics* (Parts 2 and 3); *Journal for the History of Astronomy* (Chapters 1.2 and 1.3); Kluwer Academic Publishers, publisher of *Space Science Reviews* (Chapter 4.4); the Lunar and Planetary Institute (Houston), publisher of *Origin of the Moon* (Chapter 4.2); the Smithsonian Institution, publisher of *Space Science Comes of Age* (Chapter 1.1).

Permission to quote from unpublished materials in their archives has been granted by the Carnegie Institution of Washington, Lick Observatory, the National Aeronautics and Space Administration, Princeton University (H. N. Russell Papers), the University of California, San Diego, at La Jolla (Harold C. Urey Papers), the University of Chicago (T. C. Chamberlin Papers), the University of Wisconsin, Madison, and Yerkes Observatory. I am also grateful to the following individuals who allowed me to quote documents in their possession and/or their own letters: Edward Anders, James G. Baker, A. G. W. Cameron, Bibhas De, Alan Harris, W. M. Kaula, John A. O'Keefe, Brian Marsden, A. E. Ringwood, Evgenia Ruskol, Christopher Russell, Carl Sagan, S. R. Taylor, S. J. Weidenschilling, and Donald Wise.

Finally, for all kinds of support throughout this project I am grateful to my wife, Phyllis.

PART 1
Planetesimals and stellar encounters

I.I

Introduction

Two hundred years ago there were many theories about how the world began. They were assessed by different criteria – astronomical, geological, dynamical, and theological (Miller 1803). Pierre Simon de Laplace and William Herschel proposed theories that were linked together to form the "Nebular Hypothesis," connecting the origin and development of the Solar System with the condensation of stars from diffuse matter. The development of this hypothesis in the 19th century is discussed in my book *Nebulous Earth*.

In Part 1, I describe how, after dominating cosmogonic speculation for a century, the nebular hypothesis failed the geological, dynamical, and astronomical tests of 1900 (by then, theology was no longer considered relevant to the evaluation of scientific cosmogonies). Rival theories proliferated but by 1940 most of them had also been found inadequate, though a few of their components – especially the "planetesimal" hypothesis of T. C. Chamberlin – survived.

In the meantime the foundations for a new improved Nebular Hypothesis had been constructed with the help of nuclear physics (see my *Transmuted Past*). The development of the modern theory is presented in Parts 2 and 3; special theories for the origin of Earth's Moon are discussed in Part 4. Astronomical, physical (including dynamical), and chemical criteria rule; geology is secondary. I follow the story only to about 1985; for surveys of more recent research the reader should consult works such as the compendium edited by Levy and Lunine (1993), the papers published in the November 1993 issue of *Icarus,* and the brief overview by Ahrens (1994).

1.1.1 Nebulae, planetesimals, and the big bump

At the end of the 19th century most astronomers accepted the Laplace–Herschel Nebular Hypothesis. According to Laplace, the atmosphere of the primeval Sun extended throughout the entire space now occupied by planetary orbits; it was a hot, luminous, rotating cloud of gas, similar to the nebulae that Herschel thought to be the progenitors of stars. As the nebula cooled it contracted; conservation of angular momentum then required it to rotate more rapidly. By hypothesis, the gas rotated like a rigid body in the sense that the angular velocity was the same at all distances from the center, so that the linear velocity would be greatest at the periphery. Eventually the centrifugal force on the outer portion of the nebula would exceed the gravitational attraction toward the center, and a ring of gas would separate and remain at the same distance while the inner part continued to contract. By hypothesis, again, the gas in the ring would collect into a single large sphere, which would then gradually cool and condense to a liquid or solid planet. Meanwhile the contracting nebula would spin off additional rings at regular intervals, until finally only the Sun was left at the center. Satellites could be formed by a similar process of ring separation as the protoplanetary sphere cooled down and condensed.[1]

The Nebular Hypothesis was closely connected with 19th-century geological theories, which generally presumed that the Earth had been formed as a hot fluid ball and then cooled down, solidifying on the outside first. According to the "contraction theory," the solid crust would not contract as rapidly as the fluid interior, so it would have to wrinkle in order to adjust its diameter to that of the shrinking core; in this way one could explain various geological formations.

Lord Kelvin, the most influential British physicist of the 19th century, adopted the general scheme of a cooling Earth but attacked two features that geologists relied on for their explanations. First, he estimated the time required to cool down from an initial hot fluid state and found it to be only 20 to 100 millions years compared with the hundreds of millions of years geologists assumed to have been available for slow processes like erosion to produce the observed effects. Second, Kelvin showed that the fluid interior could not be so large as to extend to within 30 or 40 miles below the surface (as the geologists thought it should); in fact he went to the other extreme and concluded that the entire Earth is now solid.[2]

The debate between Kelvin and the geologists on the age of the Earth was, of course, eventually settled by the discovery of radioactivity, which not only provided a source of heat to replace that which was lost from the original store (thereby invalidating Kelvin's conduction calculations), but also furnished a direct method for estimating the ages of some surface rocks. By 1905 Rutherford and his colleagues were proposing a time scale of billions of years, as part of the revolution that was sweeping through science (Brush 1979a).

But even before the implications of radioactivity had been generally understood, the American geologist T. C. Chamberlin challenged Kelvin's theory of the cooling Earth in an assault so successful and far-reaching that it overthrew the nebular hypothesis itself.[3] Today it is hard to appreciate what a

---

1 For further details see my *Nebulous Earth,* chapter 1.2. Hereafter, chapter and section numbers alone, without "chapter" or "section" will be used in cross-references to this volume and to *Nebulous Earth* and *Transmuted Past.*

2 See *Nebulous Earth* 2.2 and 3.3.

3 Thomas Chrowder Chamberlin (1843–1928) studied geology with Alexander Winchell at the University of Michigan, then taught science at a normal school in Wisconsin. He was appointed to the Wisconsin Geological Survey in 1873 and headed the glacial division of the U.S. Geological Survey from 1881 to 1904. By the end of the 19th century he had established a considerable reputation as a geologist and educator. Chamberlin served as president of the University of Wisconsin from 1887 to 1892, then was called to the newly established Univer-

Figure 1. Thomas Chrowder Chamberlin (*Biographical Memoirs of the National Academy of Sciences,* vol. **15**, 1934).

tremendous feat that was, because we now realize the hypothesis had some serious flaws that had already been pointed out decades earlier. Yet no one had persuaded astronomers that patching up Laplace's theory was less fruitful than looking for a fundamentally different hypothesis; no one had shown geologists that the evidence for a hot primeval Earth a few tens of millions of years ago was really quite flimsy; and no one had dared to tell Lord Kelvin that he was wrong in his basic assumptions about Earth history. Chamberlin, a geologist who ventured into the apparently more difficult and prestigious field of theoretical astronomy in his 50s, was hailed as the brash American who pulled the tail of the British lion and liberated geologists from the tyranny of the truncated time scale. At the same time he introduced into planetogony a hypothesis – accretion of cold solid particles – that, despite temporary rejection, has become an essential feature of most modern theories.

Chamberlin's original objection to the Nebular Hypothesis was based, as one

sity of Chicago, where he headed the geology department from 1892 to 1918. As founder-editor of the *Journal of Geology* (1893–) he continued to exert considerable influence on U.S. geology through the 1920s. He was also well known for his "method of multiple working hypotheses" – his 1890 article on that subject was so often cited that it was reprinted in *Science* in 1965.

The most comprehensive source of information on Chamberlin is the dissertation by Schultz (1976). See also the memoir by his son, R. T. Chamberlin (1934), and articles by Winnik (1970), Leith et al. (1929), and Willis (1929). Rainger (1993) discusses his attempt to move invertebrate paleontology into the Geology Department at Chicago.

might expect, on geological considerations. Having studied the glacial formations in North America, he examined the contemporary attempts to explain
the cause of the Ice Age, in particular the hypothesis that the Earth originally
had an atmosphere rich in carbon dioxide. A drop in the carbon dioxide
content supposedly reduced the absorption of solar heat and thus lowered
global temperatures. But when Chamberlin learned of calculations, based on
the kinetic theory of gases, showing that gases at high temperatures would
have molecular velocities great enough to escape the Earth's gravitational
field, he realized that the notion of a dense carbon dioxide atmosphere was
inconsistent with the assumption that the Earth had once been a hot fluid
ball; not only carbon dioxide but all the other gases in the atmosphere would
have escaped at temperatures high enough to melt rocks.

When Chamberlin looked into the possibility that the Earth had been
formed by accretion of cold solid particles, he found that this idea had indeed
been discussed by astronomers under the name "meteoritic hypothesis." But
it seemed to have a fatal defect: Planets formed by combining solid particles
moving in adjacent circular orbits would have retrograde rotation. The reason
is that according to Kepler's Third Law, linear velocity decreases with distance
from the Sun, so the particle in the inner orbit would be moving faster than
the one in the outer orbit just before they collided, and the combined body
would have a net backward rotation. Since it was thought that all planets
(with the possible exception of Uranus and Neptune) have direct rotation,
accretion from solid particles did not look very promising.

But the astronomers who rejected the meteoritic hypothesis on the basis of
Kepler's Third Law had forgotten to apply Kepler's other two laws. In general
the particles would move in elliptical orbits (first law), and a given particle
would move faster in the part of its orbit which is closer to the Sun (second
law). Chamberlin showed by analyzing several examples that unions of particles moving in intersecting elliptical orbits would be more likely to leave the
resulting particle with direct rotation.

Here and elsewhere Chamberlin had the assistance of a young astronomer,
F. R. Moulton, who was completing his Ph.D. research at the University of
Chicago, where Chamberlin headed the Geology Department.[4] In addition to
reviewing and working out the details of Chamberlin's ideas, Moulton put
together the objections to the Nebular Hypothesis in a comprehensive paper

4 Forest Ray Moulton (1872–1952) received his B.A. degree in 1894 from Albion College
   (Albion, Michigan), then went to the University of Chicago as a graduate student, and was
   appointed assistant in astronomy in 1896. He was awarded a Ph.D. in astronomy and mathematics in 1899, and later became professor of astronomy at Chicago. He published several
   articles and books on topics in astronomy, in addition to his work with Chamberlin on the
   origin of the Solar System. He retired from Chicago in 1926; he served as executive secretary
   of the American Association for the Advancement of Science from 1936 to 1940.
      Biographical details may be found in an anonymous article in the Albion alumni magazine
   (Achates 1947) and in the DSB article by H. S. Tropp (1974). Both date Moulton's collaboration with Chamberlin from 1898, but the documents to be cited later indicate that it had
   begun at least as early as summer 1897. Achates also confuses T. C. Chamberlin with his son
   Rollin. See also Gasteyer (1970).

Figure 2. Forest Ray Moulton (*Io Triumphe*, vol. **12**, 1947. Reproduced by permission of Albion College).

published in 1900. While his name became attached to Chamberlin's theory, Moulton's major contribution to planetogony was to convince astronomers that the Nebular Hypothesis must be abandoned.

The major defect of the Nebular Hypothesis in 1900 was its failure to explain the distribution of angular momentum in the Solar System. Laplace's spin-off of successive rings should have left the Sun with much greater rotational speed than it now has. In fact Jupiter has most of the angular momentum of the Solar System, contrary to what one would expect from any reasonable estimate based on the Nebular Hypothesis. On a smaller scale, the discovery in 1877 that Mars has a satellite (Phobos) that goes around it in only one-third of the rotation period of the planet contradicts the assumption that satellites have formed from rings spun off by the planet's nebula.

Chamberlin was primarily interested in the thermal and mechanical development of the Earth rather than the rotation of the Sun. He suggested that the accumulation of particles by the growing Earth might have been so slow that the heat released by conversion of mechanical energy would be mostly dissipated into space and would never produce a large amount of melting. Thus, he agreed with Kelvin that the Earth is now entirely solid, a view that prevailed until 1926 when the British geophysicist Harold Jeffreys established the existence of a liquid core.[5]

Nevertheless, once Chamberlin had become interested in astronomical problems he could no longer confine himself to geology. Looking at James

5 See *Nebulous Earth*, 2.3.

Keeler's photographs of spiral nebulae, he speculated that the two prominent arms belonged to two previously distinct celestial objects. From this thought, and contemplation of solar prominences, he was led to the idea that a planetary system could be generated when another star passed close to the Sun. He did not require an actual collision though this was being suggested by others. The tidal force of the intruder would cancel the gravitational force holding in the solar gases on the near and far sides of the Sun, allowing two filaments of material to flow out; the filaments would then be curved by the continued action of the intruder as it recedes. Chamberlin assumed that the filaments would eventually condense into small solid particles that would be captured into orbits around the Sun.

The Chamberlin–Moulton theory thus consisted of two distinct hypotheses: (1) close encounters of two stars, drawing filaments of gaseous material out of one of them to form a spiral nebula; (2) condensation of the gases to small solid particles, called "planetesimals" (i.e., infinitesimal planets), which accrete to form planets and satellites (Chamberlin 1905, 1906; Moulton 1905, 1906a). When it later became clear that the spiral nebulae are galaxies rather than objects that could be as small as planetary systems, Chamberlin dropped this part of the first hypothesis but retained the assumption that two stars interacted in order to release into space the material from which planets formed.

A decade after the publication of the Chamberlin–Moulton theory, Jeffreys and the British astronomer-physicist J. H. Jeans independently adopted the first hypothesis of the American theory, but rejected the second. Jeffreys (1916a, 1917a, 1918) argued that high-velocity collisions among the planetesimals would vaporize them so quickly that the material would remain gaseous until it collected into planets; thus, he proposed to return to the 19th-century assumption that the Earth was originally a hot fluid ball and has been cooling down. Jeans (1917, 1919) was more interested in developing idealized mathematical models to represent the initial ejection of material from the Sun under the tidal influence of the other star. Thus, Jeans concentrated on the astronomical side of the theory, while Jeffreys developed it from a geophysical viewpoint.

The tidal theory, whether the Chamberlin–Moulton or the Jeans–Jeffreys version, was generally accepted by astronomers until 1935, even though it was never worked out in sufficient detail to provide a convincing explanation of the quantitative properties of the Solar System. Its supporters believed that the tidal theory could overcome the major defect of the nebular hypothesis by showing at least qualitatively how most of the angular momentum could have been given to the major planets rather than to the Sun.

In the meantime, the French mathematician Henri Poincaré had demonstrated a theorem that seemed to provide another serious objection to the nebular hypothesis: If the present mass of the planets were spread out over the entire volume of the Solar System, this material would be of such a low density that it would dissipate into space before condensing. The filaments postulated in the tidal theory would not have to be spread out over such a large volume, so this difficulty would be avoided.[6]

Of course, astronomers recognized that any theory that required the encounter of two stars to form planets would entail an extremely small frequency of planetary systems in the universe. This was consistent with the failure to find any convincing evidence for nonsolar planetary systems. Jeans at one time seemed to take perverse pleasure in the idea that we are the result of a chance event that has happened only once in the universe and (because the stars are decaying and thinning out by expansion) will probably never happen again (Jeans 1932: 2–5). Later he changed his mind and postulated that stars were much larger in the past so the frequency of collisions and hence of planetary systems was correspondingly larger (Jeans 1942).

1.1.2 Astrophysics strikes back

In 1796 Laplace proposed his nebular hypothesis, not in one of his technical papers on celestial mechanics, but in a nonmathematical book on astronomy intended for the layman. Similarly, the American astronomer H. N. Russell in 1925 began to think about the origin of the Solar System when working on a textbook and presented his criticisms of the tidal theory in *Scientific American, Saturday Review,* and finally in a series of public lectures in 1934 (Russell 1935), but he never discussed the subject in an astronomical journal.[7] Did astronomers still think that cosmogony was not quite an appropriate subject for serious research, and does this attitude account for the slow rate of progress?

Russell found two major objections to the assumption that material extracted from the Sun by a passing star would condense into the planets of the present Solar System. First, theories of stellar structure developed by A. S. Eddington and others in the 1920s indicated that gases from the interior of the Sun would be at such a high temperature – on the order of a million degrees – that they would dissipate into space before they could condense into planets. Second, a simple dynamical calculation showed that it would be impossible for the tidal encounter to leave enough material with the necessary angular momentum in orbits at distances from the Sun corresponding to the giant planets.

6 Poincaré's career, his research on cosmogony, and related topics are discussed in *Nebulous Earth,* 1.7. For his own survey of selected theories see Poincaré (1913).
7 Henry Norris Russell (1877–1957) earned his Ph.D. at Princeton University in 1900 and taught astronomy there from 1905 to 1947. He was director of the University Observatory from 1912 to 1947. His research on stellar evolution was associated with the "Hertzsprung–Russell" diagram universally used by astronomers. His research on spectral analysis led to the theory of Russell–Saunders (L–S) coupling, well known in atomic physics, for atoms with more than one valence electron.

Although R. A. Lyttleton, with Russell's encouragement, attempted to rescue the encounter theory by introducing a third star, most astronomers seemed to think after 1935 that there was *no* satisfactory theory of the origin of the Solar System. Russell had refuted the encounter theory, yet the fatal objections to the Nebular Hypothesis remained, and no other alternative seemed very plausible. Jeffreys in particular continued to insist that we simply have no adequate explanation for the existence of the Solar System.

Russell indirectly helped to demolish another argument that had previously been used by planetogonists to account for the near-circularity of most planetary orbits. If the planets are initially formed with highly eccentric orbits, one must find some mechanism to reduce the eccentricity. A popular choice was the hypothetical "resisting medium," that part of the dust and gas from the original nebula that did not condense into planets. Theoretically its viscous resistance should have helped elliptical orbits to evolve into more nearly circular ones. Russell was skeptical about this mechanism, pointing out that the medium would consist mainly of hydrogen and would be accreted by the planets if they interacted with it; it would thus be hard to explain why the Earth's atmosphere and oceans contain so little hydrogen. Russell also encouraged H. P. Robertson at Princeton to look into the old claim by J. H. Poynting that the absorption and re-emission of solar radiation by small bodies in the Solar System would decrease their angular momentum and eventually cause them to fall into the Sun. There had been some dispute as to whether Poynting's result, derived from the ether theory in 1903, was consistent with Einstein's theory of relativity. Robertson (1937) showed that there is indeed a dragging effect (though Poynting's formula is not accurate) and that particles less than 1 cm in radius in the vicinity of the Earth's orbit would be swept into the Sun in less than 40 million years. Thus, the "Poynting–Robertson effect" makes it unwise to invoke a resisting medium to round up planetary orbits except under carefully defined conditions.

In addition to knocking out the encounter theories, Russell also participated in an important discovery that later removed one of the objections to the Nebular Hypothesis and substantially influenced its modern form. In the 1920s astronomers believed that the Sun had roughly the same chemical composition as the Earth; this would be consistent with the hypothesis that the Earth was formed from material drawn out of the Sun by a passing star. Thus, the Sun should contain substantial amounts of elements such as iron, silicon, and oxygen but relatively little hydrogen and helium. In 1929 Russell, confirming an earlier finding of Cecilia Payne (1925a, 1925b), showed that hydrogen is by far the most abundant element in the Sun's atmosphere, and other astrophysicists in the 1930s established that the same is true for many other stars and probably for the universe as a whole.

If one assumes that the Earth was formed from a cloud of material characterized by the typical "cosmic abundance" of elements, then most of the hydrogen originally present in this cloud – now called the "solar nebula" – must have been lost. Hence, its original mass must have been much greater than that of the present planets, and its density could have been great enough to satisfy Poincaré's criterion for condensation.

Of course, one still has to deal with the other long-standing objections to the Nebular Hypothesis, and in addition explain the chemical processes that produced planets with compositions radically different from the cosmic abundance table.

1.1.3 Intermission

In the decade following Russell's refutation of the encounter theory, no single theory was supported by more than a handful of astronomers. Nevertheless, there were some significant developments in this decade that influenced later work: (a) revival of the planetesimal hypothesis; (b) the concept of "magnetic braking" of the Sun's rotation; (c) suggestions that the Sun had encountered an interstellar cloud and captured from it the material that later formed planets; (d) claims for discovery of extrasolar planetary systems; (e) research on cosmic abundances of the elements. I will summarize these briefly.

(a) The Swedish astronomer Bertil Lindblad showed that partly inelastic collisions between particles initially moving with different speeds in eccentric orbits with different inclinations will tend to make all the particles move at similar velocities in circular orbits lying in a flat ring.[8] Collisions between the particles would then occur with small relative velocities, thereby avoiding Jeffreys's argument that collisions would vaporize the particles. Lindblad (1934, 1935) suggested that a cold particle immersed in a hot gas would tend to grow by condensing the gas on its surface. Dirk ter Haar (1944, 1948) in Leiden elaborated this idea by using the Becker–Döring kinetic theory of the formation of drops in a saturated vapor and reinforced Lindblad's proposal that solid particles could grow initially by nongravitational forces.

Jeffreys himself began to reconsider his objection to the planetesimal hypothesis and suggested that the vapor pressure of solids at very low temperatures might be below the pressure in the surrounding medium, so that condensation would outweigh the vaporizing effect of collisions (1944). Alfred Parson (1944, 1945) published an estimate of the vapor pressure of iron ($10^{-46.7}$ atm at $273°K$), which indicated that condensation would be favored in interstellar space, and Jeffreys (1948) admitted that his original objection to the planetesimal theory had thereby been answered.

The American astronomer Fred Whipple proposed in 1942 that radiation pressure acting on particles in a dust cloud would tend to push them together; each of a pair of nearby particles would shield the other from the radiation,

---

8 Bertil Lindblad (1895–1965) studied at Uppsala University, receiving his Ph.D. in 1920 with a dissertation on the theory of radiative transfer in the solar atmosphere. After a few years of research at the Lick and Mt. Wilson Observatories in the United States and at Uppsala, he was appointed director of the Stockholm Observatory in 1927 and stayed there for the rest of his career.

leaving an effective attraction.[9] This mechanism is similar to the "kinetic" explanation of gravity proposed by LeSage and others in the 18th and 19th centuries (Brush 1976: 21, 48). Whipple proposed condensation of dust particles initially as a means of star formation from the dark clouds studied by Bart Bok, but also used it as an initial stage in the formation of planetary systems (Whipple 1948a, 1948b).

(b) The Swedish plasma physicist Hannes Alfvén, whose early papers on cosmogony were communicated for publication by Lindblad, incorporated the concept of planetesimal accretion into his own theory but also added an important idea that removed a major objection to the Nebular Hypothesis.[10] He showed (1942a) that an ionized gas surrounding a rotating magnetized sphere will acquire rotation and thereby slow down the rotation of the sphere. V. C. A. Ferraro (1937) had obtained this result earlier but did not suggest its possible use in cosmogony. Alfvén (1946, 1954) proposed that the early Sun had a strong magnetic field, and that its radiation ionized a cloud of dust and gas, which then trapped the magnetic field lines and acquired most of the Sun's original angular momentum. This mechanism of "magnetic braking" was later adopted by other theorists who rejected the rest of Alfvén's cosmogony.

If a star is rotating, the Doppler effect will cause a shift in the frequencies of spectral lines for radiation emitted by those parts momentarily moving toward or away from us, and by careful analysis of stellar spectra it is possible to estimate the speeds of rotation. Otto Struve and others found that stars in later stages of evolution generally rotate more slowly than those in earlier stages. There is apparently some fairly universal process by which a star loses most of its angular momentum at an early stage of its evolution (Struve 1945; 1950: 120–53). Whether or not this process involves the formation of planets, at least one can no longer use the slow rotation of the Sun as an argument against the Nebular Hypothesis.

(c) Alfvén proposed that the Sun encountered a cloud of neutral gas that became more or less completely ionized at the distance of Jupiter; the magnetic field of the Sun prevented it from moving any closer. Another cloud, consisting of dust particles, was postulated to account for the terrestrial planets (Alfvén 1942b, 1946). At about the same time Otto Schmidt in the USSR proposed (1944) that the Sun had captured an interstellar cloud of meteorites; his theory was based on gravitational capture rather than electromagnetic ef-

---

9 Whipple (1942/1946). Spitzer (1948). Fred Whipple (b. 1906) received his Ph.D. in astronomy from the University of California, Berkeley, in 1931 and has been at Harvard College Observatory since then. He was director of the Harvard/Smithsonian Astrophysical Observatory from 1955 to 1973. He is best known for his research on comets and meteors. He retired as professor of astronomy in 1977.

10 Hannes Olof Gosta Alfvén (b. 1908) received a Ph.D. in physics at the University of Uppsala in 1934. He served as professor of electronics and plasma physics at the Royal Institute of Technology in Stockholm from 1940 to 1973; starting in 1967 he also had an appointment at the University of California, San Diego. He was awarded the Nobel Prize in Physics in 1970. On his career at Stockholm see Larsson (1993).

fects. Like Alfvén, he assumed that the Earth was formed by accretion of cold solid particles.[11] This assumption provided a common basis for the discussion of questions about the thermal history of the Earth, evolution of its core, and so forth for scientists who disagreed on whether the Sun itself was formed from this cloud or encountered it later.

(d) In 1943, two reports of extrasolar planetary systems provided a new argument against all theories that treated the origin of the Solar System as an extremely rare event. Both reports inferred the existence of an invisible third component, with mass much less than the smallest known stellar mass, from observations of a binary system (Reuyl and Holmberg 1943; Strand 1943). The headline of Russell's monthly column in *Scientific American* proclaimed: "Anthropocentrism's Demise: New Discoveries Lead to the Probability That There are Thousands of Inhabited Planets in our Galaxy" (Russell 1943).

The obituary turned out to be premature as the validity of such "discoveries" became a matter of controversy; yet many scientists wanted to find life elsewhere in the universe and would not be happy with a theory that made it unlikely (Huang 1973; Van de Kamp 1956).

(e) Research on the cosmic abundance of chemical elements, mentioned earlier in connection with Russell's interpretation of the solar spectrum, was also pursued through analysis of meteorites; much of this information was synthesized in a classic paper by V. Goldschmidt (1937) and brought up to date in a review by Harrison Brown (1949).

One way to resolve the discrepancy between the high cosmic abundance of hydrogen and its low abundance at the Earth's surface was to postulate that the Earth's core contains a large amount of hydrogen. This was the proposal of Werner Kuhn and Rittmann. They criticized the standard iron-core model of the Earth on the grounds that there was no plausible mechanism for separating the iron during the evolution of the Earth. Their model and other alternatives to the iron-core model, were eventually refuted on physical grounds.[12]

---

11 Schmidt (1958). See also Randic (1950); Levin and Brush (1995). Otto Iulevich Schmidt (1891–1956) came from a Russian family of German origin. He studied mathematics at the St. Vladimir Imperial University in Kiev, graduating in 1913 and continuing there as a graduate student (Master's degree 1916) and as a mathematics instructor. In 1917 he moved to St. Petersburg, where he became involved in the revolution and was appointed to a series of administrative positions in the new government. In 1929 he was appointed professor of mathematics at Moscow State University, and during the 1930s he was prominent in arctic research, leading the first Soviet air expedition to the North Pole in 1937. He was founding director of the Institute of Geophysics of the Soviet Academy of Science, from 1937 to 1948. But he incurred the displeasure of Josef Stalin during World War II (perhaps in part because of his German name) and was relieved of his administrative positions. As a result he had time to undertake a new research program in cosmogony, a subject that had interested him since his student years. He assembled a research group within his institute, first called the Department of the Evolution of the Earth and later known as the Laboratory of the Origin of the Earth and Planets. (The main English-language source for biographical information is the article by Aleksey Levin in Levin and Brush 1995.)

12 The debate about the chemical composition of the Earth's interior is discussed in *Nebulous Earth*, 2.4.

1.1.4 A nebula that clumps and coughs?

The postwar revival of the nebular hypothesis was due mainly to a paper by the German physicist C. F. von Weizsäcker, published in 1944.[13] Weizsäcker postulated a gaseous envelope surrounding the Sun and associated with its formation; in order to contain enough heavy elements to form the planets and at the same time have the high proportion of hydrogen and helium characteristic of the Sun, this envelope must have had about one-tenth of the mass of the Sun. If the envelope or solar nebula were concentrated in a flat disk with diameter approximating that of the orbit of Pluto, its density would be relatively high (about 10^9 g/cm^3), and it might stay together long enough to develop a regular pattern of motions.

Whereas Laplace had assumed, rather implausibly, that the gaseous nebula would rotate like a rigid solid, Weizsäcker pointed out that there would be a tendency toward differential rotation with faster motion inside and slower outside, as in Kepler orbits. But friction between adjacent streams would tend to equalize their speeds by accelerating the outer stream and decelerating the inner one.[14] This creates an instability, causing the outer stream to move further out and the inner stream to move inward, resulting in turbulent convection currents and eventually the formation of a pattern of vortex motions. Each vortex moves in a circular orbit around the Sun, and there must be an integer number of vortices in a ring. If one postulates exactly five vortices per ring then the nth ring will be at a distance $r_n = r_0 E^n$, where $E = 1.9$, giving an approximation to the Titius–Bode Law of Planetary Distances. Weizsäcker assumed that the best place to accumulate particles into planets would be the regions where adjacent vortices come into contact producing violent turbulence.

Weizsäcker's theory was initially greeted with enthusiasm, especially in the United States where it was reviewed by George Gamow and J. A. Hynek (1945), and by S. Chandrasekhar (1946). The Dutch scientist Dirk ter Haar (1948) adopted it as a basis for further work, incorporating his own mechanism for condensing dust particles. But subsequent work on turbulence theory by Werner Heisenberg (1948a, 1948b), Weizsäcker (1948, 1951), and Chandrasekhar (1949; Chandrasekhar and ter Haar 1950) indicated that the regular pattern of vortices originally postulated by Weizsäcker could not occur, but instead must be replaced by a range of eddy sizes.

13   Carl Friedrich von Weizsäcker (b. 1912) was educated at the universities of Leipzig, Berlin, and Goettingen, and held research positions at Leipzig and Berlin in the 1930s. He was a professor of physics at the University of Strassburg from 1942 to 1946 and participated in the German program to develop nuclear weapons (see the recently published "Farm Hall Transcripts" based on recordings made of the conversations of the leaders of this program while they were interned in England at the end of the war). After World War II he held academic positions at the universities of Gottingen, Hamburg, Starnberg, and Munich. See his recollections (Weizsäcker 1988).

14   This is the classical kinetic theory mechanism for gas viscosity proposed by James Clerk Maxwell (Brush 1976: 189–92).

Although Weizsäcker's theory was abandoned, it had served the important function of moving planetogonists away from the dualistic hypotheses dominant in the first third of the 20th century toward monistic models. While it was generally agreed that the planets could not have been formed from material pulled out of the Sun,[15] it was not universally accepted that the Sun and planets came from the same nebula. Thus, as noted in 1.1.3, Alfvén and Schmidt postulated that a previously formed Sun captured material from interstellar space – either a single "protoplanetary cloud," as Schmidt's followers called it, or several different clouds (Alfvén). Such theories could be called dualistic, although they ascribed a different role to the two actors in the creation drama; the planets are adopted rather than natural children of the Sun.

Among those who developed monistic nebular theories in the United States, Gerard Kuiper[16] and Harold Urey were the most influential. Kuiper had initially judged the origin of the Solar System a problem not yet soluble by direct attack so he turned instead to what he considered an easier problem: the origin of double stars. He then developed a picture of the Solar System as an "unsuccessful" double star (Kuiper 1951b). Kuiper postulated a massive solar nebula, about 0.1 $M_\odot$ (exclusive of the mass of the Sun itself), that is, about 100 times the present mass of the planets, and assumed that it would form large protoplanets by gravitational collapse (Jeans instability). After the planets formed, the excess material would be blown away by the Sun's radiation pressure.

Urey started from Kuiper's theory but soon rejected the protoplanet hypothesis, assuming instead that numerous smaller objects of asteroidal and lunar size were first formed and later accumulated into planets.[17] He was primarily interested in explaining the chemical properties of Solar System

15 The presence of significant amounts of lithium, beryllium, and boron in the Earth was considered conclusive evidence that the planets were not formed directly from stellar material, since those elements would have been destroyed by nuclear reactions (Reeves 1978a: 4–6). The theory of M. M. Woolfson is the major exception; he postulated that the planets came not from the Sun but from a passing light, diffuse star, or protostar, presumably one in which nuclear reactions had not yet begun.

16 Kuiper (1951a, 1951b, 1951c, 1956a, 1974). Gerard P. Kuiper (1905–1973) was born in the Netherlands and educated at the University of Leiden. He studied with Jan Woltjer, Willem de Sitter, and Ejnar Hertzsprung and completed his doctoral dissertation on spectroscopic binaries in 1933. Kuiper then moved to the United States and worked at the Lick Observatory in California, McDonald Observatory in Texas, and the University of Chicago's Yerkes Observatory in Wisconsin. Later he moved to the University of Arizona. For details of his career see Doel (1990), and for further comments about his impact on modern astronomy see Waldrop (1981).

17 Harold Clayton Urey (1893–1981) received his Ph.D. in physical chemistry at the University of California, Berkeley, in 1923. He was the leader of a group that discovered deuterium in 1931, for which he received the 1934 Nobel Prize in Chemistry. He participated in the Manhattan Project during World War II; in 1945 he moved to the University of Chicago. In 1951, with his graduate student Stanley Miller, he conducted an experiment on the formation of biologically significant molecules in conditions similar to those he assumed to have been present in the early Earth (in particular, a reducing atmosphere, associated with the presumed abundance of hydrogen in the early Solar System). In 1958 he accepted a professorship at the University of California, San Diego (at La Jolla), where he remained for the rest of his life.

constituents and in elaborating the consequences of his assumption that the Moon was formed before the Earth and later captured by it. Both Kuiper and Urey employed primarily qualitative or semiquantitative reasoning.

Urey (1952: 153) argued that the high abundance of hydrogen in the primeval solar nebula should be taken into account in research on the origin of life; the first organic compounds could have been formed under reducing conditions in the Earth's early atmosphere, while later stages in the process took place as the hydrogen escaped and conditions changed from reducing to oxidizing. The famous experiment by Urey's student S. L. Miller, which initiated a new epoch in research on chemical evolution, was thus indirectly inspired by the revival of the Nebular Hypothesis with the help of the Payne–Russell discovery of the high cosmic abundance of hydrogen.

Schmidt's theory was developed by V. Safronov and others throughout the 1960s and 1970s; it became primarily a model for the accumulation of solid particles from the protoplanetary cloud into planets.[18] Safronov worked out the quantitative results of the model by analytic approximations.[19] Evgenia Ruskol applied the theory to the formation of the Moon by simultaneous accretion in orbit around the Earth. Safronov's model was adopted, with some modifications, by G. W. Wetherill in the United States; he explored its consequences with the help of numerical computer calculations. The Safronov–Wetherill model is now considered the most plausible one for the formation of the terrestrial planets, though it does not yet account quantitatively for their properties.

The theory of Woolfson[20] and Dormand develops Schmidt's capture hypothesis in a different direction, leading to gaseous protoplanets rather than planetesimals as the precursors of planets (Dormand and Woolfson 1971, 1974, 1977, 1989). But while Safronov's theory discarded the capture hypothesis and became part of the dominant (monistic) paradigm in the 1970s, Woolfson and Dormand remained in the dualistic camp and their work was ignored, except in England, by others who advocated gaseous protoplanets.[21]

---

18 According to Shklovskii (1988), Schmidt's theory was promoted so aggressively in the USSR that scientists holding other views felt somewhat intimidated, which created for a few years a situation comparable to Lysenko's domination in genetics. For further discussion of this point and translations of Schmidt's ideological interpretation of planetogony, see Levin and Brush (1995).

19 Victor Sergeivitch Safronov (b. 1917) studied astronomy at Moscow State University; his graduate work was directed by Ambartsumian. He joined Schmidt's planetary cosmogony group at the Institute of Earth Physics in Moscow and eventually became the group's leader. He is married to Evgenia Ruskol.

20 Michael Mark Woolfson (b. 1927) was educated at Oxford University, completing his doctorate in physics in 1952. He held faculty positions at the University of Manchester from 1955 to 1965, when he was appointed professor of theoretical physics at the University of York.

21 Woolfson (letter to S.G.B., 1988) pointed out that no one actually criticizes the capture theory anymore – "It is simply unread and scientifically invisible." He blamed this on a "cosmogonic semiconductor" – the Atlantic Ocean – "which allows information to flow well from West to East but very poorly the other way." While it is true that most of the citations of the Dormand–Woolfson papers are by British scientists (primarily I. P. Williams and his colleagues at Queen Mary College in London), it also seems that those papers are cited mainly in connection with the gaseous protoplanet hypothesis rather than capture.

I. P. Williams is perhaps the only scientist outside of Woolfson's group who has carefully evaluated their theory; he says it "is capable of explaining most of the feature of the solar system" and that he "can see no fundamental fault in this theory" although "perhaps because of personal bias," he prefers another type of theory (Williams 1975: 49).

Of the theorists still active in 1985, Alfvén was undoubtedly the one who had pursued a cosmogonic research programme most persistently for the longest period of time, starting in 1942. His original suggestion that magnetic braking of a plasma cloud could transfer angular momentum from the Sun to the planets was adopted by Kuiper, Hoyle, and other theorists even though they rejected other aspects of his theory. Alfvén postulated that the basic structure of satellite and ring systems, as well as the Solar System as a whole, was determined by the nature of plasmas and the process by which they condense (the "critical velocity" effect and "partial co-rotation"). In addition he proposed that inelastic collisions between planetesimals in orbit around the Sun would focus them into a "jet stream," thereby promoting accretion of planets. In his work with Arrhenius he dropped the earlier assumption of separate formation of the planet-forming clouds and adopted a monistic scenario (Alfvén and Arrhenius 1973, 1975, 1976b).

1.1.5 Stars, dust, and protoplanets

The most striking new feature of the period 1956–85 was the role played by isotopic anomalies. Although these anomalies have little bearing on most of the traditional problems of planet and satellite formation, they were believed to offer important clues to the initial stages of formation and contraction of the solar nebula as related to nuclear processes in the Sun and other stars. The best-known example is the "supernova trigger" hypothesis, based in part on the excess ^{26}Mg found in the Allende meteorite; the earlier history and recent demise of this hypothesis are not so well known. Starting with the discovery of excess ^{129}Xe in the Richardton meteorite by J. A. Reynolds in 1960, theorists reasoned that a short-lived isotope (in this case ^{129}I) must have been synthesized in a supernova, ejected into the interstellar medium, and incorporated into a meteorite parent body that cooled down enough to retain xenon gas, all within a period of only about 100 million years. Since a supernova explosion also produces a shock wave that might compress rarefied clouds to densities high enough for them to become unstable against gravitational collapse, the isotopic anomalies might indicate that a supernova *caused* the Solar System to form (Cameron 1962b).

If a supernova is *necessary* to produce a planetary system, then one loses an

During the past 20 years only four papers explicitly supported the Dormand–Woolfson *capture* theory, including one from the United States (Kobrick and Kaula 1979) and one from Australia (Gingold and Monaghan 1980); an earlier paper by those authors was somewhat tentative in its conclusions. The other two (not including self-citations from Woolfson's group) were by J. Geake and D. G. Ashworth in England.

attractive feature of monistic cosmogonies, namely, the inference that the same process that forms a star generally forms a planetary system as well; hence, planets and life are widespread in the universe.

The supernova trigger hypothesis was not taken seriously until the establishment of the ^{26}Mg anomaly by Lee, Papanastassiou, and Wasserburg (1976); this was attributed to the isotope ^{26}Al, which has a half-life of only 700,000 years and thus was synthesized less than a few million years before the formation of the Solar System (Cameron and Truran 1977).

Before 1976, aside from the lack of convincing evidence, there was an alternative explanation for isotopic anomalies: They might have been produced by irradiation of planetesimals in the early Solar System. That explanation was primarily associated with the names of W. A. Fowler, J. Greenstein, and F. Hoyle (the "FGH theory").

The FGH theory was linked to Fred Hoyle's (1960) theory of the origin of the Solar System. Building on Alfvén's magnetic-braking hypothesis, Hoyle postulated that planetesimals would be formed in a disk surrounding the Sun and then pushed outward by gas flowing from the Sun as magnetic forces transfer angular momentum to the disk. The dissipation of magnetic energy in this process would accelerate protons to high speeds and they will bombard the planetesimals. Spallation reactions would produce deuterium, lithium, beryllium, and boron. The FGH theory was primarily intended to supplement another theory of element synthesis in stars (Burbidge et al., 1957), which could not account for the abundances of the light elements deuterium, lithium, beryllium, and boron. But it was also put forth as an alternative to the supernova trigger for producing ^{26}Al, ^{129}I, and other isotopes (Fowler, Greenstein, and Hoyle 1961: 403).

The FGH theory was later abandoned as an explanation of the production of light elements. As a theory of planet formation it had already come into conflict with new ideas about the early evolution of the Sun proposed by C. Hayashi (1961). Hayashi argued that before a star reaches the main sequence, it must go through a convective stage in which it will be highly luminous. This stage was thought by some theorists to be associated with the strong mass outflow (greatly enhanced solar wind) observed for T Tauri stars. The young star's emission of radiation and matter would destroy or sweep away the kinds of planetesimals postulated by Hoyle, although other theorists used the same process to get rid of excess nebular material *after* planets had been formed.

A. G. W. Cameron became the most influential North American theorist after 1960.[22] Previously an expert on nucleosynthesis in stars, he could speak authoritatively on the significance of isotopic anomalies. He could substanti-

---

22 Alastair Graham Walter Cameron (b. 1925) was born and educated in Canada; he received his Ph.D. in physics in 1952 at the University of Saskatchewan, then worked for Atomic Energy of Canada until 1961, when he moved to the Goodard Institute of Space Studies in New York. He was professor of space physics at the Belfer Graduate School of Science, Yeshiva University, from 1966 to 1973, and was then appointed professor of astronomy at Harvard.

ate Hayashi's ideas about the early evolution of the Sun with independent calculations (Ezer and Cameron 1963). Taking full advantage of the fast but cheap computers available in the 1960s and 1970s, he developed a series of numerical models for the condensation of a solar nebula, experimenting with a range of different physical assumptions. He was one of the first to discover that the mathematical collapse of a cloud does not ordinarily lead to a large central body surrounded by smaller bodies unless special processes are postulated (Cameron 1963d: 88–93). Contrary to Alfvén, Hoyle, Mestel and others, Cameron concluded (1966, 1969a) that turbulent viscosity rather than magnetic braking is primarily responsible for the transfer of angular momentum from the Sun to planets. In 1976 he revised his models to incorporate the theory of accretion disks developed by Lynden–Bell and Pringle (1974); at the same time he concluded, contrary to his earlier views, that the planets were probably formed from giant gaseous protoplanets. Cameron was also one of the major proponents of the theory that the Moon was formed by impact of a Mars-size planet on the Earth (4.4.5).

Cameron's approach stressed calculations with hydrodynamic models (even in the analysis of the impact selenogony model) and was quite compatible with the strong interest in star formation among astrophysicists. Yet in 1985 calculations of cloud collapse had not yet reached the point where they could be used as a firm basis for theories of Solar System origin (Morfill, Tscharnuter, and Völk 1985: 495).

Cameron's new model, along with the discovery of isotopic anomalies, did have a strong negative impact on another hypothesis that had been popular in the early 1970s. To explain the specific chemical and physical properties of planets and meteorites, several scientists had suggested that all of the material in the Solar System (or at least in the region of the terrestrial planets) had been completely vaporized and thoroughly mixed. This assumption was justified by evidence, found in the 1930s and 1940s, that the isotopic abundance ratios of many elements are the same in terrestrial and meteoritic samples (Manian, Urey, and Bleakney 1934; Valley and Anderson 1947, and other work cited therein). It seemed reasonable to infer that the primordial nebula had a fairly uniform composition (Brown, 1950a, 1950b). Thus, the solar system was "born again," preserving no evidence of its earlier history aside from its overall chemical and isotopic composition. As the homogeneous gas cooled down, its components would condense in a sequence determined by their thermodynamic properties and the pressure–density–temperature profile of the primordial nebula. With some additional assumptions about the relative rates of cooling and aggregation, and about the extent to which thermodynamic equilibrium prevails in the nebula, one could then calculate the chemical compositions of the solid bodies formed at different distances from the Sun.

This "condensation sequence" model was very attractive to meteoriticists. When J. S. Lewis (1972a) used the pressure–density profile from Cameron's nebular model to deduce the densities of the terrestrial planets, it appeared that a method was also available to explain the chemistry of the entire inner Solar System on the basis of a simple hypothesis about its initial state.

But the fact that meteorites with similar chemical composition varied in their isotopic ratios undermined the assumption that the primordial nebula was well mixed, and it became increasingly difficult to account for the details of their structure by simple condensation from a high-temperature gas. In the late 1970s and early 1980s meteoriticists began to favor more complex histories, including the possibility that certain components had been formed elsewhere in the galaxy and survived as interstellar grains through the formation epoch of the Solar System.

Cameron's new models reinforced this view: Whereas his calculations in the 1960s and early 1970s led to temperatures of thousands of degrees in the region of the terrestrial planets, now the temperatures were no more than a few hundred degrees, not high enough to vaporize the more refractory elements and compounds. At the same time the "Hayashi track," with its superluminous early Sun and powerful T Tauri stellar wind, vanished in the face of more accurate computations (Larson 1969). Some meteoriticists continued to report evidence for high temperatures in the solar nebula, but they could not count on astrophysics to support them – at least not until 1984, when Cameron again reversed his conclusions and suggested that small bodies might be vaporized out to the Mars region during a later stage of nebular evolution (Cameron 1984a).

Those scientists who were more interested in learning how planets were constructed than in analyzing star formation still preferred to concentrate on the condensation of planetesimals from a gas/dust cloud and the accumulation of larger bodies from planetesimals. The fundamental problem was: How can dust particles stick together to form bodies large enough to capture more particles and gas by gravitational attraction?

In 1973 Peter Goldreich and William Ward pointed out a plausible solution to this problem.[23] Gravitational instabilities may develop in the thin disk of dust that collects at the midplane of the nebula, even though the nebula as a whole is too rarefied and hot to be unstable. This will cause collapse to planetesimals with sizes up to the kilometer range, whether the dust particles are sticky or not. Although it was soon recognized that a similar phenomenon had been discussed earlier by several theorists (e.g., Gurevich and Lebedinskii 1950) and had been invoked by Safronov in his theory, it had not played an important part in Anglo-American theories. After 1973, "Goldreich–Ward instability" became an essential concept in most theories of planetary formation, though it is by no means considered a definite solution to the problem of growing kilometer-sized objects from centimeter-sized objects (Boss 1989).

23 Goldreich and Ward (1973). Peter Goldreich (b. 1939) received his Ph.D. in physics at Cornell in 1963; he has been on the faculty of Caltech since 1966. William R. Ward (b. 1944) received his Ph.D. in planetary science from Caltech in 1972, conducted research at the Harvard–Smithsonian Center for Astrophysics from 1973 to 1977, then moved to the Jet Propulsion Laboratory, where he is now senior research scientist.

In the late 1970s it was generally agreed that the terrestrial planets were formed by accretion of solid planetesimals, but this process seemed too slow to account for the outer planets; Cameron's alternative of giant gaseous proto-planets was still a viable hypothesis. The accretion calculations of Safronov and Wetherill were based on the assumption that no gas is present. Hayashi, Nakazawa, and Nakagawa (1977) developed a theory in which planetesimal accretion is accelerated by gas drag. Building on this theory, Mizuno (1980) showed that after accretion had built up a solid core of 10 or 15 Earth masses surrounded by a gaseous envelope, the envelope would collapse onto the core (see also Harris 1978a). Giant planets formed by this process would have the same size core regardless of their distance from the Sun, in agreement with planetary models developed by Hubbard, Slattery, and others.

According to Gautier and Owen (1985: 832–7), the infrared measurement of high carbon abundances in the atmospheres of Jupiter and Saturn by the *Voyager* space probe favors the Mizuno nucleation model over the gaseous protoplanet model. Gautier and Owen argued that during the formation of a core from planetesimals, accretional heating would have vaporized methane ice and enriched the gaseous envelope in carbon, whereas in the gaseous condensation model no significant deviation from solar abundance would be expected. This conclusion was disputed by Pollack (1985: 810), but if accepted it would be the first case in which planetary observations made by spacecraft (other than lunar) have clearly supported one theory of planetary formation over another.

I.2

A geologist among astronomers: The Chamberlin–Moulton theory

Although several scientists in the preceding two centuries had proposed that a collision or close encounter of the sun with another star had started the planetary formation process, this idea was not able to displace the nebular hypothesis until it was presented by T. C. Chamberlin and F. R. Moulton in 1905. Whether combined with Chamberlin's theory that the planets had accumulated from small solid particles – "planetesimals" – or with the previously accepted doctrine that the planets condensed from hot gaseous balls, the encounter theory became the most popular explanation for the origin of the Solar System in the first third of the 20th century.

1.2.1 Nineteenth-century collision/encounter theories

For nearly a century after Buffon proposed that the planets were formed from gases struck off from the Sun by a comet, there was little or no interest in collision theories for the origin of the Solar System. The first such theory to attract any attention in the 19th century was proposed by James Croll, a Scottish geologist now best known for his astronomical theory of ice ages.[1]

Croll's point of departure was the conflict between geology and physics resulting from Lord Kelvin's estimate that the Sun has been shining for no more than about 20 million years. To resolve the conflict he suggested that if two bodies, each having half the Sun's mass, collided with speeds of 476 miles per hour, the heat generated would supply the Sun at its present rate for 50 million years. Croll elaborated this hypothesis in several publications during the 1870s and 1880s, brushing aside objections that no stars were known to have such high velocities. Few scientists paid any attention to Croll's theory,

---

1 Imbrie and Imbrie (1979). Irons (1896). James Croll (1821–1890) had no formal training in science. In 1859 after failing to establish a successful business career he found a position as caretaker of Anderson's College and Museum in Glasgow. His publications on changes in the eccentricity of the Earth's orbit and the ice ages gained him a position, which he held from 1867 to 1880, as geologist with the Edinburgh Office of the Geological Survey. He engaged in several controversies about scientific topics such as the cause of ocean currents.

perhaps because it postulated a very improbable event for which there was no independent evidence.[2]

A slightly more probable hypothesis was proposed by the English scientist A. W. Bickerton.[3] He became interested in stellar collisions on reading Proctor's account "of the star which faded into star mist" but realized that, as Stoney argued, most such collisions would be grazing rather than direct (Bickerton 1911c). In a long series of papers and books published from 1878 to 1915, he suggested that "partial impact" of the Sun and another star could produce fragments that form planets. According to a biographer, these publications "attracted considerable attention and some hostility in astronomical quarters" (Kopal 1970).

Bickerton noted that the Nebular Hypothesis had recently lost much of its support among astronomers. He mentioned two major objections: "the extreme slowness of the sun's present rotation; and the irregularities in the system, such as the eccentricity of the orbits, the inclinations of the axes and orbital planes, and retrograde motions" (1880: 156). In particular, the "extreme inclination of the axis of Mercury and Venus, and the retrograde motion of Uranus, appear only explicable on the assumption that these bodies were independent satellites existing as such before the impact which gave birth to our system" (1880: 156). The meteoric hypothesis of Proctor provides a better explanation of the formation of the other planets from a nebula but does not satisfactorily account for the origin of the nebula itself.

Since his hypothesis was designed to make plausible some of the irregularities in the Solar System, Bickerton put less emphasis on explaining its regular features. He considered Bode's law "a very empirical and imperfect law" (1880: 158) and did not attempt to derive it from his theory, though he later claimed that he could explain it qualitatively (1915: 115).

In 1894 Bickerton found further evidence for his theory in observations of Nova Auriga (December 1891), which he interpreted as a collision of two stars, actually a triple system according to A. Taylor and Vogel. This agreed with his prediction that when two stars collide they will separate leaving a third between them.[4]

In 1910 Bickerton managed to get financial support for a trip to Europe, and he appeared in London on 28 December to address the British Astro-

2 See Croll (1889) and earlier works cited therein.
3 Alexander William Bickerton (1842–1909) was born in England and studied at the Royal School of Mines in London. He held teaching positions at the Hartley Institution (Southampton) and at Winchester College. His publications on the relation between heat and electricity led to his appointment as professor of physics and chemistry at Canterbury College in Christchurch, New Zealand. He was one of Ernest Rutherford's teachers there (Badash 1975: 25).
4 The "third-body" hypothesis was rejected by an anonymous reviewer of Bickerton's *Romance of the Heavens* (1901) in *Nature*, **63** 607; the reviewer stated that Bickerton's theory "has not been hospitably received by astronomers, and the more elaborate exposition now presented will probably meet no better fate. The truth seems to be that in spite of his claim to have discovered numerous facts not known to 'ordinary' astronomers, the author lacks familiarity with spectroscopic work and astronomical methods generally."

nomical Association. By this time the idea that stellar encounters may produce novae and perhaps even lead to the formation of planetary systems had become rather popular; as an early advocate of this idea Bickerton was able to get a respectful hearing. E. W. Maunder, A. C. D. Crommelin, and F. W. Henkel were still skeptical, arguing that actual collisions must be quite rare and could hardly account for all double stars. Bickerton appealed to the authority of Lord Kelvin and Sir Robert Ball, both of whom seemed to assume that such impacts must occur frequently enough to be worth considering. Moreover, Kapteyn had found that stars collect in streams moving in opposite directions, which would enhance the likelihood of collisions (Bickerton 1910: 145–6). If two stars are near each other, the chance that they will undergo a grazing encounter is increased by their gravitational attraction; this statement was confirmed by a calculation by G J. Burns (1911). Bickerton (1911c) estimated that all these factors together should raise the probability by a factor of at least 10,000 over that for random encounters (1911c: 195–6).

While Bickerton managed to stir up a number of favorable and unfavorable comments about his theory, he did not succeed in developing it into a detailed quantitative explanation of the formation of the Solar System; his attention was diverted to the interpretation of novae and Wolf–Rayet stars.[5] Conversely, other theorists were led by the observation of novae to revive collision theories for the origin of the Solar System (1.2.4).

1.2.2 T. C. Chamberlin and the overthrow of the Nebular Hypothesis

At the end of December 1905 Edwin B. Frost, recently appointed director of the Yerkes Observatory, received the following announcement from the chairman of the Geology Department of the University of Chicago:

I believe it is the practice of celestial astronomers (which should certainly be followed by telluric astronomers) to notify their colleagues of important events impending in any part of the universe. This is therefore to inform you that on and after January 1st proximo, the solar system will be run on the new hypothesis. It is not expected that the transition will be attended by any jar or other perceptible perturbation or that the change from a gaseous to a planetesimal feed will occasion any nausea. Everything is expected to work smoothly. There will be somewhat more play of the parts than heretofore and it will not be necessary to put the screws on quite so much to keep the machine running according to the rule. The planets will be allowed to rotate as fast or as slow as they please without regard to the speed of their satellites, and these will be permitted to go round forwards or backwards as they see fit without incurring the

5 W. H. S. Monck (1911). J. McCarthy (1911). M. Davidson (1911). E. H. Beattie (1912). S. N. E. O'Halloran (1912). C. W. Raffety (1912). See also anonymous notes and reviews in *Journal of the British Astronomical Association*, **21** (1911): 242; *Atheneum*, **1** (1911): 365; *Current Literature*, **51** (July 1911): 50–1; *Observatory*, **34** (1911): 95–6, *Knowledge*, **8** (August 1911): 324.

suspicion of being illegitimate members of the family. The inclination of the Sun's axis will not be regarded as a moral obliquity but merely as a frank confession that once on a time he flirted with a passing star and got twisted a bit, as was natural. The solar family will not appear as prim and precise as heretofore, and at first the neighbors may cast some reflections on it, and will take kindly to the notion of new families springing up here and there in a legitimate way without sending the whole celestial concern to smash.

Please post notice according to the rules of the game.

Yours truly,

(Signed)   T. C. Chamberlin,
Late of Yerkes Observatory
(N.B., Telluric Annex.)[6]

The occasion for Chamberlin's letter was the forthcoming publication of the second volume of his textbook on geology, containing a comprehensive exposition of his "planetesimal hypothesis" for the origin of the Solar System and its application to the formation of the Earth (Chamberlin and Salisbury 1906, **II**: ch. 1). In this section I trace the early development and subsequent fate of Chamberlin's hypothesis. As the preceding quotation suggests, the story involves not only the interaction of theory and observation, but also the personal characteristics of an unusual scientist – a man who, in his late fifties, jumped into a new field and had a major impact on it.

In the first two decades of the twentieth century, the Nebular Hypothesis for the origin of the Solar System was replaced by the hypothesis that the planets were formed as a result of a close encounter of the Sun with another star. In modern terms, a "monistic" cosmogony was replaced by a "dualistic" one (though the replacement was incomplete since the new hypothesis unlike the old one did not attempt to explain the origin of the Sun itself). Several competing theories based on the dualistic hypothesis were proposed during this period. The most comprehensive, and at first the most successful, was that of Chamberlin and his junior colleague at Chicago, the astronomer F. R. Moulton. Chamberlin and Moulton's effective critique of the Nebular Hypothesis was also in large part responsible for allowing alternative theories to receive a fair hearing, including the "tidal" theory of James Jeans and Harold Jeffreys that subsequently was adopted by astronomers as the successor to the Chamberlin—Moulton theory. But the unique and in some ways most valuable aspect of the Chamberlin—Moulton theory – that the planets and satellites were formed by accretion of small solid particles ("planetesimals") rather

6 T. C. Chamberlin (D1905a). Here "D" means an unpublished document, available at a public archive (see the list of abbreviations preceding the Reference List at the end of this book). There is a large collection of Chamberlin's papers at the University of Chicago, including copies of his own letters to other scientists as well as originals of their letters to him. There are also some letters (primarily dealing with administrative matters) at the University Archives of the University of Wisconsin, Madison; correspondence concerning research grants may be found at the Carnegie Institution of Washington.

than by condensation of gaseous or liquid material – was forgotten or rejected by most other cosmogonists.

To Chamberlin the most objectionable feature of the Nebular Hypothesis was the inference that the Earth had begun in a hot molten state and gradually cooled down while solidifying. Lord Kelvin had estimated the time required for this cooling process and found it to be a few tens of millions of years, a result that contradicted geological evidence indicating much longer periods. Moreover, Chamberlin found that the assumption of a dense primeval atmosphere rich in carbon dioxide, whose variation was frequently invoked to explain major climatic changes and glaciation, was untenable in the light of deductions from the kinetic theory of gases about the escape of atmospheric gases at high temperatures.

In addition to its geological implications the Nebular Hypothesis had an indirect influence on 19th-century biological theories, and its overthrow by Chamberlin and Moulton was associated with a major change in the scientific worldview. Laplace's cosmogony could be considered a paradigm of evolutionary naturalism, a plausible prehistory for theories of organic development. In England, early evolutionists like Robert Chambers and Herbert Spencer constructed schemes beginning with the formation of a nebula in space and running through the evolution of living forms including man.[7] In the United States, according to one historian, the debate about astronomical evolution helped prepare intellectuals to accept Darwin's theory of natural selection (Numbers 1977).

Since the formation of a star involved the formation of planets, the Nebular Hypothesis suggested a plurality of inhabited worlds throughout the universe. Every planet should pass through a series of similar stages as it cooled down, and every civilization or race should pass through the same stages of cultural and scientific attainment as it developed, just as 19th-century anthropologists described all terrestrial societies in terms of their degree of advancement along a one-dimensional path, the highest point being that reached by white European males (Harris 1968: 108–216; Stocking 1968: 69–132).

As historians of the biological and social sciences have shown, evolutionary naturalism suffered a sharp setback in the early 20th century at the same time mechanistic determinism was being discredited in the physical sciences.[8] As in cosmogony, continuous predictable evolution in one direction was replaced by a discontinuous random process going in many directions (Brush 1979a). The Chamberlin–Moulton theory is not, however, typical of 20th-century theories that emphasize mathematical structure at the expense of intuitive comprehensibility; the Jeans cosmogony is more "modern" in this respect. Chamberlin's personal achievement was against the trend of the times; he demonstrated the power of what he called a "naturalistic" method of science and showed that the qualitative reasoning of the geologist could contribute something in

7 See Nebulous Earth, 1.4.
8 Rádl (1930: ch. 33). Harris (1968). Stocking (1968: 133–269). Cravens (1971). Cravens and Burnham (1971).

the domain of an exact quantitative science such as astronomy. Nevertheless, the demise of this theory resulted from its failure to satisfy quantitative criteria.

As late as 1906 a popular British writer on astronomy proclaimed that "the nebular theory, modified by subsequent research, seems destined to hold its own against all comers" (MacPherson 1906: 233). But the American astronomer Simon Newcomb warned that "cautious and conservative minds will want some further proof of the theory before they regard it as absolutely established" even though many facts support it (Newcomb 1903: 106). Most scientists at the end of the 19th century favored the Nebular Hypothesis but with reservations – they believed that the Solar System had formed from parts of a whirling cloud of hot gas, but they did not think that Laplace's original theory gave a completely satisfactory explanation of how this process worked.[9]

1.2.3 The assault on the Nebular Hypothesis

The overthrow of the Nebular Hypothesis was primarily due to the efforts of two Americans, the geologist T. C. Chamberlin and the astronomer F. R. Moulton.[10] Chamberlin was clearly the dominant partner who generated most of the ideas, but Moulton's contribution was essential in making the ideas consistent with celestial mechanics and thus acceptable to astronomers.

Chamberlin began working on his theory of the Earth's origin in 1892, the year he came to the University of Chicago to head its Geology Department. His initial reason for doubting the validity of the Nebular Hypothesis had nothing to do with any of the well-known astronomical problems discussed in the preceding section. His studies in glacial geology led him to investigate theories of the Ice Age, in particular the suggestion that it was caused by a decrease in the amount of carbon dioxide in the atmosphere (Chamberlin 1891, 1899b). He then undertook a more general study of the early atmosphere, starting from the usual assumption that the Earth was formerly molten, with a dense atmosphere rich in carbon dioxide.

At this point a paper by Irish scientist G. Johnstone Stoney exerted a crucial influence on Chamberlin's thinking.[11] In 1892 Stoney wrote a paper entitled "On the Cause of the Absence of Hydrogen from the Earth's Atmosphere

9 Ball (1901: 205–6). Becker (1898). Berget (1904: 45–51). Berry (1898/1961: 409). Cortie (1899). G. H. Darwin (1898: 338). Delaunay (1905: 705). Fison (1898: ch. 1). Geikie (1894: 227–8). Howe (1896: 327–33). Klein (1903). Mascart (1902). Mooser (1904). Morehouse (1898: 39, 56). Moulton (1896). Ritter (1906: 64–78; 1909: 64–78). See (1896, I: 1–3). Seemann (1900). Shaler (1898: 34–8). Sollas (1900). Todd (1899: 245–8). Von Braumüller (1898). Young (1900: 566–74).

10 For biographical notes see 1.1, notes 3 and 4.

11 George Johnstone Stoney (1826–1911) was the uncle of the physicist G. F. FitzGerald. He graduated from Trinity College Dublin in 1848, was professor of natural philosophy at Queen's College Galway for five years, then in 1857 returned to Dublin as secretary to the Queen's University. He remained there until 1882, when he moved to London. Stoney had been one of the first to take up the kinetic theory of gases and used it to estimate the size of an atom; he is now known mainly for having invented the word "electron" (O'Hara 1975; Brush 1976: sect. 5.4).

and of Air and Water from the Moon" in which he pointed out that gaseous molecules whose velocities exceed the "escape velocity" of a planet or satellite would be lost from its atmosphere.[12] According to the kinetic theory of gases the average molecular speed varies as the square root of the absolute temperature divided by the molecular weight ($mv^2 \propto kT$) so the lighter gases such as hydrogen are most likely to escape.

If the atmosphere were in thermal equilibrium at all heights, the number of molecules having velocities high enough to escape could be calculated from Maxwell's Velocity Distribution Law. Stoney did not believe this law would apply to planetary atmospheres. Instead he argued in a more detailed paper in 1897 (of which Chamberlin obtained an advance copy through G. E. Hale) that the Earth's atmosphere has retained water vapor but not helium; hence, there will be a significant loss of a gas when the velocity is less than about 10 times its average molecular speed, but not when it is 20 times or more.[13]

Chamberlin used Stoney's arguments in a paper on climatic change, presented at a meeting of the British Association for the Advancement of Science in Toronto in August 1897. He began by challenging the "time-honored conception of an exceedingly extensive, dense, warm and moist atmosphere" of the primeval Earth, and the assumption that this atmosphere contained the huge amount of carbon dioxide that is now locked up in mineral deposits on the Earth's surface (Chamberlin 1897: 656). He noted that the idea of a "vast original atmosphere" is based primarily on the assumption of the Earth's "supposed original molten state," which in turn is based on the Nebular Hypothesis (Chamberlin 1897: 656–67). But, in his first published remark on cosmogony, Chamberlin stated:

There is still some ground to doubt the nebular hypothesis and to entertain some of the various phases of the meteoroidal hypotheses. The nebular hypothesis correlates a wonderful array of remarkable facts and has gained a profound hold upon the convictions of the scientific world, yet some of its great pillars of support have recently weakened or have fallen away entirely. Of the 5000 unknown nebulae to which we naturally look for analogy very few, if indeed any, strictly interpreted, exemplify in a clear and decisive manner the systematic annular evolution postulated by Laplace. The photographs of the nebula of Andromeda, that were hailed with such delight on their first appearance as exemplifying the Laplacean hypothesis, appear upon more critical study to support it only in vague and general terms, if indeed they lend it support at all. The Saturnian rings, the trite source of illustration and analogy, prove under the test of the spectroscope to be formed of discrete solid particles, and not of gas, and the investigations of Roche have put a new phase on their theory. While in their form

12 I am unable to locate a publication with this exact title; it is cited by Chamberlin (1897: 658) and Moulton (1900: 111). Chamberlin's letter to G. J. Stoney, 6 September 1897, requesting a copy of this paper, indicates some bibliographical confusion (D1897a).

13 Stoney (1897, 1898). Chamberlin wrote to Stoney on 10 December 1897, thanking him for a copy of the paper and stating that he had seen an advance copy sent by G. E. Hale (Chamberlin D1897b). In a letter to C. R. Van Hise, 30 November 1897, he said he saw this paper "yesterday" (Chamberlin D1897c).

they tally with the annular hypothesis they do not support its gaseous phase, if indeed they lend it any support at all. (Chamberlin 1897: 657)

Turning to the supposed molten state of the primeval Earth, Chamberlin raised the question of whether such an Earth could have retained gases such as oxygen and carbon dioxide long enough for them to form an atmosphere. To answer this question he presented tables of the escape velocity (called "parabolic velocity" here) at various heights above the Earth's surface, computed for him by F. R. Moulton. The escape velocity varies from 11181 m/sec (= 6.9 miles/sec) at the Earth's surface ($r = 6.38 \times 10^6$ m from the center) to 565 m/sec at a distance 25×10^8m from the Earth's center. But when the centrifugal force due to the Earth's rotation is taken into account, the escape velocity decreases rapidly above that height and becomes zero at 304×10^8m from the Earth's center.[14] He then gave a table of average molecular velocities of helium, water vapor, carbon dioxide, oxygen, and nitrogen, computed by A. W. Whitney.[15] Another table shows the proportion of molecules having speeds of various multiples of the average speed, for temperatures up to 4000°C; this table is based on the Maxwell Velocity Distribution Law with no mention of Stoney's reservations about the applicability of that law to atmospheres.[16] Using Maxwell's estimates of the frequency of molecular collisions, Chamberlin then computed the times required for all the molecules to acquire the escape velocity.

Chamberlin recognized the difficulty of drawing definite conclusions in a situation where there are so many unknown or conjectural factors, but the main line of his argument was fairly convincing. If the Earth were molten, its temperature must have been at least 4000°C during its condensation, and this high temperature would have been shared by the gases in the upper atmosphere where the escape velocity is relatively small. Moreover, if one assumes that the Moon was created from the primeval Earth, either by condensation from a secondary Laplacian ring or by fission as in G. H. Darwin's hypothesis (Darwin 1879), then the rotational speed of the Earth at that time must have been much greater than now; the period of rotation may have been as short as 1 hour 24 minutes, corresponding to equality of centrifugal force and gravitational force at the equatorial surface. Under such conditions the escape velocity would be even further reduced; "Indeed it is difficult to see how the Moon could have separated from the Earth without carrying away the atmosphere" (Chamberlin 1897: 666). In any case, the numerical estimates from kinetic theory suggested that if the temperature were higher than 3000°C,

14 Chamberlin (1897: 660). This assumes a period of rotation of 23 hours 56.067 minutes. Another table gave the escape velocities for a rotation period of 1 hour 24 minutes (see upcoming text). See Kerz (1884) for a similar objection to the original nebular hypothesis.
15 Chamberlin (1897: 661). Letters from Chamberlin to Moulton, 10 September 1897, and to Whitney, 9 October 1897, indicate that there was some disagreement between Whitney and Moulton on the calculation of centrifugal force effects (Chamberlin D1897d, D1897e).
16 Chamberlin's source of information about kinetic theory was a recent book by Allan Douglas Risteen (1895).

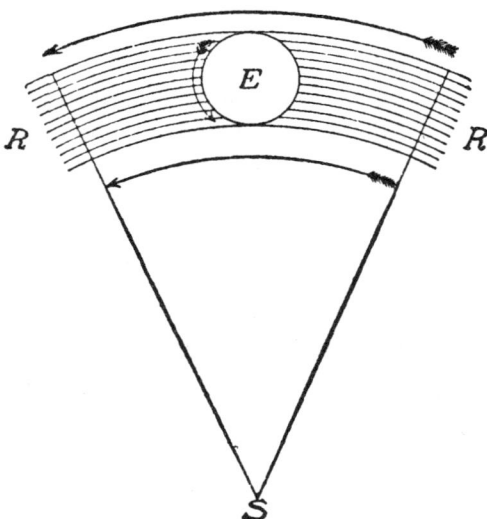

Figure 3. Chamberlin's illustration of the explanation of direct rotation on the Nebular Hypothesis: "RR represents a ring of gas moving as a unit and hence the outer portion the faster. If converted into a spheroid, E, centrally located, the rotation is forward, as shown by the arrow" (T. C. Chamberlin, *Origin of the Earth*, University of Chicago Press, 1916, p. 91).

there was no hope of retaining much more than a small residue of the atmosphere, and the only question was whether the gases would disappear in a few seconds or a few hundred years. The same arguments would apply with even greater force to the earlier stage of the Nebular Hypothesis, the supposed gaseous rings from which the Earth condensed; only by asuming that such a ring rapidly cooled and crystallized into small solid particles could one avoid the conclusion that its material would have been dissipated into space.[17] The only way out of the difficulty, if one still accepted the Nebular Hypothesis, would be to assume that the original Earth–Moon ring was not gaseous but composed of solid particles.

But that assumption, known in the 19th century as the meteoritic hypothesis, ran into other difficulties, as Chamberlin soon discovered when he began to read the literature on cosmogony. According to a recent English book, citing Faye's authority, planets formed by aggregation of solid particles would be expected to have retrograde rotation.[18]

Chamberlin detected a fallacy in the argument: It was based on the assumption that the particles move in circular orbits so that in an encounter of

17 For further remarks about the bearing of the kinetic theory of gases on the origin of the Earth, see Chamberlin (1916: ch. 1; D1899).
18 Chamberlin (D1895). The "recent English work" may be a book by Gore (1893), which includes an extensive discussion of Faye's work.

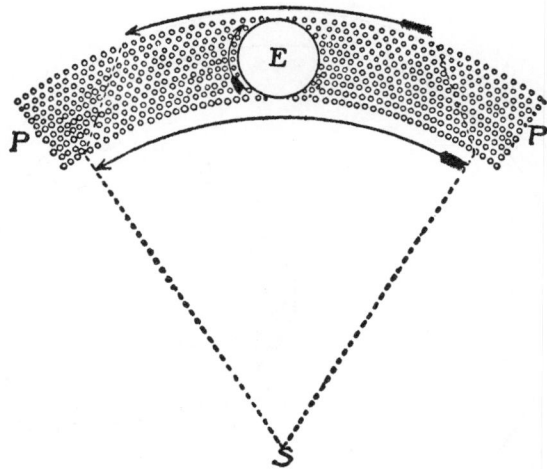

Figure 4. Chamberlin's illustration of the rotation produced by combining particles moving in adjacent circular orbits, with linear speeds decreasing with distance from the Sun according to Kepler's Third Law. "PP represents a belt of planetesimals revolving concentrically about the center, S. If these collect about the central point of the belt into a spheroid, E, by the enlargement of the inner orbits or the reduction of the outer ones, the concentric arrangement remaining, the rotation will be retrograde, as shown by the arrow" (T. C. Chamberlin, *Origin of the Earth*, University of Chicago Press, 1916, p. 92).

two particles, the outer one will be moving more slowly. But in reality collisions occur only if and when the orbits of two particles intersect, and this implies that at least one of them moves in an ellipse. The geologist, applying Kepler's Second Law, noticed a point that seemed to have escaped the attention of the astronomers who had previously written on this subject: The orbits are likely to intersect at a place where the aphelion of the inner corresponds to the perihelion of the outer orbit; at this place, the outer body will move faster than the inner one.

Puzzled that the eminent Faye and other cosmogonists could have overlooked such a simple argument, Chamberlin wrote to his former colleague George C. Comstock, director of the Washburn Observatory at the University of Wisconsin, asking him to verify the reasoning (Chamberlin D1895). Comstock replied that while Chamberlin's conclusion about speeds of particles in elliptical orbits was correct, retrograde rotation had not been a strong argument against the meteoritic theory since one could account for the change to direct rotation by invoking various other causes such as tidal friction.[19]

19 G. C. Comstock (D1895). For an account of Faye's theory and an explanation of how direct rotation might be produced he referred Chamberlin to C. A. Young (1895: 518–19).

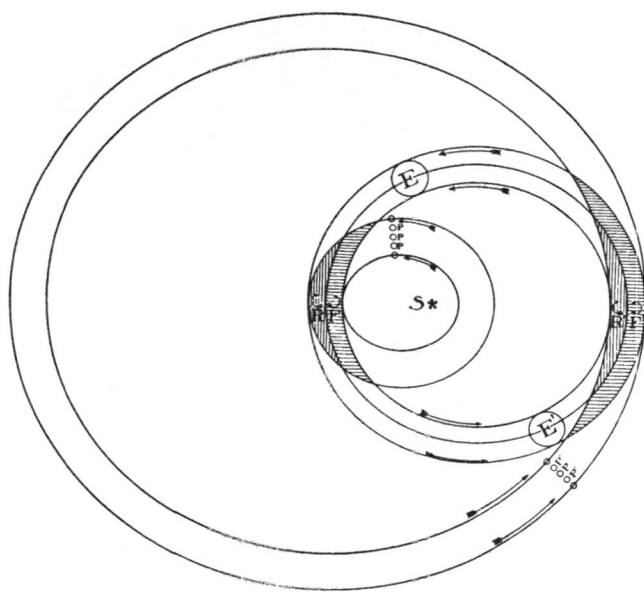

Figure 5. Chamberlin's explanation of the direct rotation of a planet produced by accretion of planetesimals in circular orbits onto a nucleus in an
elliptical orbit: "A diagram intended to illustrate the proportion of normal cases of collision that will tend to give *forward* rotation as against
retrograde rotation, a uniform distribution of planetesimals being assumed.
To avoid unnecessary complication, a belt of planetesimals in the minimum normal orbits that permit collision, each having the breadth of the
planetary nucleus, are chosen, as these are the cases of greatest differential
velocities, and hence the most effective in producing rotation. The intermediate cases are not only less effective because of less differential velocity, but because the collisions of the two classes are more nearly equal in
number and more nearly neutralize one another. In inspecting, let it be
noted that the nucleus E, at the left, is moving faster than the planetesimals in the area of possible collision, and hence that the planetesimals
which it overtakes on its inner side tend to cause forward rotation, while
those which it overtakes on its outer side tend to produce retrograde
rotation. The collisions in the larger area, F, favoring forward rotation, are
much more numerous than those in the small area, R, favoring retrograde rotation. At the right, the planetesimals are moving faster than the
nucleus E', and those that overtake it on the outside in the area F' tend
to forward rotation, while those that overtake it on the inner side in the
area R' tend to retrograde rotation. In both cases, forward rotation is
favored. If belts between these limiting belts be drawn, the difference
between the two classes of areas will be less, but of the same phases"
(T. C. Chamberlin and R. D. Salisbury, *Geology*, Holt, 1906, vol. 2, 76).

In spite of Comstock's warning that it was not decisive, Chamberlin incorporated the argument about collisions producing direct rotation in his early publications on cosmogony, beginning with a footnote in his 1897 paper (Chamberlin 1897: 668–9). The argument has not received much attention from astronomers, and it is not clear that it is valid in all cases (Darwin 1909: 930). However, calculations by R. T. Giuli in 1968 and A. W. Harris in 1977 indicated that it is probably correct.[20]

By the end of 1895 Chamberlin had begun to study other astronomical phenomena involving rings that might be comparable to those postulated in the Nebular Hypothesis. In the letter to Comstock mentioned earlier he inquired about the nature of Saturn's rings. He was aware that the rings were generally believed to be composed of separate particles (as proposed on theoretical grounds by James Clerk Maxwell in 1857 and supported by James Keeler's spectroscopic observations a few months earlier in 1895) but was not sure if this meant that the particles each moved in separate orbits around the planets. If so, and if they were subsequently to condense into a satellite, would that satellite have a backward or forward rotation? Comstock replied that he thought the perturbations between particles did not cause "any radical deviation from approximately circular motion." But there was no chance of testing the theory of the direction of rotation of aggregated bodies in this case, since Edouard Roche had proved that the rings could never condense into a satellite; they are inside the "Roche limit" (as it is now called) so tidal forces would prevent any large body from being in hydrostatic equilibrium.[21]

In a lecture on 27 January 1896, Chamberlin explored the geological implications of the Nebular Hypothesis and analyzed some of the astronomical evidence often given for it. His contact with Washington geologists, resulting from association with the U.S. Geological Survey, had led him to consider the hypothesis that the Earth's crust is mobile. That hypothesis, based earlier on

20 R. T. Giuli (1968a, 1969b). A. W. Harris (1977). Dr. Harris kindly sent me the following comments on the Chamberlin–Moulton theory (letter of 15 July 1977): "It seems to me that they correctly point out that the retrograde rotation paradox is resolved by the effect of eccentricity: the simple cases with both planet and planetesimals in circular orbits as shown in their Figures 27 and 28 do not apply because there is no means for planetesimals to reach the planet embryo from significantly far away from the planet orbit. In their solution, they propose eccentric orbits for the planet embryos, while retaining circular orbits for the planetesimals. More recent works (e.g., V. S. Safronov [1969/1972]) favor the reverse situation: the planetesimals travel in more eccentric orbits than the planets. However, exactly the same geometry applies, and their Figure 30 is an excellent illustration of the effect which I attempt to illustrate in my Figure 1 and develop into an analytical theory in my paper. My result is as they suggest, that 'the resulting rotation is likely to be relatively low, though the total force of impact be great.' In fact, I find that this mechanism produces rotation rates that are about an order of magnitude less than observed rotation rates, if one assumes eccentricities of planetesimals sufficient to deliver them to the planet orbits from distances midway in between. Satisfactory results can be obtained from this mechanism by assuming smaller (but non-zero) eccentricities, as was done by Giuli in his numerical theory." See also Artemév and Radzievsky (1965), Artemév (1969), Lissauer and Safronov (1991).

21 For theories of Saturn's rings see *Nebulous Earth*, 1.5. On 6 March 1903, Chamberlin wrote to Comstock asking for the loan of Roche's work, which was apparently not available at Chicago; see Chamberlin (D1903a). See also Chamberlin (D1903b).

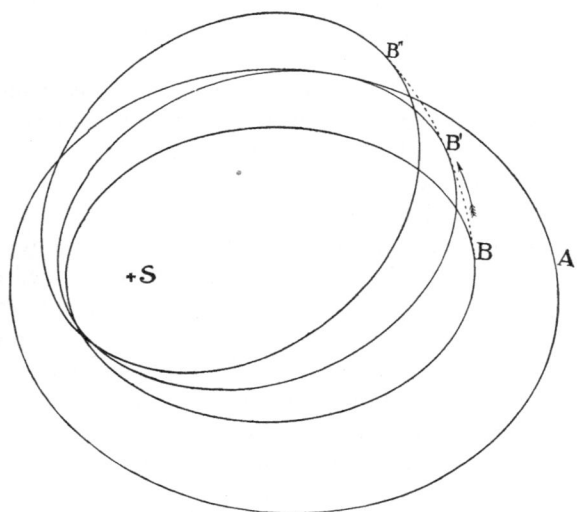

Figure 6. Chamberlin's explanation of why a perturbation of the elliptical orbits of planetesimals, causing them to intersect, will lead to accreting collisions more likely to produce a body with forward rotation: "Diagram showing that in the shifting of orbits, the first contingencies of collision favor forward rotation. B represents a smaller elliptical orbit within a larger one, A. If B be shifted progressively in the direction B', B", it will first come into possible collisional relations with A on its inner side, and at this point a body in the orbit A is moving faster than a body in the orbit B, as shown by the larger orbit the former describes, and the collision being on the inner side, forward rotation is favored" (T. C. Chamberlin and R. D. Salisbury, *Geology*, Holt, 1906, vol. 2, p. 77).

the assumption that there is a fluid layer below the crust, had been attacked by physicists like Lord Kelvin who argued that the entire Earth must now be solid, since otherwise one would expect tidal effects different from those observed. While many geologists had bowed to Kelvin's authority, Chamberlin (D1896) reported that John Wesley Powell and Grove Karl Gilbert still believed in liquidity, while Clarence Dutton insisted on "mobility" without using terms such as "fluidity" or "plasticity" to account for it. Chamberlin was willing to consider these ideas, but differed from Dutton in one respect: He thought there was evidence for the differential motion of parts of the crust (e.g., the raising of mountain chains), while Dutton denied differential motion. If the Earth had been formed according to the Nebular Hypothesis, it would have had a liquid layer under the crust and differential motion would have been possible. But he felt uneasy about drawing such an important geological conclusion without first examining with a critical eye the fundamental hypothesis on which it seemed to depend. At first sight, the rings of Saturn seemed to provide striking support for the Nebular Hypothesis. So did the photographs of nebulae showing a ringlike structure. But then he realized that only a few of the thousands of known nebulae really had anything like a ring structure, and on closer examination those looked more like spirals than rings.

About halfway through this lecture, according to the notes that have been preserved, Chamberlin confessed his attraction to the meteoric cosmogony, though not yet to the exclusion of the Nebular Hypothesis. If a nebulous mass, extending far out into space, throws off a ring, this ring would be very attenuated and would rapidly lose heat; probably it would solidify into separate particles, such as those in Saturn's ring, before condensing to a planet. If the Earth were formed by aggregation of solid particles, heat would be generated during its formation, and the age of the Earth might be considerably greater than that estimated by assuming it had simply cooled down from a molten or gaseous state. Thus, the geologists might be able to escape the restrictive time limits imposed by physicists.

These arguments were elaborated in Chamberlin's 1897 paper at Toronto. If the Earth was formed by the slow accretion of solid particles it may have remained solid and cold throughout its history, or until it grew to such a size that gravitational contraction generated internal heat as in the Helmholtz solar theory.[22] The gases trapped inside would then be expelled to the surface, which would still be cold enough to hold most of them in an atmosphere. Conditions favorable to life could thus be present early enough "to satisfy the most strenuous demands of theoretical biology. Most of the restrictive arguments of Lord Kelvin and others lose their application under this hypothesis" (Chamberlin 1897: 676).

Chamberlin did not want to reject the Nebular Hypothesis without giving it a fair chance to overcome its difficulties. In November 1897, Moulton presented him with calculations indicating discrepancies between the observed rate of rotation of the Sun and that expected from Laplace's theory. Chamberlin suggested that "at the time of formation of Neptune, the material of the Sun was so concentrated toward the center that the faster moving outer portion was just able to accelerate the central or axial slow-moving portion, so as to give the present Sun's energy of rotation. Then suppose that previous to the separation of each subsequent planet the central concentration had increased relatively and the density of the peripheral portion had diminished relatively, so that in each instance the same energy of rotation was present (neglecting the discharged portions), and yet the ratio between the slow-moving central portion and the fast-moving outer portion was such as to give in each the present solar rotation" (Chamberlin D1897i). Moulton replied that while this might be the only possible explanation of the slow rotation of the Sun on the assumption of contraction from a large homogeneous mass, he

22 Chamberlin (1897: 668–70). There is a letter to Comstock, dated 30 November 1897, thanking him for criticisms and suggestions on this point in the paper (Chamberlin D1897f). Earlier Chamberlin had asked Moulton to calculate the heat that would be generated in the Moon by its self-compression; see Chamberlin (D1897g). See also Chamberlin (D1897h).

preferred to consider the question more carefully before reaching a definite conclusion (Moulton D1897).

But Chamberlin had moved too far from the Nebular Hypothesis to invest any further effort in patching it up; he was already beginning to realize the enormous changes in geological theory that would result from adopting the meteoric theory instead. In January 1898 he warned his Wisconsin colleague Charles Van Hise, who was about to publish a major paper on crustal shortening in Chamberlin's *Journal of Geology*, that quantitative estimates of forces determining the structure of the Earth's crust might well be affected by the choice of a theory of the Earth's origin (Chamberlin D1898a). A few days later he wrote to the publisher Henry Holt, to whom he had promised a comprehensive new textbook on geology, asking for an extension of time in order to work out the consequences of his new interpretation of the history of the Earth (Chamberlin D1898b; see also Chamberlin D1898c). The extension was granted; it was not until 1904 that the first volume of *Geology* appeared, coauthored by Rollin D. Salisbury, head of the Geography Department at Chicago (Chamberlin and Salisbury 1904).

Astronomy might also be affected by taking seriously the hypothesis that a nebula is composed of cold solid particles. Previously it had been claimed that the light emitted by nebulae is conclusive evidence that they are composed of hot gases, and this result was also considered favorable to the Nebular Hypothesis.[23] But, Chamberlin asked E. E. Barnard at the Yerkes Observatory, might it not be shown spectroscopically that the luminosity of nebulae is actually due to "an electrical or phosphorescent condition" rather than to high temperature? (Chamberlin D1898d).

In May 1899, *Science* reprinted Lord Kelvin's address on the age of the Earth, delivered in London two years earlier (Kelvin 1897/1899). Chamberlin quickly prepared a sharp critique, which appeared in the issue of 30 June.[24] This was quite a memorable event for his biographers: Finally a geologist dared to attack the time limits imposed by calculations of heat conduction in the Earth, striking at the foundations of Kelvin's argument rather than haggling about the numerical details. According to Carroll and Mildred Fenton, Chamberlin "tore the restrictions to shreds while the more conservative Britishers gasped at New World temerity" (Fenton and Fenton 1945: 512–13). His son Rollin Chamberlin recalled in 1934: "Not a few British scientists expressed surprise that anybody should have the temerity even to reply to Lord Kelvin, let alone to attempt to unhorse him in his own field" (R. T. Chamberlin 1934: 351).

Chamberlin attacked Kelvin's "very sure assumption that the material of our present solid Earth all-round its surface was at one time a white-hot liquid." Though widely accepted, this assumption does not rest on "any conclusive geological evidence." He noted that Kelvin himself had at times fa-

---

23 See *Nebulous Earth*, 1.3.11.
24 Chamberlin (1899a). The manuscript was reviewed by Moulton before publication; see Moulton (D1899a).

vored the idea that the Earth was built up by the infall of meteorites; the main
difference between them seems to be whether the process was so rapid that
the material vaporized by heat of impact, or whether the heat was generated
more slowly (Chamberlin 1899a: 892–3).

In addition to his argument that a hot fluid Earth could not have retained
water and atmospheric gases, Chamberlin also seems to have had some more
philosophical reasons for his belief in a slow process of accretion. As Winnik
(1970) has noted, Chamberlin thought of natural processes as resulting from
purposeful design, allowing progressive opportunities for man to participate in
shaping his future. With this attitude he could not accept the pessimistic cosmology leading to the "heat death," which Kelvin and others had inferred
from the Nebular Hypothesis and the Second Law of Thermodynamics. In
his 1899 critique of Kelvin, Chamberlin gave a remarkable comparison of the
two hypotheses for the Earth's history, his own and the accepted one of a
cooling contracting ball:

The slow-built meteoric Earth delayed the exercise of thermal agencies until the life
era and gradually brought them into play when they were serviceable in the prolonga-
tion of the life history, whereas the liquid Earth exhausted these possibilities at a time
of excessive conversion of energy into heat and thus squandered its energies when
they were not only of no service to the life history of the Earth, but delayed its
inauguration until their excesses were spent. (Chamberlin 1899a: 897)

Implicit in Chamberlin's theory was a radically new philosophy of natural
science. The 19th-century worldview was evolutionary in a very limited
sense, especially when applied to the history of the Earth: a simple cooling,
solidification, and contraction, giving little scope for fluctuations in temperature (as needed to account for glacial epochs) or for the hundreds of millions
of years of temperate conditions postulated by Uniformitarian geologists and
evolutionary biologists. Now at last with Chamberlin's theory, a more complex type of evolutionary development of the Earth was being introduced,
compatible with modern theories of the development of living organisms on
the Earth's surface. The evolutionary process could yield complexity and nov-
elty rather than degeneration and death as 19th-century scientists had supposed. With Chamberlin's view of geological processes involving rearrangement of the solid materials of the Earth's interior, lubricated by temporary
melting and driven by the heat of gravitational contraction, one might have
expected a more favorable climate for hypotheses such as continental drift.
But in fact Chamberlin did not establish a new worldview and did not support Wegener's continental drift hypothesis in spite of his remarks in the 1896
lecture about differential motion of the crust. What he did was important
enough without having to be exaggerated: He liberated astronomers and geologists from the restrictions of the Nebular Hypothesis and the assumption
of a uniformly cooling, solidifying Earth.

Early in 1900 the Scottish geologist James Geikie reviewed the geological objections to Kelvin's theory of the cooling of the Earth from a molten state. He quoted at length from Chamberlin's "very interesting essay" on Kelvin's theory, concluding that Chamberlin's concept of a solid Earth built by slow accretion of meteors, with heat generated over a long period of time by consolidation and rearrangement of materials, is just as plausible as Kelvin's on astronomical grounds and much more satisfactory from a geological viewpoint (J. Geikie 1900).

Chamberlin wrote immediately to Geikie, thanking him for his support, and telling him that in a forthcoming paper on the Nebular Hypothesis, "I have been rash enough to try to carry the war into Africa, if I may speak so slightingly of our mathematical and physical friends. I may come home on a stretcher, but at any rate it seems worth while to direct our friends' attention to the p's and q's of their own domain. In any case the war is likely to be quite as wholesome for being less purely on the defensive on our part" (Chamberlin D1900a).

In order to "carry the war into Africa" and defeat the Nebular Hypothesis on the astronomical battlefield, Moulton's efforts were essential. But in the summer of 1899 Moulton had no assurance that he would be able to remain at Chicago to collaborate with Chamberlin after receiving his Ph.D. He expressed his financial anxieties in letters to George Ellery Hale at the Yerkes Observatory. (Hale was also on Moulton's examining committee at Chicago.) Moulton complained that President (William Rainey) Harper of Chicago would not pay him enough to live on; he was now in debt and intended to leave Chicago as soon as he could find a suitable position elsewhere (Moulton D1899b). Hale replied that Harper had told him he was pleased with Moulton's work, though he could not yet promise a promotion (Hale D1899a). Shortly after that the situation seemed to improve.[25]

Moulton completed his Ph.D. requirements at the end of 1899; on 3 January 1900 he wrote to Hale outlining his results on the Nebular Hypothesis, with a view toward possible publication in Hale's *Astrophysical Journal* (Moulton D1900a). One result had already been announced at the first meeting of the Astronomical and Astrophysical Society, held at the Yerkes Observatory the previous September: "If we supposed that the condensation towards a planet had gone on to a considerable extent the chances are against its completing itself."[26]

Hale encouraged Moulton to send his critique of the Nebular Hypothesis for publication in the *Astrophysical Journal*. In another letter Moulton stated that "I have been associated with Prof. Chamberlin in attacking the problem, and I have obtained many valuable ideas from him, but of course all the mathematical features are mine" (Moulton D1900b). The paper was accepted for publication and appeared as the first article in the issue of March 1900.[27]

---

25 "I am glad to hear that there is some prospect of your remaining at the University. If you decide to do so I shall be glad to recommend you for promotion next year" (Hale D1899b).
26 Only the title was published (Moulton 1899/1910).
27 "Your paper on the Nebular Hypothesis seemed to be all right and I therefore did not call. I

Most of Moulton's arguments against the Nebular Hypothesis had already been made by other writers, but his presentation was exceptionally forceful and comprehensive; this paper is probably his most important contribution to cosmogony. He gave some rather convincing new arguments, on the basis of the Lagrangian solution of the three-body problem and Roche's theory of stability, that a Laplacian ring could not condense into a single planet. He claimed that the angular momentum objection is so simple and certain that

it seems to render the nebular hypothesis in the simple form in which it has usually been accepted absolutely untenable unless some fundamental postulates now generally accepted are radically erroneous. It seems a necessary inference from the results of the discussion that the solar nebula was heterogeneous to a degree not heretofore considered as being probable, and that it may have been in a state more like that exhibited in the remarkable photographs of spiral nebulae recently made by Professor Keeler.[28]

In an account published 35 years later, Moulton seemed to have forgotten these first tentative steps toward the new theory; he wrote that in 1900 Chamberlin and Moulton "occupied the almost unique position in science of abandoning a theory because of its own weaknesses rather than because of the superior merits of a rival hypothesis." Implying a complete absence of any preconceptions about an alternative theory at that time, Moulton said they felt like Adam and Eve on leaving the Garden of Eden – "The world was all before them" (Moulton 1935: 143–44).

The spiral nebulae were clearly regarded as the key to an alternative theory. On 30 January 1900, Chamberlin wrote to James Keeler that he and Moulton were considering alternative theories of the origin of the Earth, including "an origin from a spiral nebula." For this purpose he asked Keeler to send him copies of the photographs of spiral nebulae currently being made at the Lick Observatory (Chamberlin D1900c). The request was granted; a month later Chamberlin wrote to express his thanks and to promise Keeler a copy of his forthcoming paper. "I sincerely hope," Chamberlin told Keeler, "you may be able to come to the relief of geologists in the dilemma in which we seem to be finding ourselves as the result of the apparent breaking down of the accepted form of the nebular hypothesis" (Chamberlin D1900d).

The paper just mentioned was "An Attempt to Test the Nebular Hypothesis by the Relations of Masses and Momenta," which Chamberlin published in the January–February (1900) issue of his own *Journal of Geology*. The opening paragraph pointed out the link between geology and astronomy: Recent work in glacial geology

hope you will not make any more changes in the galley proof than are really necessary as it is very expensive" (Hale D1900).

28 Moulton (1900: 130). For Chamberlin's comments on the manuscript of this paper (chiefly pertaining to the discussion of G. H. Darwin's explanation of the motion of Mars and Phobos), see Chamberlin (D1900b).

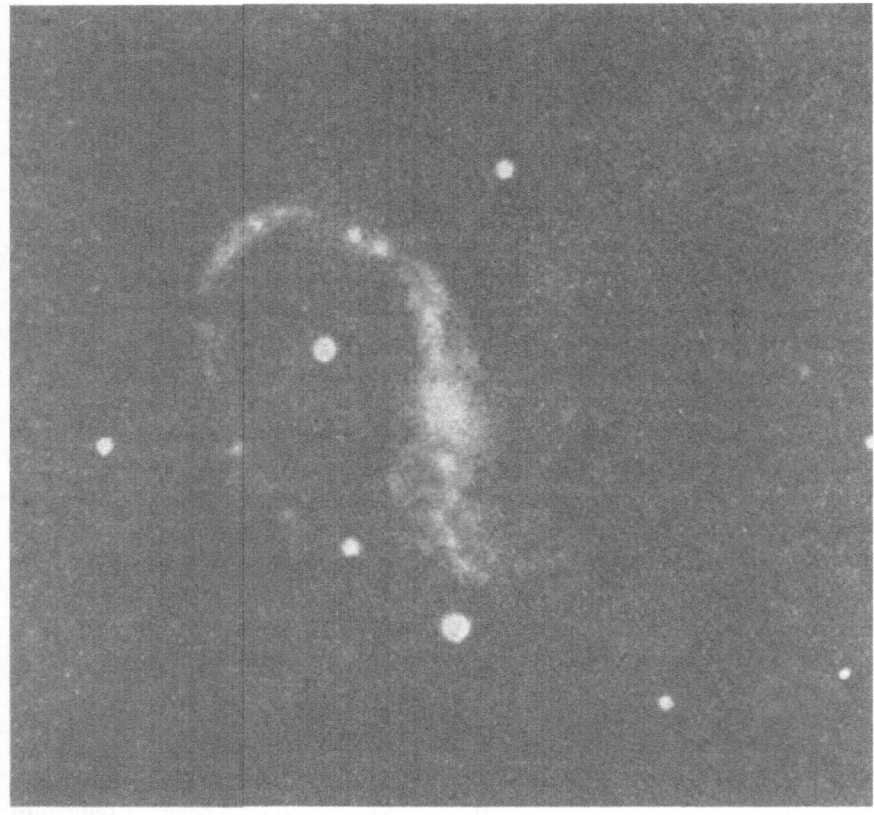

Figure 7. Chamberlin's example of the kind of nebula from which a planetary system might evolve: "A spiral nebula in which the two arms are especially distinct, HI 55 Pegasi" (T. C. Chamberlin, *Origin of the Earth*, University of Chicago Press, 1916, p. 125).

had made it imperative to seriously consider inherited views relative to the nature of the Earth's early atmospheres, and this in turn forced an inquiry into the current postulate of a primitive, vast, gaseous envelope exceptionally rich in carbon dioxide; for the special heat-absorbing qualities of this constituent render it doubtful whether its presence in large amount is compatible with glaciation.

But that postulate now seemed inconsistent with the Laplacian hypothesis that the Earth had been formed from a hot gaseous ring, in the light of calculations of molecular velocities based on the kinetic theory of gases (Chamberlin 1900: 58). As he had suggested earlier, carbon dioxide and other atmospheric gases including water vapor would largely have escaped from the primeval Earth, if it was so hot that its mineral components were in a liquid state.

Possibly having in mind current disputes about the applicability of the kinetic theory of gases to atmospheres,[29] Chamberlin pointed out that his conclusions did not depend on the precise validity of his computations of molecular velocities. The qualitative results of kinetic theory were in any case completely in agreement with the known absence of atmospheres from all satellites and asteroids, the thin atmosphere of Mars, the absence of the lightest gases (hydrogen and helium) from the Earth's atmosphere, and the "vast and deep atmospheres" of Jupiter and Saturn – "in short, there is a general correspondence between the mass of the atmosphere and the gravitative competency of the body" (Chamberlin 1900: 59).

The major astronomical test of the Nebular Hypothesis was a comparison of calculated and observed distributions of angular momentum among the planets and the Sun. The ancestral nebula was assumed to be a sphere just extending to the orbit of Neptune, rotating with the present angular velocity of that planet. This assumption would lead to a lower limit for the angular momentum, since the equatorial bulge of a rotating mass would increase its angular momentum for a given speed of rotation, and since it was likely that the nebula actually extended some distance beyond the orbit of Neptune; hence, as will be seen, it was favorable to the hypothesis being tested. The distribution of density found by G. H. Darwin[30] was used; the details of the calculation are attributed to Moulton. The result was that the ancestral nebula must have had an angular momentum at least 213 times that of the present Solar System, according to the Nebular Hypothesis.[31]

The discrepancies between hypothesis and astronomical observation became even more striking if the outer planets were omitted and the hypothesis was applied only to the formation of the inner planets and the Sun. For example, if the Sun, Mercury, Venus, and Earth were formed from a nebula rotating with the present angular velocity of the Earth, that nebula must have had angular momentum 28 times as great as the Sun for those bodies, even making the most favorable assumption that the Sun is homogeneous in density rather than condensed toward the center.

In the next section Chamberlin pointed out that the ratios of masses to

29 Bryan (1899, 1900). Cook (1899, 1900). Stoney (1900).
30 Darwin (1888). Moulton (1900) stated that he used Darwin's "Isothermal—Adiabatic" sphere given on p. 25. A year later Anne Sewall Young (niece of the astronomer Charles Augustus Young) attacked the problem from the other end, using the present distribution of angular momentum to find the law of density of the initial nebula and showing that this leads to impossible results. See A. S. Young (1901). Her work was apparently started under the direction of Moulton, who submitted it to Hale on 26 March 1901 with a letter saying that "in the light of these results many of Darwin's cease to have importance, and in my opinion, this might have been pointed out, but Miss Young felt timid about doing so" (Moulton D1901).
31 Another way to express the discrepancy was suggested by Chamberlin: "If the matter of the solar system were expanded to some point beyond the orbit of Neptune in conformity to the laws of hydrodynamic equilibrium, and given the moment of momentum of the present solar system and allowed to contract by secular cooling, the centrifugal force in its equatorial portion would not become equal to the centrifugal force [and hence allow separation of a ring] until *after it had contracted far inside the orbit of Mercury*" (Chamberlin D1900e).

momenta of bodies in the Solar System are highly irregular; for example, Jupiter, having less than one-tenth of 1 percent of the total mass of the nebula, carried away 95 percent of its angular momentum. While this circumstance does not seem to be forbidden by the Nebular Hypothesis, Chamberlin implied that it was a fact that must weigh against the regular development usually associated with that hypothesis.

Chamberlin concluded by stating that the relationships of mass and momentum "are seemingly altogether incompatible with an evolution of the solar system from a gaseous spheroid controlled by the laws of hydrodynamic equilibrium and developing by secular cooling. The argument is equally cogent against an evolution from a meteoroidal spheroid controlled by the laws of convective equilibrium" (as in G. H. Darwin's theory) (Chamberlin 1900: 71–2).

He then gave a hint of how he might develop his own theory. It was necessary to explain how

the peripheral portion of the system acquired all but a trivial part of the moment of momentum, while it possesses but a trivial part of the mass. The first suggestion of these conclusions was the possible formation of the system by the collision of a small nebula upon the outer portion of a large one. . . . Following a purely naturalistic and inductive method, it would seem that the spiral nebulae, whose abundance is attested by the recent notable success of Professor Keeler in photographing numerous small ones, offer the greatest inherent presumption of being the ancestral form. (Chamberlin 1900: 72–73)

Shortly after the publication of Chamberlin's and Moulton's separate papers in 1900, James Keeler gave an extended discussion of his photographs of spiral nebulae. He concluded:

The idea at once suggests itself that the solar system has been evolved from a spiral nebula, while the photographs show that the spiral nebular is not, as a rule, characterized by the simplicity attributed to the contracting mass in the nebular hypothesis. This is a question which has already been taken up by Professor Chamberlin and Mr. Moulton. (Keeler 1900: 348)

A few months later Chamberlin and Moulton, in a joint article in *Science*, agreed with Keeler that the consideration of spiral nebulae should have "precedence in attempts to find analogies for the origin of our system . . . in the absence of any knowledge of the origin of spiral nebulae, it is possible to conjecture that they arose from peripheral collisions of antecendent nebulae" (Chamberlin and Moulton 1900: 208). This hypothesis was explored but eventually dropped; according to a memorandum written in 1906, published by Chamberlin and Moulton in 1909, the investigation of collisions between "nebulous bodies" as a mode of origin of spiral nebulae was "futile . . . no escape was found from the high probability, amounting almost to certainty,

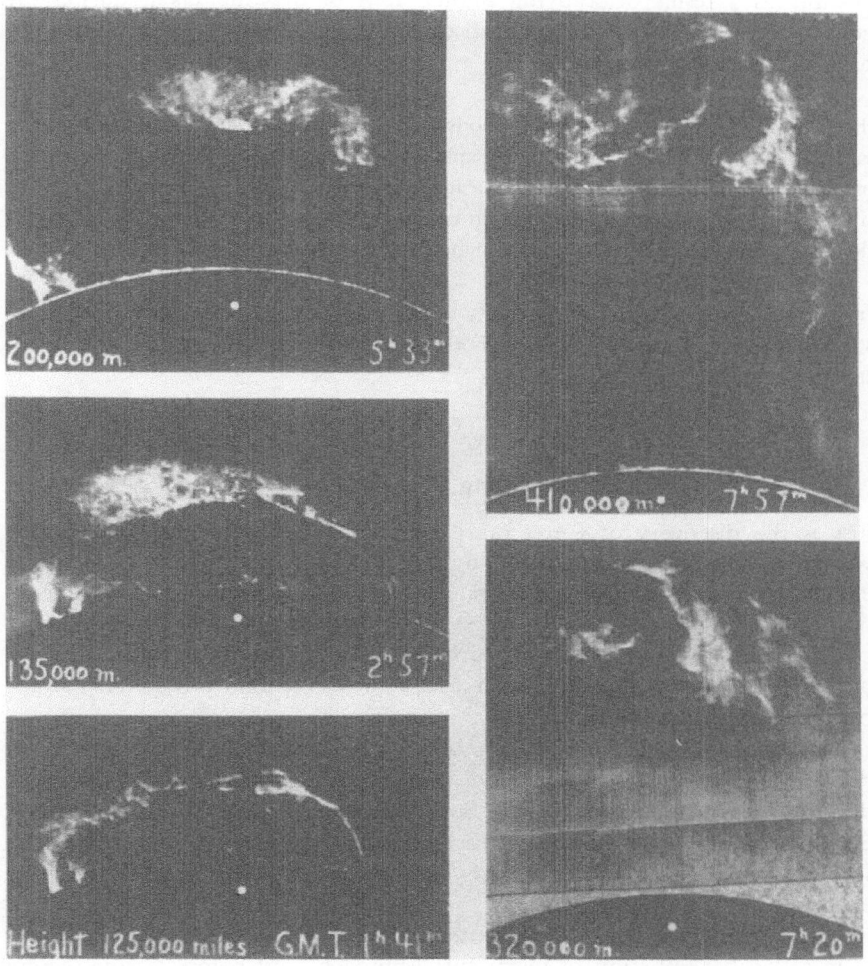

Figure 8. Chamberlin thought that solar prominences indicated the enormous tendency of solar gases to erupt into space, just barely restrained by
gravity under ordinary conditions, and capable of ejecting a planet-forming filament when the Sun's gravity is briefly neutralized by the tidal
force of a passing star. These photographs show the rise of a very high
prominence of 29 May 1919; the dot represents the size of the Earth
(H. N. Russell, *The Solar System and Its Origin*, Macmillan, 1925, facing
p. 101; photographs taken at Yerkes Observatory).

that the resulting orbits would be too eccentric to fit the case of the solar
system in any instance that was likely to occur" (Chamberlin and Moulton
1909; see also Chamberlin D19XX).

In the meantime Chamberlin had become interested in solar eruptions, possibly as a result of studying photographs of the Sun taken during the eclipse of 28 May 1900.[32] He began to study the problem of "disruptive approach," in particular

the projective effect developed in a body of enormous elasticity already under high pressure and affected by violent local explosions which were subject to intensification by the changes of gravity brought to bear on them by a passing body . . . the explosive projections from suns under the influence of the passing body gave a reason for the two-armed feature of most spiral nebulae. (Chamberlin and Moulton 1909: 644)

Just as he was completing an article on this subject, the idea of a catastrophic origin of spiral nebulae received unexpected support from an astronomical event.

1.2.4 Competing theories, 1901–1912

Just as the discovery of Ceres on 1 January 1801 ushered in a century in which knowledge of the smaller members of the solar system was greatly extended, the observation of Nova Persei on 22 February 1901 heralded a period of discovery for stellar astronomy. The *Astronomisches Jahresbericht* recorded 228 publications dealing with Nova Persei or novae in general in 1901, 97 in 1902, 27 in 1903, and 13 in 1904. (Nova Geminorum stimulated another 40 items in 1903.)

Many astronomers attributed novae to the collision of a star with other bodies;[33] the new observations indicated that the nova might subsequently develop into a nebula. Since the Solar System was still thought by almost everyone to have evolved from some kind of nebula, the conclusion drawn by several writers was that the Solar System originated in a stellar collision or encounter of some kind.[34]

Chamberlin had apparently written the following sentence in his paper "On a Possible Function of Disruptive Approach" just before learning about Nova Persei:

That disruptions or explosions of some kind actually take place in the heavens, and that not uncommonly, seems to be implied by the sudden appearance of new stars, often with great brilliancy, followed by rapid decline to obscurity or extinction. (Chamberlin 1901: 21)

32 Tropp (1974). He had earlier remarked on the high speed of gases ejected in solar prominences (Moulton D1899b).
33 Monck (1885). Huggins (1892). Bickerton (1894, 1904). Shaler (1898: 42). Clerke (1902: 397–8). Turner (1903/1970). Campbell (1905: 308). MacPherson (1906: 195). Gregory (1912: 29). Clifford (1925). Waterfield (1938: 84–9).
34 Fiske (1876: 11–14). Gore (1902). Tornow (1902). Campbell (1905). Halm (1905a, 1905b). Meyer (1906: 43–5). Snyder (1907: 421–3). S. Arrhenius (1908: 152). Bickerton (1911a, 1911b). H. Poincaré (1913: 63). See also the collision/encounter theories discussed earlier, 1.2.1.

A footnote to this sentence adds: "A fact which has become very familiar and impressive, since this was written, by the appearance of *Nova Persei*." As Chamberlin pointed out, the probability that two stars actually collide would seem to be much too small to account for the observed frequency of novae; five had been reported in the decade before 1901. But, he argued, collisions are not necessary to produce such an explosion; if the distance of closest approach in an encounter of two bodies is less than the Roche limit, tidal forces will tear apart at least one of them.[35] The fragments will form comets, or in some cases a spiral nebula.

Chamberlin considered several possible types of disruptive encounter between astronomical objects (including both gaseous stars and solid bodies) without explicitly proposing any one of them as a theory of the origin of the Solar System.[36] He presented a more definitive suggestion in a talk to the geology and geography section of the American Association for the Advancement of Science at its meeting in Denver during the last week of August 1901. The talk, "Report on some Studies Relative to Primal Questions in Geology," was not published, but a brief and rather confusing summary of it by E. O. Hovey appeared a month later in the *Scientific American Supplement*. The main point seemed to be that if one believes the Earth was formed from meteorites, it is necessary to explain the fact that some of them contain hydrocarbons and therefore can never have been subjected to the extremely high temperatures one would expect if they originated in the Sun. The solution was to postulate that a small body (comet or planet) passed near a large gaseous or nebular body and was disrupted by the encounter without being subjected to great heat. But the nebula itself would also explode, and its fragments would form a spiral shape (Hovey 1905).

Spiral nebulae now seemed more than ever to be the necessary foundation for the new geology Chamberlin hoped to develop. On 30 September 1901, he wrote to W. W. Campbell, director of the Lick Observatory, asking to see Keeler's latest photographs of nebulae as soon as possible (Chamberlin D1901b). (Keeler had died 12 August 1900; his work with the Crossley reflector was being completed by C. D. Perrine, and was not published in full until 1908.) Campbell's response was prompt and apparently so generous that Chamberlin felt compelled to explain in some detail his need for the photographs. Now that the Nebular Hypothesis had collapsed as a basis for geological reasoning, it was imperative to develop a satisfactory alternative hypothesis, especially since the geological interpretations had been so greatly hampered in the past few decades by deductions based on the Laplacian theory. "I am not a little like the poor heathen Goths of old, who, driven by the Huns and Vandals behind, could not well avoid invading the precincts of the sacred city, however reprehensible it was in itself."[37]

---

35 In a memorandum dated 17 October 1901, Chamberlin (D1901a) noted that he had found somewhat similar ideas expressed in T. J. J. See's book (See 1896, **I**: 258).

36 Nevertheless, it is often stated that the Chamberlin—Moulton theory was first published in this paper.

37 Chamberlin (D1901c). Further letters to Campbell and H. D. Curtis, requesting photographs of nebulae, are in Chamberlin (D1901d).

The guardians of the sacred city of astronomy did not seem particularly offended by Chamberlin's incursion. One of the earliest responses came from George Ellery Hale, who wrote in *Scientific Monthly* for February 1902 that the Nebular Hypothesis

has encountered fresh attacks on the part of Chamberlin and Moulton, and it now seems doubtful whether it will be possible to overcome their criticisms, which are based on dynamical considerations. It may prove to be sufficient, however, to forsake the lenticular mass of vapor predicted by Laplace in favor of the spiral form which Keeler has shown to characterize so many nebulae.[38]

But the same evidence that led some scientists to give favorable consideration to the Chamberlin–Moulton theory also inspired them to think of their own hypotheses. Since I now want to survey some of the alternative theories proposed in the early 20th century before returning to the development of the Chamberlin–Moulton theory, it is appropriate to mention here another piece of evidence that emerged in 1904 to help discredit the Nebular Hypothesis by destroying the regularity of satellite motions usually cited in support of the hypothesis.

Before 1904 it was believed that all satellites of a planet must move in the same direction that the planet rotates. Phoebe, the ninth satellite of Saturn, was discovered by W. H. Pickering in 1899, but he was not able to analyze its orbit until five years later. He found that Phoebe moves in the retrograde direction, unlike the eight satellites discovered earlier (Pickering 1899, 1904b, 1905).

A. C. D. Crommelin, who came to the same conclusion on the basis of a smaller number of observations of Phoebe, stated that the retrograde motion

undoubtedly is unfavourable to the nebular hypothesis, and I anticipate that upholders of that hypothesis will say that the satellite is not an original member of the Saturnian family, but a later capture, which is, of course, possible, though the chances of such a capture are slender indeed.[39]

Retrograde satellites of Jupiter were discovered a few years later. While it could be argued that these discoveries, added to all the other objections, were fatal to the Nebular Hypothesis (Meyer 1910: 60; Moreux 1926: 134–5), they did not have as much impact as one might have expected. Pickering himself thought the apparent irregularities of satellite rotations could be explained by a "tidal inversion" mechanism that would change most but not all spins from

38 Hale (1902: 292). This is also quoted at the end of a memorandum (Chamberlin D19XY) submitted to the Carnegie Institution of Washington to support Chamberlin's application for a research grant. Another astronomer thought that recent nova observations provided evidence for Chamberlin's theory of tidal disruption (Very 1903: 55).

39 Crommelin (1904). One of Crommelin's colleagues immediately fulfilled his prophecy by suggesting a capture explanation: W. T. Lynn (1904).

retrograde to direct as the primary contracted, in agreement with the Nebular Hypothesis.[40]

It seems odd that Chamberlin and Moulton are usually credited with the first 20th-century encounter theory, since two eminent scientists had suggested such a theory, though rather vaguely, before the Chamberlin–Moulton publications of 1905. The Swedish physical chemist Svante Arrhenius in 1901, and the American astronomer Percival Lowell in 1903, proposed that two stellar bodies could collide, forming a nebula from which a planetary system could develop.

Arrhenius[41] developed his ideas on cosmogony in a short article in the *Archives Néerlandaises* (1901) and at greater length in his *Lehrbuch der Kosmischen Physik* in 1903. His approach was somewhat similar to that of Chamberlin; he was interested in the role of atmospheric carbon dioxide in terrestrial climates and glacial epochs,[42] and cited Stoney's kinetic theory calculations to show that the Earth could not have retained atmospheric gases if it had been formed as a hot gaseous ball. He mentioned the possibility that two stars could collide to form a nebula and discussed briefly the formation of planets at condensation points in a spiral nebula. He suggested that the planets will be heated by their contraction, then cool by radiation (Arrhenius 1901; 1903: 221–30).

In 1906 Arrhenius developed his theory in more detail in *Världarnas Utveckling*, soon translated into German and then into English as *Worlds in the Making*. He emphasized the cosmic role of radiation pressure, predicted by Maxwell's electromagnetic theory, recently discovered experimentally by P. N. Lebedev in Russia, and confirmed by E. F. Nichols and G. F. Hull in the United States. According to Arrhenius, radiation pressure pushes dust particles away from stars. The material collects in nebulae, where it is reorganized in an entropy-reducing process similar to that effected by Maxwell's Demon. In this way the universe can be rejuvenated in cycles rather than succumbing to the heat death. Stars, which gradually cool off, will form crusts that keep their energy bottled up inside for billions of years until released by collisions, thus producing a new spiral nebula and a new planetary system.[43]

40 Pickering (1893a, 1893b, 1901, 1904a, 1905/1910). Stratton (1906). Redman (1919). In reply to a criticism by Moulton, Pickering insisted that his theory is not the same as Daniel Kirkwood's (*Nebulous Earth*, 1.3.8). See Pickering (1905), with Moulton's reply and Pickering's rejoinder on subsequent pages.

41 Svante August Arrhenius (1859–1927) was professor at the technical high school in Stockholm, later director of the Physical Chemistry Department at the Nobel Institute in Stockholm. He received the 1903 Nobel Prize in Chemistry for his theory of electrolytic dissociation in solutions. See *Nebulous Earth*, 2.2.8, for his views on the internal state of the Earth.

42 In a long article, T. C. Chamberlin (1899b) made frequent use of the results presented by Arrhenius (1896). But later he protested to C. G. Abbot that he had never favored Arrhenius's theory of glaciation and had been misled by Arrhenius's results (Chamberlin D1913).

43 Arrhenius (1908: 148, 193–4, 206–10). His cosmogony is not even mentioned by many of his biographers, e.g., J. Walker (1929), H. A. M. Snelders (1970). The account by his grandson makes it appear that his theory was monistic; see G. O. S. Arrhenius (1962).

For supporters of the Arrhenius theory see Snyder (1907: 423), Amaftunsky (1909), Henkel (1909), Kaempffert (1909), Musson (1909), Busco (1912: 69–70), Becher (1915).

Lowell was probably the best-known planetary astronomer in the United States at the turn of the century; his speculations about life on Mars caught the public fancy, and his initials were commemorated in the name and symbol of the new planet whose discovery he persistently sought.[44] In 1903, in a book entitled *The Solar System,* he asserted that the Nebular Hypothesis is erroneous because the different parts of a gaseous nebula would not all revolve at the same angular velocity; hence, Laplace's explanation for the common direction of revolution and rotation of the planets fails. Lowell suggested that the angular-momentum distribution might be explained by assuming that the mass of the nebula is concentrated at its center – "This would be the case if the present system had been formed by the collision of two bodies." The collision would scatter meteorites through a large space, and from their distribution Lowell suggested one might explain the retrograde motion of the outer planets.[45]

Lowell's biographers have stated that he supported the Chamberlin–Moulton theory, apparently misled by the usual statement that this theory was published in 1901 (A. L. Lowell 1935: 128; Marsden 1973: 520–3). In fact Lowell attacked the Chamberlin–Moulton theory in 1909, while at the same time elaborating his own somewhat similar theory; he rejected the Chamberlin–Moulton explanation of the production of direct rotation by particle impacts, relying instead on tidal forces to reverse the initially retrograde rotations.[46] Reviewers of his book missed the distinction and thought he was simply presenting the Chamberlin–Moulton theory.[47] Apparently Lowell never advanced any claim for his own priority.

By 1910 several other cosmogonies involving collisions of stars and/or nebulae had been proposed. Only one of them received much notice and that was primarily because of the persistence of its author in publicizing it. In 1897 the American astronomer T. J. J. See pointed out that the Nebular Hypothesis should be revised in the light of Darwin and Poincaré's (and his own) results; nebulae do not shed rings, but split into globular masses (See 1897).[48] See was

44 Percival Lowell (1855–1916) came from a prominent New England family; his younger brother and biographer, Abbot Lawrence Lowell, was president of Harvard. Percival Lowell graduated from Harvard in 1876, traveled in Europe and Asia for several years, then started astronomical observations in the 1890s. He established his own observatory at Flagstaff, Arizona, in 1894, where observing conditions were superior to others he considered in the Western hemisphere. In addition to his controversial observations on Mars, he instigated a search for a planet beyond Neptune. Pluto was finally spotted by Clyde Tombaugh in 1930 at the observatory Lowell had founded. See A. L. Lowell (1935); Strauss (1993) discusses the efforts by Lowell and Edward Morse to promote "planetology."

45 Lowell (1903: 118–30). For earlier remarks favorable to the Nebular Hypothesis see Lowell (1895: 4).

46 Lowell (1909a). The controversy between Lowell, See, and Moulton is summarized by Hetherington (1975). See also Lowell's refutation (1909b) of the planetesimal hypothesis applied to satellite motions. His own theory was presented in Lowell (1909c: 24–30, 142–51) and in a popular article (Lowell 1909d). For its detailed application to individual planets see Lowell (1916).

47 Rolston (1910). Poor (1910). "Dr. Lowell's Cosmogony" (1910).

48 Thomas Jefferson Jackson See (1866–1962) graduated from the University of Missouri in 1889, then earned his doctorate at the University of Berlin in 1892 with a dissertation on the theory of binary stars. This work, based on an extension of G. H. Darwin's tidal theory

responsible for popularizing the phrase "Babinet's criterion" in connection with the angular-momentum objection to the Nebular Hypothesis. I suspect he first learned about Babinet's paper from Agnes Clerke's book on cosmogony, which he reviewed in 1906.[49]

Previously skeptical of theories of the Solar System based on analogies with spiral nebulae (See 1906b), See proposed in 1909 that "our system was formed by the unsymmetrical meeting of two streams of nebulosity or by the mere gravitational settling of a single nebula of curved and unsymmetrical figure, but without hydrostatic pressure as imagined by Laplace." As an example he pointed to a photograph of H. V. 2 Virginis, "a spiral nebula just beginning to coil up and form a system . . . as the mass whirls and condenses under resistance, it will necessarily retain and draw down most of the nebulosity into the plane of motion" (See 1909a; see also 1909b, 1909c). Having previously claimed to have proved that the Moon was captured by the Earth rather than being spun off by it (as G. H. Darwin and others believed), See now argued that the planets were all formed independently in the nebula, then were captured by the Sun into nearly circular orbits with the help of a resisting medium. The planets were gradually built up to their present size over a period of up to a billion years by accretion of dust and meteoric matter.[50]

One other theory must be noted because, though generally ignored in the decade after it was proposed, it was later recognized as a precursor of Alfvén's cosmogony. Kristian Birkeland (1867−1917), a Norwegian geophysicist, studied cathode ray discharges from magnetized globes in connection with the aurora and other solar−terrestrial electromagnetic phenomena. In 1912 he proposed that the Sun emits charged particles into space; some of them cluster into orbits determined by the solar magnetic field and eventually form planets (Birkeland 1912, 1913a, 1913b).

1.2.5 The planetesimal hypothesis

Chamberlin first introduced the term "planetesimal" in a paper presented to the Washington meeting of the Geological Society of America on 1 January 1903. In the published abstract he distinguished his theory from "the Lapla-

of rotating liquid masses, established See's reputation. He conducted astronomical research at the University of Chicago and the Lowell Observatory; in 1899 he was appointed professor of mathematics in the U.S. Navy and served most of his career as director of the Observatory at Mare Island, California. He failed to attain the more prestigious posts to which he aspired and later became known as a crackpot for his attacks on Einstein and for his own unorthodox theories. There is an extensive account of his early career by Webb (1913) and a short article by Ashbrook (1962); a comprehensive biography is being prepared by Charles J. Peterson at the University of Missouri, Columbia.

49 Babinet (1861). Clerke (1905: 44). See (1906a). André (1912). Jeans (1919: 272, 275). For further details on Babinet see *Nebulous Earth*, 1.3.10.

50 See (1909d, 1909e). A comprehensive exposition of See's theory may be found in his book: See (1910a). Qualitatively similar theories had been proposed earlier by F. B. Taylor (1898) and A. Gareis (1901).

     See's theory was supported by Henkel (1910a, 1910b), Kerfoot (1912, 1665−76), Mumford (1912), Dean (1913), Wallace (D1913), MacPherson (1921).

cian and other gaseous hypotheses, and from the meteoroidal hypothesis as set forth by Lockyer and Darwin." In those theories, "the aggregation is the simple work of gravity," whereas "in the planetesimal hypothesis the aggregation is dependent on orbital conjunction" and is relatively slow, involving much lower temperatures. "Planetesimals" were defined simply as "infinitesimal planetoids" from which the planets grew by accretion (Chamberlin 1903a).

As might be expected the discussion of Chamberlin's theory on this occasion centered on its geological aspects.[51] Harry Fielding Reid (later known for his "elastic-rebound" theory of earthquakes, and a severe critic of the Chamberlin–Moulton theory 21 years later) asked friendly questions about possible evidence from rocks and from the study of other planets. George Perkins Merrill, an expert on meteorites as well as geology, objected that acid rocks in the Earth's crust did not have the same character as meteorites, to which Chamberlin replied that "the hypothesis was not meteoroidal, but nebular. That he considered meteoroidal material a negligible quantity."[52] Grove Karl Gilbert, another prominent geologist, "cited the results of his study of the Moon as showing the effects of the impact of masses falling upon it, and supporting in this way the hypothesis."[53]

By this time Chamberlin realized that he needed more help in developing his theory than Moulton could give, and was beginning to call on other Chicago scientists for assistance. He wrote to Robert A. Millikan on 19 December 1902, asking about a problem involving the cooling of the Earth, and received a quick reply citing Lord Kelvin's solution of the problem in 1862 (Millikan D1903). But the people he really needed to work on his research programme were for the most part occupied with their own projects; presumably most graduate students in the Geology Department did not have the required expertise in mathematics, physics, and astronomy to solve the special problems involved in working out the planetesimal hypothesis.

Fortunately the means for organizing an interdisciplinary research team were at hand. The Carnegie Institution of Washington had recently been established to finance scientific research, and one of its trustees was Chamberlin's friend Charles D. Walcott, director of the U.S. Geological Survey. On 23 January 1902, Chamberlin wrote to Walcott, alluding to an earlier conversation – "Since you were kind enough at Rochester to mention what you had in mind for me in connection with the Carnegie Institution" – and requesting a grant that would allow him to reduce his commitment to teaching and administrative duties as well as to hire research associates.[54] Enclosed with the

---

51 The discussion was reported following the account of Chamberlin's .talk, under the title "The Origin of Ocean Basins on the Planetessimal [sic] Hypothesis" (Chamberlin 1903b). Chamberlin soon dropped the extra "s"; as he explained in a letter to H. L. Fairchild, "The analogy of 'infinitesimal' permits the omission of one 's,' and this is in the line of simplification and progress" (Chamberlin D1903c).

52 Quoted from *Science*, **17** (1903b): 301. For Merrill's later views on this subject see his (1909).

53 Quoted from *Science*, **17** (1903b): 301.

54 Chamberlin (D1902c). A copy of Walcott's reply and other correspondence may be found in the file of Chamberlin correspondence at the Carnegie Institution of Washington (CIW).

letter was a detailed memorandum, outlining a number of related problems in geophysics and astronomy.[55]

Walcott's response was favorable, though the negotiations proceeded rather slowly (detailed procedures for handling grant applications had not yet been worked out), and it was not until 2 January 1903 that Chamberlin submitted a formal application to the Carnegie Institution for a grant of $6000 (Chamberlin D1902b). Walcott immediately approved the application, and Chamberlin informed him about how the money was to be spent in a letter dated 10 January 1903.[56] Moulton was to work on the project half-time at $1000 per year; L. M. Hoskins of Stanford would "undertake certain computations relative to the competency of certain shells of the Earth to *accumulate stresses;*"[57] C. S. Slichter of the University of Wisconsin would study the "changes of form of the Earth due to supposed changes in the rate of rotation and the geological consequences of this,"[58] and A. C. Lunn of the Mathematics Department at Chicago planned to investigate the "development and distribution of the heat of the Earth's interior."[59] Hoskins, Slichter, Lunn, and others recruited later were paid on a per diem basis. Chamberlin himself received $3500, half of his annual salary, from the grant.[60]

The relationship between Chamberlin and the Carnegie Institution was mutually beneficial though not without some friction and misunderstanding. By 1910 Chamberlin had received a total of $34,000, more than any other grantee; the next highest amount was $33,000 to Simon Newcomb (Woodward D1910a). But Chamberlin frequently resented the need to make an annual plea for funds, feeling that once the importance of his work had been recognized he should be able to count on automatic renewal of his grant without more paperwork (Chamberlin D1910a, 1910b; Woodward D1910a, 1910b). The Carnegie Institution gained the public credit for having supported one of the major U. S. scientific achievements of the early 20th century; the first detailed statement of the complete Chamberlin—Moulton theory was published in the institution's *Yearbook* for 1904 (Chamberlin 1905).

55 This is probably the undated "Memorandum" (Chamberlin D19XY) mentioned earlier; however, that memorandum contains a quotation from the February 1902 issue of *Popular Science Monthly* (Hale 1902), so one would have to assume that this article was available to Chamberlin before the nominal publication date in order to accept this identification.

56 Chamberlin (D1903e). The letter begins, "Yours of the 6th announcing that my application for a grant has been approved, is at hand."

57 Chamberlin (D1903e: 1). Correspondence on Hoskins's work for the project may be found in Letterbooks XVI and XVII, and in Addenda Box II, folder 4, C-UC. See also Chamberlin (D1903f). Leander Miller Hoskins (1860–1937) had taught civil engineering and mechanics at Wisconsin from 1885 to 1892 and was professor of applied mathematics at Stanford from 1892 to 1925.

58 Correspondence with Slichter on this problem may be found in Letterbooks XIV, XVI, and XVII, and Addenda Box III, Folder 5, C-UC. Charles Sumner Slichter (1864–1946) was educated at Northwestern University. He taught mathematics at the University of Wisconsin, with a special interest in the motion of underground water.

59 Lunn had been mentioned in Chamberlin's earlier letter to Walcott (D1902a) as "a rising young mathematician of remarkable promise." Arthur Constant Lunn (1877–1949) taught applied mathematics at Chicago starting in 1903; he was promoted to professor in 1923.

60 Walcott (D1902b). Chamberlin (D1903g, D1908c, D19XZ).

Chamberlin also served on the Geology Advisory Committee and gave the institution the benefit of his long experience in conducting and administering scientific research; much of his advice is recorded in correspondence with Robert S. Woodward, president of the Carnegie Institution from 1904 to 1920.[61] However, Chamberlin was not able to prevent a major change in policy that eventually denied Carnegie funds to most academic scientists; Woodward apparently found that the aggravation of dealing with individual applicants for grants was too burdensome and decided that the endowment should be used primarily to support a small number of major laboratories directly administered by or affiliated with the institution.[62] The grants to Chamberlin and Moulton were continued after this policy change had been set, though on at least two occasions permission to hire relatives as assistants was denied,[63] and in another case the existence of a large unexpended balance from a previous grant was used to justify the denial of a new allotment, much to Chamberlin's dismay.[64]

At the same time that he was contracting for geophysical calculations to support his planetesimal theory, Chamberlin continued to pursue the astronomical side with Moulton. On 12 January 1903, he wrote:

On mulling over our problems, I am wondering whether the best step to take first may not be to prepare a preliminary discussion of *the possible methods of nebular aggregation and their limitations*. The purpose would be to clear up the ground and present some general fundamental statements worked out in mathematical form which would serve as an introduction and groundwork to the more special problems to be taken up afterwards. The proposition would include a more elaborate treatment of the limitations of gaseous and quasi-gaseous aggregation which appeared in your paper on the nebular hypothesis . . . it would be an extension of celestial dynamics in a field so far, I think, much neglected.

There would, I think, be two rather distinct classes of aggregations to be considered, first, those in which the constituents were moving in essentially parallel concentric orbits and were only influenced by the differential attraction of the central body (and slightly of other bodies). Second, those in which there was kinetic action between the constituents within the aggregation. . . .

The proposed discussion might well perhaps include a statement of the planetoidal or planetesimal method of aggregation, with certain general mathematical determinations relative to its applicability and limitations. I do not think there has ever appeared

---

61 Walcott (D1902a). Chamberlin (D1902b). Walcott (D1902c) (on committee appointments). Chamberlin (D1902c).
62 Woodward (D1906). Chamberlin (D1906b). For detailed discussion of the Carnegie Institution's change in policy see Reingold (1979). Reingold's interpretation is that Woodward wanted the Carnegie Institution to be not merely a "disbursing agency" but a real participant in research.
63 Moulton (D1905) (on hiring Moulton's brother). Chamberlin (D1909a, D1909b). Woodward (1909a, 1909b) (on hiring Chamberlin's son Rollin).
64 Woodward (D1910b, D1910c). Chamberlin (D1910a) (the denial has "wrecked my plans" to recruit a new collaborator).

in print a full and explicit statement of this doctrine. I have never seen any statement of it in print, except those which I have myself made, and the fact that Fay[e] in his discussion of the origin of the Earth entirely ignores it and drew his conclusions without regard to it, and the fact that these conclusions seem to have been generally accepted, or at least to have escaped criticism among astronomers, leads me to think that the doctrine is essentially new. (Chamberlin D1903h)

By the end of 1903 Chamberlin had consolidated his ideas about the initial disruptive encounter of the Sun with another star, and the shapes of the orbits into which the fragments torn from the Sun would be thrown. Moulton was in poor health during this time, spent the winter of 1903–4 in New Mexico, and thus was not able to contribute as much as might be expected to developing this part of the theory; but he reviewed all of Chamberlin's ideas and drafts by correspondence and thus helped prevent serious technical errors (Chamberlin D1903i; Moulton D1903, D1904a).

In February 1904 Chamberlin was invited to present an account of his theory to the Washington Academy of Science (Chamberlin D1904a, D1904b). Chamberlin later considered this to have been the "original announcement of the planetesimal hypothesis" and recalled that it was understood that his theory would be discussed "by Simon Newcomb, who was then regarded as the foremost mathematical astronomer in the country, by Grove Carl Gilbert, one of the most trusted geologists in the country, and by George E. Becker, who stood for a man using mathematics to a considerable extent" (Chamberlin D1928a: 1).

The talk at the Washington Academy was given on 29 March 1904. Chamberlin was somewhat disappointed that the three experts did not completely understand what he had done, even though he had sent them each an advance copy of the exposition of his theory in his forthcoming geology text (Chamberlin D1904c, 1904d, 1904e, 1904f). He reported to Moulton:

While the presentation appeared to be satisfactory and to awaken a good deal of interest, the discussion did not amount to a great deal.

Dr. Newcomb had informed me beforehand that he could not undertake to seriously discuss the subject because the pressure of his duties would not permit him to give the previous consideration which was required. He was however present, and was good enough to say that he thought our studies were an advance, and that perhaps our views were the best now available. He discussed some minor points of no special moment. (Chamberlin D1904g)

Becker had written out his criticisms, pertaining mainly to the change in angular momentum of the system; he also stated that the Chamberlin–Moulton theory had been anticipated by Buffon in the 18th century. Chamberlin thought he could dispose of the first objection without difficulty, though the details have not been preserved; and he protested that his own theory was fundamentally different from Buffon's (Chamberlin D1904h).

Becker was an exponent of Kant's ideas (Becker 1898), and in his address to the Congress of Arts and Science at St. Louis on 21 September 1904, he asserted that "every attempt to devise an essentially different hypothesis [from that of Kant and Laplace] has failed." Chamberlin's assumption that the accretion of the Earth was slow enough to avoid fusion was mentioned briefly but quickly dismissed (Becker 1904).

As for the third expert at the Washington meeting:

Mr. Gilbert's discussion lay chiefly on geological lines, the leading point being that, as he conceived it, the hypothesis of a molten Earth permitted considerable heterogeneity in the distribution of density, while he thought that the accretion hypothesis would give rise to uniformity. Precisely the opposite seems to me to be the case. . . .

On the whole, therefore, the discussion did not contribute much that was helpful to me. Of course the qualified endorsement of Professor Newcomb was gratifying. Many pleasant things were said privately. (Chamberlin D1904g; see also Chamberlin D1904i)

Chamberlin's theory now contained two new features. First, the "ancestral Sun" was assumed to have had a family of planets, swept away by the encounter; second, the other body was several times as massive as the Sun. The connection between these two features is not made clear in published expositions of the theory, but is suggested in Chamberlin's letter to Moulton of 2 March 1904:

The very small ratio of the mass of the planets to the mass of the Sun seems to make it evident either that the Sun did not go anywhere near the Roche limit of the disturbing body, or else that the disturbing body was small. As the postulated ancestral Sun may have had a family of planets, and as there are now no signs of these in the Sun's sphere of influence, it seems best to suppose that the disturbing body was massive enough to have stripped these away, while at the same time it caused the eruption of enough nebulous matter to form a new family.

The preferable case, therefore, as I see it, is one which supposes that the ancestral Sun passed some distance outside of the Roche limit of a rather large body, say some few times as massive as the Sun, and that protuberances to the amount of one, two or three percent were shot forth on the opposite sides of the Sun in the line of differential attraction, with velocities for the outer parts or ends of the arms sufficient to carry them beyond the orbit of Neptune. (Chamberlin (D1904j)

In the report in the Carnegie *Yearbook* for 1904, the two assumptions are explained separately; (1) "The inference that a spiral nebula is formed by a combined outward and rotatory movement implies a preexisting body that embraced the whole mass. In harmony with this, an ancestral solar system has been postulated." (2) The other body is assumed to be "several times the mass of the Sun, since it is regarded as a small star."[65]

These two features were in turn connected to a third assumption, explained

65 Chamberlin (1905). Similar remarks are found in the text by Chamberlin and Salisbury (1906, **II**: 51, 55).

in a letter to Moulton on 16 May 1904. Instead of assuming, as he had in his 1901 paper on disruptive approach, that the tidal force due to the other body was the *cause* of matter being ejected from the Sun, Chamberlin now emphasized that the ejection of matter was a natural consequence of the Sun's normal activity; the only function of the other body was to neutralize the restraining effects of the Sun's gravity at two particular locations on its surface (toward and away from the other body), thereby allowing intensified eruptions at those places:

I am dealing with protuberances which are not primarily dependent upon the disturbance of another body, so far as we know, for example, the protuberances which are constantly taking place in the Sun. These are supposed to be true explosions in the Sun, dependent upon conditions not understood. These, in the case of close approach, are supposed to be greatly augmented, but still to remain essentially the same in nature. It is merely their increased magnitude and localization which is due to the change in pressure, not their fundamental production. I was led to this special view by considering the magnitude of the protusions that seem to be required in the case of the solar nebula. As the planets are now only $\frac{1}{700}$ of the whole mass, and as I could not see good reasons for supposing that many times as much of the nebular matter went back to the Sun as went into the make-up of the planets, it seemed that the best working hypothesis was one in which only a very small percentage of the Sun was shot out into the nebulous form. Now this could hardly be the case with general tidal deformation such as I discussed in the *Astrophysical Journal*. That would be applicable to those nebulae in which a large part of the whole mass was dispersed, and a comparatively small portion remained in the central nucleus, as is conspicuously true in some cases, judging from the photographs.

. . . it is obvious that the solar case is one of *distant* approach. The Sun could not have gone anywhere near the Roche limit of the disturbing body; otherwise, a very much larger proportion of the Sun's mass would have been shot out into the nebular form. The solar case must have been one in which the Sun's normal protuberances were merely intensified and localized. (Chamberlin D1904k)

A further consequence of the smallness of the disruption was that the matter ejected first – and presumably shot farthest from the Sun – would originate in the surface layers of the Sun, while the later ejecta would come from the lower portions. If one assumes that elements of higher specific gravity have sunk toward the center of the Sun, this would offer a possible explanation for the fact that the outer planets have lower densities than the inner ones (Chamberlin 1905; Chamberlin and Salisbury 1906: II, 58).

But the real difficulty still remained: What would happen to the ejected matter after the perturbing body had left the scene? Would all or most of it simply fall back into the Sun, or would it go into orbits around the Sun? It seemed clear that the action of the perturbing body on the ejected matter would impart some kind of rotation, simply because the perturbing body would be moving past the Sun and thus its gravitational force at a later time would have a component transverse to the line of the original tidal action.

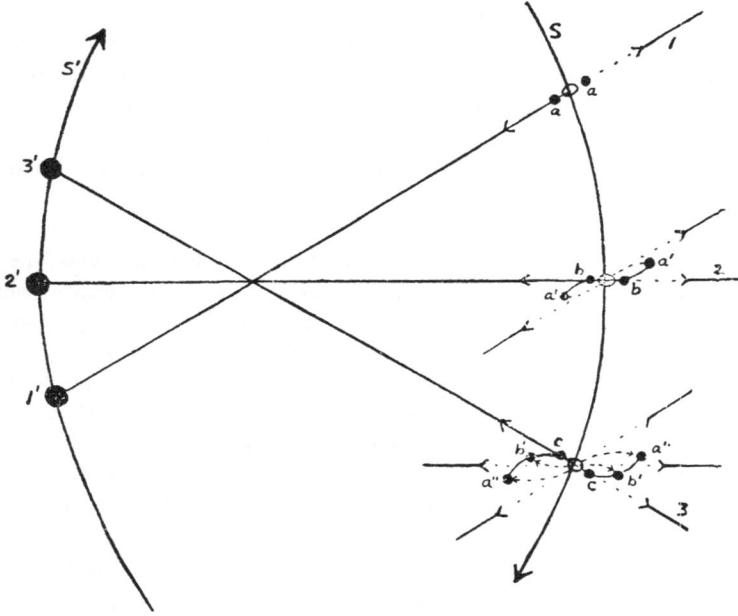

Figure 9. Chamberlin's diagram showing the development of a spiral nebula from the action of intruder star S' on the Sun S, at successive times 1, 2, 3 (T. C. Chamberlin, *Origin of the Earth*, University of Chicago Press, 1916, p. 122).

Moulton (D1904b) explained how the orbit could be computed by stepwise approximation, but admitted that in problems of this kind "the results are often the opposite of what one would naturally expect" (D1904c).

At the end of September 1904 Chamberlin submitted to the Carnegie Institution his first comprehensive report on the complete Chamberlin–Moulton theory (Chamberlin 1904l). The outline of events envisaged in the theory was now as follows: (1) the ancestral Sun is approached by another star (Moulton later calculated that it needs to come only as close as the orbit of Jupiter); (2) the tidal force of the intruder neutralizes the gravitational force of the Sun at two opposite points on its surface, allowing internal expansive forces to push out material in two protuberances; (3) as the intruder star swings around the Sun and goes off into space it draws these protuberances out into two spiral arms; (4) the arms contain "knots" due to irregular pulsations in the process of emission from the Sun; these form nuclei for condensation of material into planets and satellites; (5) rapid cooling of the ejected matter by radiation from its surface reduces its refractory components to the liquid or solid state (planetesimals); (6) knots and planetesimals move in Kepler orbits around the Sun; (7) planetesimals are gradually captured by the knots; (8) growth of Earth and

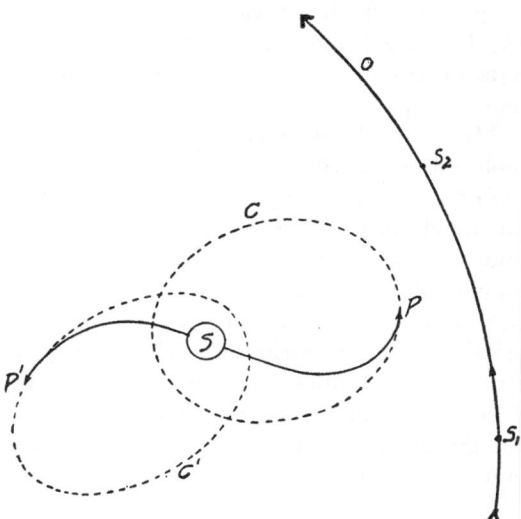

Figure 10. Chamberlin's diagram showing development of orbits of portions of material in the filament left near the Sun S, under the gravitational influence of the receding star at successive positions S_1, S_2 (T. C. Chamberlin, *Origin of the Earth*, University of Chicago Press, 1916, p. 123).

other planets proceeds as described earlier, with slow generation of heat by gravitational contraction.

Robert S. Woodward, who was about to succeed D. C. Gilman as president of the Carnegie Institution, expressed the first reaction to the theory in a letter to Charles D. Walcott on 2 November 1904, after reading the proofs of Chamberlin's report for the Carnegie *Yearbook:*

Naturally such work must be, in its inception, somewhat speculative, and hence lacking in the definiteness to be expected in a completed investigation. However, the work foreshadowed in the report is not only very interesting but also very suggestive and far-reaching in its scope. It must rank, I think, with the work of Laplace, Roche, Croll, Kelvin, and G. H. Darwin in the domain of cosmogony. (Woodward D1904)

Woodward also praised the work of Chamberlin's associates Moulton and Lunn: "They come nearest to being successors to our distinguished George W. Hill of any men in America at the present time."[66]

66 Woodward (D1904). George William Hill (1838–1914) was the leading mathematical astronomer in the United States at the time, equalled only by Simon Newcomb; he was appointed to the Rutherford chair of astronomy at Columbia (where Woodward was dean of the College of Pure Science) but resigned because there were so few qualified students to teach.

Chamberlin's report was published early in 1905 in the *Carnegie Institution Yearbook for 1904*.[67] It was a handsome dividend for the foundation that was investing heavily in his research. This form of publication called attention to the Carnegie sponsorship; curiously (from the modern viewpoint) neither Chamberlin nor Moulton acknowledged the source of financial support for their work in publications elsewhere.

Continuing the policy of separate publication, Moulton sent a paper under his own name to the *Astrophysical Journal* on 30 August 1905; it appeared in the October issue. He presented a brief discussion of the origin of spiral nebulae in encounters of two stars, not mentioning Chamberlin's assumption that the other star has a mass several times that of the Sun (see earlier), but promising numerical treatment of various special cases in a future publication.[68] Scheiner's observation of dark lines in the spectrum of Andromeda (Scheiner 1899) was cited in support of the assumed condition of nebulae, but some uncertainty about the alleged relation between spiral nebulae and planetary systems had begun to creep in:

Doubtless those spirals which have been photographed are immensely larger than the one from which our system may have developed, and as a rule have relatively less massive centers. (Moulton 1905: 169)

Most of Moulton's paper dealt with the orbits and rotations or the planets formed from the ejected matter. He argued, on the basis of approximate calculations, that the absorption of small particles by planetary nuclei would in most cases decrease the eccentricities of their orbits. The larger planets would thus be expected to have more nearly circular orbits. The Nebular Hypothesis, according to Moulton, predicts that the planets nearest the Sun would have the most nearly circular orbits because "friction would have destroyed most irregularities" (Moulton 1905: 172).

The eccentricity test proposed by Moulton does not seem to me to be very conclusive. He says that "the orbits of the terrestrial planets average more than twice as eccentric as those of the great planets" (Moulton 1905: 172). Using the data in Chamberlin and Salisbury (1906: 79), I find the average to be 0.079 for the terrestrial planets, compared with 0.04 for the great planets.

67 Chamberlin (1905). The fact that this is called the "Yearbook for 1904" has obscured the fact that Chamberlin's paper was not actually published until 1905.
68 Moulton (1905: 168). The details of these calculations were never published. He mentioned the difficulty of the work in a letter to Woodward, 3 August 1905 (Moulton D1905). The results were said to be qualitatively favorable to the hypothesis in Moulton's "Report" (1906b); see also Chamberlin (D1905b). In his report for 1906, Moulton stated that he had computed the orbits of material ejected from the Sun in 48 cases, following them for about five years each time; in most cases the disturbing star's closest approach to the Sun was 5 AU. See Moulton (1907: 168–9). But three years later Moulton reported that his work on this problem was not yet complete (1910: 225–6). I have not made a systematic search for possible later publications since Moulton himself told H. N. Russell that "in nearly 30 years I have written only three times on the Planetesimal Hypothesis, and in two of those three times the material appeared in my 'Introduction to Astronomy' " (Moulton D1929a).

Whereas Moulton points to Mercury, the smallest planet, as having the greatest eccentricity, and points out that some of the asteroids have even more eccentric orbits than Mercury, he does not mention the anomalously small (on his theory) eccentricity of Venus, or the anomalously large eccentricity of Jupiter and Saturn. On the other hand, the predicted correlation between eccentricity and distance from the Sun that he attributed to the Nebular Hypothesis was not generally considered a consequence of that hypothesis by its supporters, though Moulton had suggested this in 1900 (Moulton 1900: 108).

Moulton repeated Chamberlin's argument that inelastic collisions of particles moving in Kepler orbits would tend in most cases to give a forward rotation to the resulting planets; moreover, one would expect to find the larger planets rotating more rapidly because there have been more such collisions. He did not indicate whether this expectation was fulfilled, perhaps because of the uncertainty of existing data; modern data indicates that there is a rough correlation of this type, if one ignores the *direction* of rotation.[69]

In the Chamberlin—Moulton theory, satellites were formed by accretion of planetesimals by secondary nuclei that could have revolved in any way around the primary nucleus that formed the planet. However, the orbits of retrograde secondary nuclei would tend to become *more* eccentric and would thus tend to fall into the primary nucleus unless they moved at a considerable distance from it. Any surviving retrograde satellites would therefore be expected to be on the outside of the system and to have highly eccentric orbits; this was indeed true of Phoebe (Saturn IX) and the recently discovered seventh satellite of Jupiter. On the other hand, the behavior of Phobos, which was an anomaly for the Nebular Hypothesis, presented no difficulty: "There is no reason why a satellite may not revolve more rapidly than its primary rotates, especially as its period continually diminishes as the mass of the planet increases."[70]

Moulton devoted the last part of his paper to an attack on W. H. Pickering's attempt to explain the retrograde rotation of Phoebe in a manner consistent with the Nebular Hypothesis (1.2.4). But he misunderstood Pickering's theory, thinking it to depend on a tidal slowing and inversion mechanism like the one suggested originally by Kirkwood (1869), whereas what Pickering had in mind was a simple tipping over of the axis of rotation. Pickering's theory survived Moulton's attack, at least for a few years, and thus held out some hope of rescuing the Nebular Hypothesis.[71]

Moulton concluded his 1905 paper by suggesting some philosophical consequences of what he called the "spiral theory." Previously it was thought that

69 The rotation periods for the planets in order of mass are: Jupiter 9 hours 50 minutes 30 seconds; Saturn 10 hours 14 minutes; Neptune 16 hours; Uranus 11 hours (retrograde); Earth 23 hours 56 minutes 4 seconds; Venus 243 days (retrograde); Pluto 6 days 9 hours; Mars 24 hours 37 minutes 23 seconds; Mercury 59 days.

70 Moulton (1905: 175). According to Newtonian gravitational theory the period of revolution of a satellite varies inversely as the square root of the mass of the primary.

71 Turner (1909). A. S. Eddington (1909/1970) regarded the discovery of a retrograde eighth satellite of Jupiter as confirmation of Pickering's theory.

cosmic processes tended to aggregate matter into larger bodies while dispersing energy; but we now recognize tendencies toward dispersion of matter. This suggests that the evolution of celestial bodies may have a cyclic character, though this would require the discovery of new types of energy sources, possibly involving atomic processes, to supplement the energy available to the stars from Helmholtzian contraction (Moulton 1905: 181).

1.2.6 Reception of the planetesimal hypothesis

The announcement quoted at the beginning of 1.2.2 suggests that Chamberlin felt he had reached a milestone in his research program by the end of 1905. He sent a copy not only to Frost at Yerkes but also to Agnes Clerke, the British historian of astronomy, together with an account of the development of the Chamberlin–Moulton theory (Chamberlin D1906c). He also prepared a memorandum specifying the contributions of Moulton and himself to each stage of this development. As he recalled 22 years later, he "thought there might be some question of Who's Who" and wanted to be sure the record was clear.[72]

Not that Chamberlin thought, at age 62, that his life work was complete and was preparing to retire from science. On the contrary, he told R. S. Woodward, "I feel an intellectual vigor and spontaneity that I have rarely felt even in younger days and problems open out with wonderful vividness and fascination" – and he was in much better health than his younger collaborators.[73] But he did seem to think that the astronomical side of his theory had been placed on a sufficiently firm footing that he could turn his efforts to developing the geological side again. It is therefore appropriate now to consider how astronomers regarded the theory during the first decade or so of its existence.

As early as February 1904 Chamberlin told a correspondent that the planetesimal hypothesis was getting to be a fad, and there is no doubt that it enjoyed some popularity among those who had a general interest in science.[74] Early in 1906 he reported to Woodward that Charles D. Perrine and Agnes Clerke supported his theory.[75]

72 Chamberlin (D1928b); see also Chamberlin and Moulton (1909).
73 "As implied in our reports, the planetesimal hypothesis was worked out by myself in essential independence of Dr. Moulton, as he was compelled, on account of his health, to spend a year in the west, and I hesitated to tax him under the circumstances . . . the health of Dr. Lunn during the year has been such as to unfit him for the very difficult mathematical inquiry on which he had made good progress last year. He has been on the verge of a nervous breakdown growing out of overstrain in connection with an educational experiment under adverse auspices" (Chamberlin D1905b).
74 Chamberlin (D1904m). Upham (1905). Snyder (D1906). One of the most enthusiastic early suporters of the planetesimal hypothesis was the American geologist H. L. Fairchild; in a letter to Moulton, Chamberlin (D1904j) credited Fairchild with stimulating interest in the theory.
75 Perrine wrote, according to Chamberlin, "I think you have hit the pay-streak," and, Cham-

It must have been a source of some irritation to Chamberlin that E. B. Frost, whose advice he had sought and to some degree accepted,[76] refused to accept his theory. In 1904 Frost wrote:

I am not converted to the planetesimal theory, but it seems to me that your case is very strongly put; and so safely that your opponents would have still to drive you out from a whole series of "earth works" after, or if, they had rendered your position untenable in some of them.

The greatest difficulty to me is that any form of planetesimals, except molecules, seem rather to be products of disintegration than of integration. (Frost D1905)

In 1911 Frost denied that the Nebular Hypothesis had been discarded or supplanted, though he admitted "this view is shared by many of my friends whose opinion I value." While it needed some modification, "no adequate substitute has been proposed."[77]

In 1906 the magazine *Popular Astronomy* (edited by W. W. Payne and H. C. Wilson) solicited the views of several astronomers on the Chamberlin—Moulton theory. Four responses (by G. C. Comstock, A. Hall, S. Newcomb, and C. A. Young) were published; they seem to provide a good sample of the view of the U.S. astronomical establishment.[78]

berlin noted, "what is perhaps more significant, asks advice relative to lines of investigation"; Clerke "is very favorably impressed by the new views" (Chamberlin D1906a). But Clerke failed to mention the planetesimal hypothesis in her book *Modern Cosmogonies* (1905); Chamberlin suggested that the publisher should paste "Recent" over the word "Modern" on the title page (Chamberlin D1906c: 8). She did remark briefly in another book on Chamberlin's statement that explosive projection could produce the spiral form of nebulae, granting that it may have some truth in it but noting that it involves events on a much smaller scale than those we must ascribe to spiral nebulae (1903: 445).

76 Chamberlin (D1904n); for Chamberlin's comments on Frost's corrections see (D1904o).
77 Frost (1911: 467). In his memoirs Frost noted his early interest in the Nebular Hypothesis and recalled that Chamberlin had been an enthusiastic member of a faculty club to which both belonged, but did not mention the Chamberlin—Moulton theory (Frost 1933: 16, 45, 220).
78 *Popular Astronomy*, 14(1906): 570–2. George Cary Comstock (1855–1934) was educated at the University of Michigan and in 1879 was hired as an assistant to James C. Watson at the Washburn Observatory of the University of Wisconsin. He spent most of his career at Wisconsin, teaching astronomy and writing textbooks; his research was on the improvement of the precision of position measurements by taking account of atmospheric refraction and aberration, and on double stars.
    Asaph Hall (1829–1907) studied at Central College (McGrawville, New York) and in 1862 obtained a staff position at the U.S. Naval Observatory. In 1877 he used the 26 inch Clark telescope there to discover the two satellites of Mars. He observed other planetary satellites and double stars. After his retirement in 1891 he taught celestial mechanics at Harvard.
    Simon Newcomb (1835–1909) was born in Canada and came to the United States at age 18. Self-educated, he joined the Nautical Almanac Office in Cambridge in 1857, as a computer, and studied mathematics with Benjamin Peirce at Harvard. In 1861 he was appointed to the U.S. Naval Observatory, where he organized obvservations and undertook theoretical work on the Moon and planetary perturbations. In 1877 he was appointed superintendent of the Nautical Almanac Office (which had moved to Washington).
    Charles Augustus Young (1834–1908) graduated first in his class at Dartmouth College at the age of 18. In 1857 he changed his original plans to become a missionary and obtained a position as professor of mathematics and natural philosophy at Western Reserve College in

1. G. C. Comstock "regards it as containing very valuable material for the improvement of our theories of cosmogony. He understands that authors do not regard the theory as having taken its definite form, and that it stands now substantially as a report of progress upon their problems. As such he thinks it is extremely interesting and valuable."

2. Asaph Hall "thinks the genesis of the solar system a vast subject. Not even Laplace knew much about it, as he himself says."

3. Simon Newcomb "says this subject is one lying so far outside his sphere of work that he has not given any study to it. In scientific standing and ability, the authors are of the best, and what they say is worthy of serious consideration; but Professor Newcomb still retains a little incredulity as to our power in the present state of science to reach even a high degree of probability in cosmogony."

4. C. A. Young "has given some consideration to the 'Planetesimal hypothesis' of the genesis of our solar system and he is disposed to regard it favorably though he has not given it careful study."

Three other American astronomers were somewhat less cautious. Robert G. Aitken supported the new theory; Charles Lane Poor thought that the planetesimal hypothesis avoided many of the difficulties of the Nebular Hypothesis and is "undoubtedly the most satisfactory theory yet advanced"; S. D. Townley judged Laplace's theory "no longer tenable" in the light of the Chamberlin–Moulton work.[79]

George Ellery Hale, whose opinion Chamberlin valued highly, accepted as conclusive the Chamberlin–Moulton criticism of the Nebular Hypothesis; he gave considerable publicity to the planetesimal hypothesis though without granting it an explicit endorsement.[80] J. M. Schaeberle and T. J. J. See were skeptical about Chamberlin's interpretation of spiral nebulae.[81] Chamberlin

Hudson, Ohio. He returned in 1866 to Dartmouth to occupy the chair previously held by his father and grandfather, then in 1877 moved to the College of New Jersey (now Princeton University) where he stayed until his retirement in 1905. He conducted research on the solar spectrum and wrote several widely used astronomy textbooks.

79 Townley (1905). Aitken (1906). Poor (1908: 304).

80 Hale (1908: 182–6 and ch. 21). Chamberlin thanked Hale for his favorable account in a letter (D1908a). See also Hetherington (1994). George Ellery Hale (1868–1938), son of the man who manufactured "the hydraulic elevators that would make possible the tall buildings of the new Chicago" (DSB), studied solar prominences at the Harvard College Observatory while earning his B.S. at MIT. In 1892 he was appointed associate professor of astrophysics at the new University of Chicago and in 1897 founded the university's Yerkes Observatory at Williams Bay, Wisconsin. In 1895, with James Keeler, he founded the Astrophysical Journal. In 1904 he obtained funds from the Carnegie Institution of Washington to establish the Mount Wilson Solar Observatory near Pasadena, California. The 60 inch telescope installed there in 1908 allowed research on stellar spectra; it was followed by the 100 inch telescope erected in 1917. Thus, Hale had a major impact on astronomy by providing the tools for others to do important research; each of the three observatories he founded (Yerkes, Mt. Wilson, and Palomar) was "in its time the greatest in the world" (DSB). He also helped to transform Throop Polytechnic Institute into the California Institute of Technology.

81 Schaeberle (1906); he thought Chamberlin's invocation of the action of another body to account for the ejection of streams of matter from stars was unnecessary. T. J. J. See (1906b);

was not unhappy that See eventually decided to reject the planetesimal hypothesis in favor of his own "capture" theory,[82] apparently feeling that it would be safer to have such an eccentric thinker identified as an opponent rather than as a supporter (Chamberlin D1906d, D1906e). On the other hand, Percival Lowell, who voiced strong objections to Moulton's work in 1909, was sometimes regarded as a supporter of the Chamberlin—Moulton theory because of the similarity of some of his own ideas (1.2.5).

The Chamberlin—Moulton theory probably achieved its highest degree of acceptance among U.S. astronomers around 1915, as indicated by a review article published by W. W. Campbell. Campbell agreed with Chamberlin that the 7° inclination of the Sun's plane of rotation from the principal plane of planetary orbits is a fatal objection to the Nebular Hypothesis (1915: 192). He favored the planetesimal hypothesis for the origin of the Solar System, though he did not think the Chamberlin—Moulton theory could account for most spiral nebulae (1915: 238–44). Articles and books by H. N. Russell, Harlow Shapley, J. C. Duncan, and others supported the planetesimal hypothesis into the 1930s.[83]

In Britain the reception of the Chamberlin—Moulton theory was less favorable but by no means entirely hostile. G. H. Darwin, whose views on cosmogony undoubtedly commanded great respect, mentioned Chamberlin's *Carnegie Yearbook* paper briefly in his Presidential Address to the British Association in 1905. He admitted that the Nebular Hypothesis had encountered serious difficulties, but still thought many of them could be resolved by combining it with the meteoric conceptions he had presented earlier (Darwin 1888). He agreed with Chamberlin that

it is difficult to understand how a swarm of meteorites moving indiscriminately in every direction could ever have come into existence. But my paper may have served to some extent to suggest to Chamberlin his recent modification of the Nebular Hypothesis, in which he seeks to reconcile Laplace's view with a meteoritic origin of the solar system.[84]

Chamberlin, who had recently been going through Darwin's 1889 paper on meteoric cosmogony with a critical eye (D1904p), and was to question the validity of some of Darwin's work on tides (Chamberlin D1907a, D1908b, D1908d, D1908e), must have been indignant about this misrepresentation of his work, but he made no public protest. In fact Chamberlin was unusually

See criticized speculations on the condensation of spiral nebulae to systems of stars, an idea he considered "quite unsound" though it is "even given place in one treatise of Geology."

82 See 1.2, note 46; also See (1909f, 1909g, 1909h, 1910b, 1910c, 1911a, 1911b, 1912, 1914).
83 Kippax (1914: 260). Jacoby (1915: 358–60). Buchanan (1916). Lewis (1919: 14–15). McNairn (1919). Todd (1922). Russell (1924). Duncan (1926: 370–5). Russell, Dugan, and Stewart (1926: 370–5; 1930; 1935). Shapley (1926: 100). Wood (1927: 247). Mather (1928: 18–21).
84 Darwin (1905: 20). This passage was omitted from the version Darwin published in *Science*, **22** (1905): 257–67. Reprints, extracts, and translation of Darwin's address appeared in several other journals.

deferential to Darwin, both in public and in private.[85] If he was confident that the eminent British mathematician would eventually see the light, his patience was rewarded, for in 1909 Darwin published an article on cosmogony presenting a much more accurate and generally favorable account of the planetesimal hypothesis. Darwin concluded that article by saying that Chamberlin and Moulton "are to be congratulated on having advanced views of extraordinary interest, and whether the theory be sound or not in all its parts they have made a contribution to cosmogony of great importance" (1909: 932).

Several British authors of popular works on astronomy welcomed the Chamberlin–Moulton theory. Walter W. Bryant wrote, "The weight of evidence has steadily accumulated in favour of the planetesimal hypothesis" and that it is "likely to hold the field in the near future" (1907: 128, 246). Cecil Dolmage supported it in a survey of *Astronomy To-Day*.[86] J. Ellard Gore greeted the theory with enthusiasm; he thought it "far superior to Laplace's Nebular Hypothesis which should now be definitely abandoned and consigned to the limbo of disproved theories."[87] Hector MacPherson, Jr. (1909) thought the planetesimal hypothesis contained "valuable truths," though the Nebular Hypothesis should not be completely abandoned.

The other professional British astronomer who wrote on the Chamberlin–Moulton theory soon after its publication, F. J. M. Stratton, judged that the U.S. team was responsible in part for "the general abandonment" of the Nebular Hypothesis "in anything like its original form by most astronomers of the present day." But he was not enthusiastic about the planetesimal hypothesis:

In many respects it gives a general *qualitative* agreement with observed facts, while its supporters are criticising older theories on the ground that they lack at times a close *quantitative* agreement with observed facts. It remains to be seen whether the new theory will come up to the standard by which the older theories are being judged. (Stratton 1909: 102–3)

In a review article (1910) Stratton wrote that the Chamberlin–Moulton theory had been applied

with but moderate success, to explain the equatorial acceleration of the Sun, the eccentricity of retrograde orbits, and other phenomena. Of necessity these applications are somewhat vague, and further evidence in favour of the hypothesis seems badly wanted. Spectroscopic evidence from the study of actual spiral nebulae, of such motions as are indicated by this hypothesis, would give valuable quantitative support to a theory which rests mainly on a basis of qualitative constructive reasoning and quantitative destructive reasoning.

85 Chamberlin (D1908b; 1916), see references to Darwin given in index.
86 Dolmage (1909: 335–6). According to the preface, dated 1908, the author died before this edition was published.
87 Gore (1906); quotation from Gore (1907: 334).

Arthur Holmes, the British geologist later known for his research on the age of the Earth and on continental drift, was an early supporter of the planetesimal hypothesis, calling it the most successful "of all attempts to grapple with the fundamental problem of the genesis of the solar system" (1913: 29–30; see also Holmes 1915: 103–4).

The most thorough critique of the Chamberlin—Moulton theory was that of the German astronomer Friedrich Nölke, in his book on cosmogony published in 1908. Nölke went systematically through each section of Moulton's 1905 paper and attempted to refute most of the arguments. He denied that stellar encounters could be frequent enough to produce the large number of spiral nebulae observed, or that such encounters could actually produce a symmetrical double-armed nebula as Moulton claimed. (If the intruder star S' passed close enough to the Sun to produce tidal disruption, that is, about 2.44 times the Sun's radius according to the Roche theory, it would have a much greater effect on the material ejected from the near side of the Sun than on that from the far side.) Moreover, such a nebula could not retain its spiral form very long since the material in the orbit of Mercury goes around the Sun in 88 days, while that in Neptune's orbit takes 165 years to complete a revolution; no such rapid changes have ever been observed in any spiral nebula (Nölke 1908: 63–9).

Apart from the difficulty of identifying the features predicted by Moulton's theory in observations of spiral nebulae, Nölke denied that the theory could account for the properties of the Solar System, such as the inclination of the planes of planetary orbits to the Sun's equator. He disputed Moulton's explanation of the effect of the intruder star on the Sun's rotation (arguing that any such effect would be negligible under the stated circumstances). He attacked Moulton's conclusion that the accretion of planetesimals by a planetary nucleus would reduce the orbital eccentricity, on the grounds that various important factors such as attractive forces and nonaccretional collisions had been unjustifiably neglected. He thought that Moulton's version of Chamberlin's argument, that accretion would tend to produce planets with direct rotation, was erroneous, because one category of collisions leading to retrograde rotation would be as numerous as another category leading to direct rotation.[88] Finally, he disputed Moulton's theory of the evolution of satellite motions. Almost the only favorable remark Nölke made about the work of Chamberlin and Moulton was that they have definitely proved the untenability of the nebular hypothesis by their 1900 analysis of angular momenta and kinetic energies in the Solar System, and that Moulton's critique of Pickering's explanation of the retrograde motion of Saturn IX is valid (Nölke 1908: 64, 78–9).

88 Nölke (1908: 75). Referring to Moulton's fig. 3 in his 1905 paper, Nölke pointed out that while particles whose aphelia lie between c and b tend to give the nucleus a forward rotation, those whose aphelia lie between c and a tend to produce retrograde rotation when they unite with the nucleus; and he thinks there should be as many of the second category as of the first. Moulton's discussion of this point is in fact defective, but Chamberlin's makes it at least plausible that there will be more collisions of the nucleus with particles of the first kind than those of the second. See Chamberlin and Salisbury (1906: 64–81, figs. 29, 30, 31).

Chamberlin and Moulton did not reply to Nölke's criticism, and I have found no evidence that they were even aware of it. It probably had little impact on cosmogonical opinion outside of Germany.[89]

More damaging than Nölke's explicit criticism was the fact that Henri Poincaré, the highest authority on theoretical astronomy in France, completely ignored the Chamberlin–Moulton theory in his lectures on cosmogonical hypotheses published in 1911. This book became a classic reference work for later theorists, and French writers on cosmogony during the following decades usually discussed only the theories that Poincaré had mentioned.[90] Poincaré recognized the appeal of cyclic collision theories such as that of Arrhenius but doubted that the latter's nebular Maxwell Demon could work.[91] He devoted considerable space to an analysis of the modifications of Laplace's theory proposed by Roche, Faye, and du Ligondes. Nebular collision theories of See and Belot were criticized. While Poincaré did not give strong support to any single theory, the net impression left by his book was that the Nebular Hypothesis was not yet dead even though it might require substantial revision. At the same time he derived stability conditions for the Laplacian rings that seemed to be almost impossible to satisfy with any plausible density distribution in the nebula, thus giving Jeans and other cosmogonists additional ammunition to shoot at the Nebular Hypothesis.[92]

To summarize, Chamberlin and Moulton were generally given credit for having refuted the Nebular Hypothesis and thus opening the way for other theories of the origin of the Solar System; but their own alternative theory was not widely accepted except in the United States.

But even this was a substantial achievement, for as Chamberlin himself recognized (D1906a) the real test of the value of a new hypothesis is not merely whether people believe it is true, but whether they can use it as a fruitful basis for further progress; and the United States had become the leading country in astronomical research (Brush 1979b). In particular, George Ellery Hale's giant telescopes in California offered the greatest hope for new discoveries, and Chamberlin had some reason to think that observations of spiral nebulae might be directed toward exploring the consequences of his own theory. In 1907 he corresponded with Hale about testable predictions of the planetesimal hypothesis; for example, the motion of the matter in spiral arms of a nebula should be transverse to the axes of the arms rather than either outward or inward along them as usually believed. The motion might be measured either by spectroscopic determinations of motion in the line of

---

89 A reviewer in Nature, 78 (1908) 474, suggested that though Nölke found it easy enough to demolish the theories of his predecessors, his own theory would probably not survive an equally critical examination. Stratton (1910: 368) said that Nölke's "claim to have explained many of the as yet unsolved problems of cosmogony is hardly substantiated by his work" but did not discuss his criticisms of the Chamberlin–Moulton theory.

90 Veronnet (1914). Sageret (1931: 302–12).

91 Poincaré (1911; 1913: lxiii, lxvi–lxix, ch. 9).

92 Poincaré (1913: 22–3). Jeans (1919: 147–53); 1929a: 264). Kuiper (1956b: 109). Struve (1958).

sight, or by comparisons of photographs taken at intervals. Of course, it would be necessary to select a very small, very near nebula; the more familiar spirals are probably several orders of magnitude greater than planetary systems (Chamberlin D1907b).

Efforts to test the Chamberlin—Moulton theory did have some effect on nebular astronomy. At the Lowell Observatory, V. M. Slipher attempted to measure nebular rotation spectroscopically and at one point thought his results favored the Chamberlin—Moulton theory.[93] One of the less fortunate efforts was the attempt of Adriaan van Maanen to detect internal motions in spiral nebulae by comparing photographs taken at different times. In 1916 he claimed to have detected such motions, though not of exactly the kind predicted by the Chamberlin—Moulton theory.[94] Although van Maanen's results were not confirmed by other astronomers and are now believed to be incorrect, they had some effect on the debate about the nature of spiral nebulae around 1920. Estimates of the size and distance of spiral nebulae derived from van Maanen's measurements suggested that they were only a few hundred light years away, not "island universes" comparable to the Milky Way galaxy. Eventually, thanks to the work of E. P. Hubble and others, it was decided that the nebulae are indeed huge distant galaxies.[95]

The progress of nebular research gradually removed one of the foundations of the Chamberlin—Moulton theory. Even though Chamberlin realized as early as 1907, in his letter to Hale summarized earlier, that most spirals are probably too large to be progenitors of planetary systems, he still thought it might be possible to find a few small nearby ones to illustrate his theory. Later he stated that "nebulae small enough to make planetary systems like ours could not be seen at stellar distances" (1924a: 258; see also 1928: 135). It was also pointed out by Campbell and others that the spiral nebulae are most frequently found in just those parts of the sky most thinly populated by stars, contrary to what one would expect if nebulae are the products of stellar collisions.[96] Eventually it was necessary to drop altogether the idea that the Solar System could have evolved from any nebula of the kind observed in the heavens.[97]

93 Slipher (D1914). I am indebted to Miss Helen S. Horstman for supplying a copy of this letter, and to Dr. Norriss S. Hetherington for informing me of its existence. See also Hetherington (1988) for a discussion of the relation between the Chamberlin—Moulton theory and Slipher's observations. There are brief remarks on this point in Berendzen, Hart, and Seeley (1976: 105) and Smith (1990).

94 A. van Maanen (1916b). The connection with the Chamberlin—Moulton theory was mentioned at the end of a preliminary announcement with similar title (1916a). See also Hetherington (1972; 1974; 1975; 1988: ch. 3), Berendzen and Hart (1973), Berendzen, Hart, and Seeley (1976: sect. 3).

95 Fernie (1970). Berendzen, Hart, and Seeley (1976). Smith (1982).

96 Campbell (1915: 244—5). For Chamberlin's reply to this argument see his (1924a: 258).

97 For Moulton the final blow to the long-prevailing view that nebulae are "world-stuff in an early stage of evolution" was Bowen's discovery that gaseous nebulae contain the heavier elements found in the Sun, in the same proportions, rather than consisting purely of lighter elements from which the heavier ones might be manufactured (Moulton 1939).

I.3

Jeans, Jeffreys, and the decline of encounter theories

There is a superficial similarity between the American pair, Chamberlin and Moulton, and the British pair, Jeffreys and Jeans. James Hopwood Jeans, like Moulton, was primarily interested in the mathematical aspects of astronomy and physics, but pursued original research for only about 25 years.[1] Harold Jeffreys, like Chamberlin, was more concerned with applications to the history and present state of the Earth, and was active in science into his eighties.[2] But the resemblance ends there, for Jeans and Jeffreys were not collaborators, and their ideas coincided for only a brief period. Jeffreys's approach to geophysics was radically different from that of Chamberlin – mathematical rather than "naturalistic" – although both rejected continental drift, the most important geophysical theory of the 20th century. Jeans ranged over a wider field of stellar astronomy, cosmology, and atomic physics in contrast to Moulton's concentration on planetary orbits (though both worked on the problem of stability of a rotating fluid). Jeans and Jeffreys "improved" the Chamberlin–Moulton theory by putting it on what appeared to be a superior astronomical and physical foundation; they retained the dualistic feature (subsequently abandoned) while discarding the planetesimal feature (later revived).

I.3.1 Jeans

In 1902 James Jeans, following up some comments by his mentor G. H. Darwin (1888), considered the problem of the stability of a rotating mass of gas as a basis for the Nebular Hypothesis. As he pointed out, the problem of a gaseous nebula of finite size does not lend itself to any simple mathematical

1 James Hopwood Jeans (1877–1946) was educated at Cambridge University; he was Second Wrangler in the mathematics tripos in 1898, and Smith's Prizeman in 1900. He was elected Fellow of Trinity College in 1901 and served as professor of applied mathematics at Princeton University from 1905 to 1909. He returned to Cambridge as lecturer in applied mathematics from 1910 to 1912; since he had married a wealthy American woman he decided to retire from university duties so he could devote himself to research and writing. He is known to historians of physics for his early work on the kinetic theory of gases and the law of black-body radiation as derived from classical assumptions ("Rayleigh–Jeans law").
2 Harold Jeffreys (1891–1989) studied at Cambridge University and remained on its faculty for his entire career. He established the fluidity of the Earth's core and was a long-time opponent of the theory of continental drift. See *Nebulous Earth*, 2.3.4, note 11.

treatment, because the ordinary gas equations break down over the outermost part of the nebula where the density is too small to justify statistical treatment of the atoms. One must assume either that there is an external pressure so that the gas density is finite at the boundary or that the nebula is infinite in size.

Jeans found that a uniform distribution of gas throughout infinite space would be gravitationally unstable and would tend to form separate nebulae (condensations of gas around points of maximum density).[3] The linear scale in which these nebulae are formed will depend on three quantities: γ the gravitational constant, ρ the mean density, and λT the mean elasticity; the only combination of these quantities having the dimension of length is $(\lambda T / \gamma \rho)^{\frac{1}{2}}$. Assuming the primitive temperature to be on the order of $1000°$ and estimating the density on the assumption that stars with density comparable to the Sun are at an average parallactic distance of $0.5''$ apart, he found the basic length to be of the order of $10^{19.5}$ cm. The distance corresponding to a parallax of $0.5''$ is about $10^{18.6}$ cm. Thus, the nebulae would be substantially larger than typical interstellar distances (Jeans 1902: 51).

Jeans did not draw any definite conclusions about the validity of the Nebular Hypothesis from these results, except that rotational instability is not the only mechanism by which a nebula could break up; gravitational instability must also be considered. Thus, the behavior of a gaseous nebula is "of so much more general a kind than those usually inferred from the analogy of a liquid mass that no difficulty need be experienced in referring existent planetary systems to a nebular or meteoric origin, on the ground that the configurations of these systems are not such as could have originated out of a rotating mass of liquid" (Jeans 1902: 53).[4]

In 1905 Jeans published a similar numerical estimate, using Lord Kelvin's (1901) estimate of star density corresponding to an average parallax of $0.62''$ for adjacent stars. The average mass of a nebula from which a planetary system is formed would be $10^{34.9}$, which he considered to be in good agreement with the mass of our own system, $10^{33.3}$, "considering the extreme vagueness of the data" (Jeans 1905: 99). Moreover, the agreement can be "vastly improved on taking an average parallactic distance greater than $0.62''$."

In 1917, returning to astronomy after a decade of work on radiation and quantum theory, Jeans collected his cosmogonical investigations in a classic monograph, which gained the Adams Prize at Cambridge University (Jeans 1919). Here he investigated the possibility that the Solar System had been formed by the encounter of the Sun with another star.[5] It was primarily his

---

3 See Lang and Gingerich (1979: 196) for the relation between his equations and the modern formula for the "Jeans stability criterion."

4 Jeans presented some further discussion of the application of his theory to the Nebular Hypothesis in a 1903 paper, again without reaching definite conclusions.

5 Although he later claimed to have been "the first, in 1901, to consider the possibility of the second body not colliding with the Sun, but producing planets by tidal action" (1931: 433), his early papers contain only a few vague sentences on this possibility (1901: 454–5; 1902: 52; 1905: 98, 102).

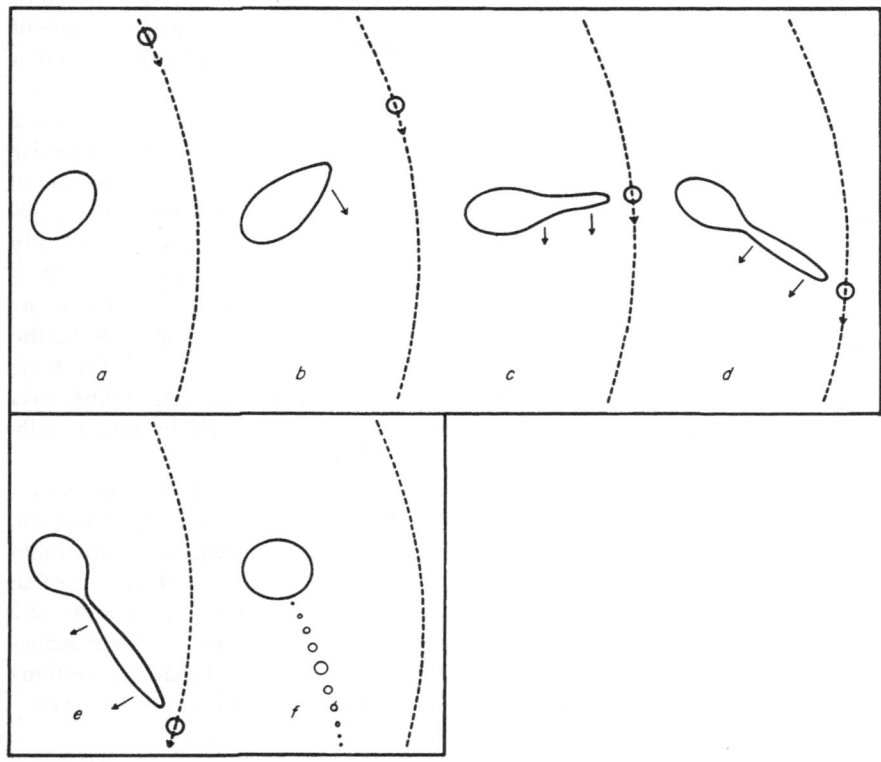

Figure 11. Jeans's tidal theory. "This sequence of six diagrams (a, b, c, d, e, f) is intended to suggest the course of events in the sun, or other star, when its planets were coming into existence. For simplicity the second star is represented throughout by a small circle, although actually this would, of course, also experience deformation and possible break-up" (J. H. Jeans, "The evolution of the solar system," *Endeavour*, **2** [January 1943], 3–11).

quantitative treatment of several definite situations that gave authority and credibility to Jeans's work.

Jeans emphasized the extremely low probability of stellar encounters of the type that could conceivably lead to the formation of planetary systems. Hence, spiral nebulae, being rather common, could have nothing to do with planetary cosmogony. Jeans demanded "the abandonment of all tidal theories, including the planetesimal, as explanations of *normal* cosmogonic processes" since "only a small proportion of the stars in the universe are likely to have been broken up in this way." He suggested that the tenability of the Chamberlin–Moulton theory had been weakened by its association with spiral nebulae (Jeans 1919: 17, 231).

1.3.2 Jeffreys

Jeffreys published his first papers on cosmogony and geophysics in 1915 and made substantial contributions to these subjects during the next 60 years. H. N. Russell called him "one of the ablest mathematicians who has ever devoted serious attention to cosmogony" (D1932a). In a paper presented to the Royal Astronomical Society in December 1916, he admitted that the Chamberlin–Moulton theory is "open to fewer objections than the other theories" but argued that if two planetesimals collide at a relative speed of a few kilometers per second and unite, the energy liberated will be sufficient to volatilize them. Chamberlin and Moulton had assumed that nearly all collisions will be overtakes, with low relative velocity, but Jeffreys argued that perturbations due to the planetary nuclei would change the orbits so much that after about 100,000 years the planetesimals would be moving in all directions. He therefore suggested abandoning what Chamberlin had considered his fundamental postulate – formation of planets by accretion of solid particles – and returning to the assumption of a gaseous nebula.[6]

Jeffreys argued in later papers that the thermal data available could best be explained by Arthur Holmes's model, which assumed that the Earth was initially fluid.[7] He proposed to revive on this basis the theory that mountains have been formed by contraction of the crust of a cooling Earth (*Nebulous Earth*, 2.2.1), with appropriate revisions of the time scale to take account of radioactive heat generation. Having contradicted Chamberlin's ideas in the areas of greatest interest to a geologist, Jeffreys went on to maintain that they were really not far apart in cosmogony and that both held slightly different versions of the "tidal theory" (Jeffreys 1916b, 1921, 1925b).

Jeffreys relied to a large extent on Jeans's calculations of the results of tidal interactions between stars, adding to them his own geophysical arguments; thus, the combination of Jeans plus Jeffreys was rather potent. It would have been irresistible if the geologists had not been inspired by Chamberlin's example to defend the legitimacy of their own arguments. There was in fact something of a reaction against "mathematicians" – a reaction that Jeffreys later protested (1929a: app C; 1973).

1.3.3 Chamberlin's reaction to Jeans and Jeffreys

Chamberlin went through Jeans's book and prepared a 7½-page memorandum of his comments and criticisms:

6 Jeffreys (1916a). In 1969 Jeffreys wrote the following note for the reprint of this paper: "The argument on volatilization of planetesimals needs modification on account of the result of A. L. Parson that the volatilized matter would recondense. But this makes the eccentricities of the orbits harder to understand than ever" (Jeffreys 1974: 105).

7 Jeffreys (1917a, 1924). A. Holmes (1915). Holmes abandoned the hypothesis of a steadily cooling Earth in 1925 (see Holmes 1931: 563).

Jeans distinctly misrepresents the planetesimal theory by calling it simply a tidal theory. It has, as distinctly stated again and again, three vital factors: (1) *eruptivity of the Sun* (*which Jeans entirely ignores,* and which is the main source from which the energy of expulsion comes), (2) *tidal stimulus* (*trigger* action rather than *cartridge* action), and (3) *transverse pull* (giving revolution). The first and third are most essential, the second forms the easiest and most natural *modus operandi* of the combination (see MS. of my 2nd paper).[8]

Chamberlin disliked Jeans's method, based on the construction of simplified models that could be treated mathematically, in contrast to his own "naturalistic" approach. For example, Jeans made the unrealistic assumption that the gas laws applied to his matter. But Chamberlin also complained that Jeans was stealing some of his ideas without proper acknowledgment:

. . . Absorbs the two-arm feature of the spiral nebulae without the slightest recognition of those who discovered it and urged its importance long ago – and this immediately after having slyly taken over in an occluded form the very principle of formation which has long been urged as an explanation.
. . . My bottom thought is that Jeans is trying to push the Planetesimal theory aside and put a substitute of his own in its place, much as See and Lowell did earlier, but not in their crude fashion. I shall expect him to come around later to something of the same nature in essence after he has done "an immense amount of mathematical work" on it, etc., etc.
. . . The most interesting feature of the whole book is the fact that after intimating at the outset that our theory would get hard knocks in the end and manifesting abundant disposition to give them, he comes as near to shutting out everything else and giving it the right of way as he well could for one working on the gas basis. (Chamberlin D1920: 5–4, 7, 8)

Chamberlin also drafted but did not publish (as he could have done in his *Journal of Geology*) an editorial titled "Our New Responsibility," pointing out that since Jeans had concluded that the stellar universe is only 630,000,000 years old whereas radioactive dating indicates that rocks on the Earth's surface are at least 1,500,000,000 years old, "we geologists . . . must now assume responsibility for the history of a great lapse of time before the universe was born."[9]

In 1924 Jeffreys published the first edition of his monograph *The Earth* (the sixth edition appeared in 1976). He adopted the Jeans "tidal theory," stating:

Its modern development was due in the first place to Chamberlin and Moulton; but several features of their formulation have been found unsatisfactory by the present

8 Chamberlin (D1920a). The memorandum appears to be addressed to one of his colleagues, other than Moulton, probably W. D. MacMillan, who did in fact write a critical review (1920) of Jeans's book including some of the points Chamberlin mentioned in this memorandum.
9 Chamberlin (D1920b). Jeans estimated that about 560 million years may have elapsed since the primeval rotating nebula began to break up into stars (1919: 286).

writer, who has therefore attempted to reconstruct the theory entirely from the time of the rupture.[10]

The Chamberlin–Moulton theory, Jeffreys continued, "has obtained the support of a considerable number of geologists," although astronomers have given it much less attention.[11] Among other objections to the Chamberlin–Moulton theory, Jeffreys argued that the planetesimals could not have condensed to solid particles before the formation of planets, and therefore the Earth could not have been initially solid. Jeffreys developed his own theory of the Earth's history on the assumption that it was originally molten.

In his review of *The Earth*, Chamberlin noted that while Jeffreys accepted Jeans's views to a large extent, he gave credit more explicitly to the earlier work of Chamberlin and Moulton than Jeans had done, and "recognizes more fully the actualities of the case, though not with adequate fullness. . . . There is much less shrouding of the vulgar facts of nature in the celestial atmosphere of the mathematical Parnassus, but still there is quite much of it" (1924b: 699).

The main points at issue between Chamberlin and Jeffreys were (1) whether the "eruptive power" of the Sun, attributed by Chamberlin to "convective turbulence," must be taken into account in analyzing its ejection of matter under the tidal forces of the other star; (2) whether the ejected matter would behave as a gaseous-resisting medium during the process of condensation of planets, or would quickly condense to small solid particles; (3) whether the Earth was initially molten or solid.

On the first point, Jeffreys denied the importance of convection in the Sun and asserted that the prominences cited by Chamberlin are a "purely superficial phenomenon" (1925b: 397–8).

Chamberlin's position on the second point was not as strong as it had been earlier, since he had recently adopted the view that planetesimals can be individual molecules (1920: 666, 668, 679).

But Chamberlin was more insistent than ever that the Earth was now and always has been solid (1916: 181, 184; 1922: 269; 1928: x). Jeffreys defended the Holmes model of an initially molten Earth but did not exploit the recent seismic evidence showing that a substantial part of the Earth's interior is now liquid or at least has a homogeneous layered structure indicating differentiation in a previous state of fusion, contrary to Chamberlin's geological theories.[12]

During the 1920s the Jeans–Jeffreys cosmogony gained ground at the expense of the Chamberlin–Moulton theory. Astronomers, especially outside

10 Jeffreys (1924: 3). Jeffreys gave a detailed critique of the planetesimal hypothesis in an appendix (pp. 250–6), including an attempt to refute Chamberlin's argument that a hot fluid Earth could not have retained water vapor and atmospheric gases. In another appendix (pp. 257–9) Jeffreys stated that the main difference between his theory and that of Jeans was that the latter assumed that the Sun was so large at the time of encounter that it occupied the orbit of Neptune. Jeffreys considered this improbable and argued that even for a much smaller Sun, the probability of encounter would not be impossibly small.
11 Jeffreys (1924: 17). This remark is elaborated in his article in *Science* (1929b: 246).
12 Oldham (1918). Knott (1919: 184–6). Jones (1922). It was later suggested that this evidence may not count decisively against the planetesimal hypothesis; see Mather (1939).

the United States, most frequently referred to the Jeans theory while occasionally recognizing the Chamberlin–Moulton theory as an earlier but cruder version of it.[13] In reviewing Chamberlin's last book, *The Two Solar Families* (published just before the author's death in 1928), Moulton complained that in their recent publications Jeans, Jeffreys, and Eddington had failed to give proper credit to Chamberlin and himself as founders of the planetesimal hypothesis and had made it appear that cosmogony is a purely British subject. He criticized these men rather harshly for overstating the validity of their mathematical calculations while rapidly changing their physical assumptions from one year to the next; this approach was to be contrasted with Chamberlin's naturalistic method, which has produced results of permanent value (Moulton 1928).

In reply, Jeffreys protested that he had given sufficient acknowledgment to Chamberlin and Moulton in his publications, and that Chamberlin himself had admitted this in reviewing *The Earth*. Jeffreys then complained that his own work in cosmogony and geophysics had been ignored by geologists for 10 years "by preconceptions based on Chamberlin's views; this was fostered by Chamberlin's own refusal even to acknowledge the existence of criticism or of an alternative theory until 1924" (Jeffreys 1929b: 245–6). Moulton insisted that Chamberlin had been concerned about "a tendency on the part of certain British scientists to adopt important essentials of the planetesimal hypothesis as their own, though under another name, and he suggested that I should undertake to clarify the history of the theory."[14]

Jeans made no public response to Moulton's criticism until the anonymous writer of a column in the British magazine *Observatory* stated that Moulton "has shown good reason for complaint" ("From an Oxford notebook" 1929). Jeans then protested "that my Tidal Theory was in no way dependent on the American Theory, and that I gave very generous recognition (but not of course acknowledgment) to the Planetesimal Theory." Moreover, Jeans claimed that he had himself thought of and published the idea of second-body influence "long before Professor Chamberlin" (Jeans 1929b: 172–3). At the same time Jeffreys (1929c) stated that the planetesimal hypothesis was similar to earlier ideas of Bickerton, Buffon, and Proctor.

The British astrophysicist Herbert Dingle, writing over the initials "H. D." in *Nature*, rejected Moulton's charge of nationalistic bias and turned it around by presenting a count of references to authors of various nationalities in the major works of Chamberlin, Jeans, and Jeffreys; he found that 65.6 percent of Chamberlin's references were to Americans, while only 58.6 percent of Jeffreys's and 22.8 percent of Jeans's were to Britons.[15]

---

13 On this, see, e.g., Hobson (1921–22/1968: 309), anonymous review of Hobbs (1922) in *Nature*, 110 (1922): 270–2, Jones (1922: 380), Crommelin (1923: 80), van Maanen (1923: 276), Abbot (1925: 231), Gheury de Bray (1925), Smart (1928: 283–5), Menzel (1931: 87).
14 Moulton (1929). For evidence of Chamberlin's resentment see the end of his "Reminiscences" (D1928b) and his (D1920a).
15 H. D[ingle] (1929). Dingle acknowledged his authorship of this note in his (1932), in which

1.3.4 Decline of the encounter theories

Modern textbooks and review articles on the origin of the Solar System state that the Chamberlin–Moulton theory had two fatal defects (in common with the Jeans–Jeffreys tidal theory): First, it failed to account for the present distribution of angular momentum within the Solar System, an objection generally attributed to H. N. Russell (1935). Second, because of the high temperature of the interior of the Sun, a gaseous filament drawn out of it by the action of another star would rapidly dissipate into space before it could condense; this argument is usually credited to Lyman Spitzer Jr. (1939). A third objection, admittedly more metaphysical in nature, was that the estimated probability of stellar encounters was so low that planetary systems produced in this way would be extremely rare, and it was difficult to accept the conclusion that we may be the only life in the galaxy or the entire universe.[16] These are strikingly similar to the reasons for rejecting the Nebular Hypothesis in 1900: failure to account for the angular momentum distribution; inability of a hot primeval Earth to hold an atmosphere (or of gaseous Laplacian rings to condense into planets); and Kelvin's estimated age of the Earth (based on a presumed initial molten state), which was too short to allow geological and biological evolution by natural means.

The paradox, that a theory is rejected for the same reasons as its predecessor, dissolves as soon as we look at the situation more closely. First it should be noted that the planetesimal hypothesis for the formation of the Earth, which Chamberlin originally considered the most important part of his theory, is not now considered obsolete, although it was abandoned by astronomers for several years on the strength of Jeffreys's argument that planetesimals would be vaporized by collisions. Geologists remained loyal to the planetesimal hypothesis through the 1930s, though they did not always accept Chamberlin's inference that the Earth had remained cold and mostly solid throughout its history. The only significant criticism was a 1924 paper by Harry Fielding Reid, who estimated the effect of accretion of planetesimals by a planetary or satellite nucleus and found that it was much too small to account for satellite orbits. Reid's critique was published in the *American Journal of Science* and remained unknown to most astronomers, though Russell gave a rather vague reference to it in 1935.[17]

---

a chronological review of theories was presented. Here Dingle quoted a letter from Jeans admitting that the first proposal of a "tidal origin for the solar system" was that published by W. F. Sedgwick (1898). Sedgwick's paper was unknown to other cosmogonists until he sent a copy of it to the editor of *Nature,* who asked Dingle to look into the priority question.

16 Jeans took this objection quite seriously but was ambivalent about whether it was really an argument against the theory. In his book *The Mysterious Universe* (1932) he seemed to be rather pleased with the idea that life on Earth is the accidental result of an extremely unlikely stellar encounter, and that we may be unique in the universe. Yet he also argued that the probability was not so small after all, since stellar encounters would have been more frequent in an earlier stage of the expanding universe, and the Sun would have been much larger at the time of such an encounter. See also Jeans (1919: 286; 1942).

17 Russell (1935: 104): "Moulton concludes that under certain circumstances, this rotation

Reid had assumed on the basis of Chamberlin's earlier expositions that the accretion of planetesimals was the most important feature of the Chamberlin–Moulton theory, and that the encounter of the Sun with another star was "not considered as a necessary part of the hypothesis" (1924: 63). Thus, he gave little attention to this part of the theory. But in his reply Chamberlin insisted that "the close approach of a star (or its equivalent) to the Sun is the very soul of the planetesimal hypothesis" (1924a: 244). Hence, Chamberlin argued, Reid's criticism was misdirected, as well as being a purely mathematical calculation based on an approximate model that did not come to grips with the naturalistic style of reasoning.

Having (rather surprisingly) declared that the encounter of two stars was the "basal hypothesis" of his theory, Chamberlin was no longer in a strong position to protest the assimilation of his theory to the Jeans–Jeffreys "tidal theory" – nor would his protests have been effective in any case as long as Jeffreys's argument about the dissipation of planetesimals by collisions was accepted. The result was that the fate of the Chamberlin–Moulton theory now depended on what happened to the tidal theory.

Moulton's failure to reply to Reid's criticism, even after some pointed warnings by H. N. Russell (D1925), is rather surprising. "I read the paper of Professor Reid in manuscript form," he told Russell, "and it was considerably revised by him as a consequence of our correspondence on the subject, though in my opinion the conclusion is not final" (Moulton D1926a). Russell assumed Moulton was thereby admitting that the planetesimal hypothesis "required correction" and suggested that Moulton ought to have published some statement to that effect (Russell D1929b). Moulton protested that he had been misunderstood, that Reid's work "was based on a number of assumptions which are more or less uncertain, and does not begin to cover the possibilities of the case" and concluded that "I know of nothing which has yet been published which has convinced me that the Planetesimal Hypothesis requires correction" (Moulton D1929b). But he did nothing to defend it.

Friedrich Nölke, one of the most persistent critics of all cosmogonical hypotheses, continued to attack the Chamberlin–Moulton theory on various points. For example, he argued that the tidal mechanism could not produce the present Solar System because the material drawn out of the Sun would either fall back or be dissipated into space. His criticisms began to have some effect, at least after they had been reinforced by other astronomers.[18]

[due to infall of planetesimals on planetary nuclei] would have been direct and rapid; but his postulates have been criticized by H. F. Reid, who concludes that the rotation would be slow" (no reference given).

Harry Fielding Reid (1859–1944) received his Ph.D. from Johns Hopkins University in 1885, then taught mathematics and physics at the Case School of Applied Science, returning to Johns Hopkins as professor in 1894. He is best known for his "elastic rebound" theory of earthquakes, developed as a result of his service on a committee appointed by the governor of California to investigate the earthquake of 1906.

18   Nölke (1930: 181–95). Russell (1935: 104, 108, 110). Luyten and Hill (1937). Page (1948). Kuiper (1956b).

Figure 12. Henry Norris Russell (American Institute of Physics, Niels Bohr Library, Margaret Russell Edmondson Collection).

But a more damaging blow against the tidal theory was struck by one of its erstwhile supporters. In 1929, Jeffreys announced that he had abandoned the hypothesis of a close encounter on the grounds that it could not account for the rotation of the large planets; he proposed instead an actual collision between the Sun and another star.[19]

Russell had become interested in planetary cosmogony around 1925, when he prepared a revision of Young's textbook in collaboration with R. S. Dugan and J. Q. Stewart (Russell D1925, D1927). In that text he credited the Chamberlin–Moulton theory with "the great advantage that it accounts for the angular momentum of the planetary motions" but qualified this statement by

19 Jeffreys (1929d). For further development of the collision theory see (1929e, 1931a).

expressing some doubt that matter ejected from the Sun could have the great amount of angular momentum now possessed by the outer planets (Russell, Dugan, and Stewart 1926: I, 465–8). He told Moulton that the objections of Reid and Jeffreys should be taken seriously even if they were not conclusive; he agreed with Jeffreys that planetesimals would collide with each other so frequently that they would be smashed to dust (though not to separate gas molecules as Jeffreys thought). The result would be a medium of particles moving in circular orbits, which would be effective in decreasing the eccentricities of the planetary orbits but would also diminish their forward rotation (Russell D1925, D1926a).

Russell then confronted Moulton, in the same letter (D1926a), with rudimentary versions of the two objections that ultimately proved fatal to the tidal theory. First, the filaments of material drawn out from the Sun would be at a temperature of at least a million degrees, if Eddington's new theory of stellar constitution was correct.[20] Second, a particle whose ratio of angular momentum to mass is the same as that of an outer planet, say Neptune, cannot have had an original perihelion distance less than one-half of that of the planet into which it is eventually accreted; it was difficult to imagine how material ejected from the Sun could have been placed in such an orbit. Nevertheless, Russell still thought the Chamberlin–Moulton theory was the best available at the time (D1926a: 8).

Moulton, instead of trying to answer these objections, accused Russell of giving Jeans and Jeffreys most of the credit for the ideas in the planetesimal hypothesis (D1926b). Russell denied that he had done so and emphasized again that more work needed to be done on the theory: "I hope you will get at it again" (D1926b). But Moulton by this time seemed to have abandoned active research, complaining that – as a result of unpleasant experiences with people like Jeffreys who tried to steal the fruits of his labors – "my early illusions as to the high ideals and the objectivity of scientific work are pretty largely shattered. For this reason I have not published much recently except in book form, which of course no one takes the trouble to read except the coming generation" (D1926a).

Early in 1927 Russell was invited to participate in a project to prepare a report on the internal constitution of the Earth for the National Research Council. In accepting, he told L. H. Adams that the report should include a section on the origin of the Solar System, and outlined his views on current theories. The planetesimal theory meets with "grave objections, which have been ably stated by Jeffreys." But the Jeans–Jeffreys theory is also subject to "grave difficulties." "The most serious of these, so far as I am aware, has never

20 Eddington (1926). His model was based on the assumption that heat transfer within the star takes place entirely by radiation. More recent models that allow for some convection lead to lower estimates for the temperature near the surface but do not invalidate the argument. These models also take account of thermonuclear reactions, which rapidly destroy deuterium, and thus indicate that the Earth and other Solar System bodies that contain significant amounts of deuterium could not have come from a mature Sun (Burbidge et al. 1957: 618; Fowler, Greenstein, and Hoyle 1961; Cameron 1965: 557).

been mentioned in print, and has to deal with the temperature of the material ejected from the Sun in the original cataclysm." This temperature he estimated as "well up towards a million degrees" at the time of ejection, contrary to the Jeans–Jeffreys estimate of a few thousand degrees. It is difficult to understand how more than a small fraction of such material could condense into planets. But he was not yet prepared to throw out the tidal theory on these grounds (Russell D1927).

When Chamberlin's last book appeared in 1928, Russell told the readers of *Saturday Review* that "many competent authorities in physics, astronomy and geology find themselves obliged to dissent from some or many of the conclusions of this Chicago school of cosmogony – and their dissent is supported by arguments of weight." He mentioned only one such argument: It is "very hard to understand" how dust particles or molecules ejected from the Sun could condense to visible grains or chondrulites from which planets could form (Russell 1929). In a letter to Moulton, he called the theory "natural history" not dynamics; he could not reconcile with dynamic principles Chamberlin's description of the motion of particles ejected from the Sun. But the most important difficulty was how material at a temperature of a million degrees could condense to planetesimals or planets; again he urged Moulton to work on improving the theory (Russell D1929a).

Russell reviewed Jeffreys's new collision theory together with the earlier tidal-encounter theories in his monthly astronomy column for *Scientific American* in 1931. He pointed out that all these theories failed to explain how the masses to be assembled into planets could have been removed far enough from the Sun to be launched into the required elliptical orbits, during the short time when they were under the influence of the other star:

It is easy to show that the shortest time in which a body can be taken from the Sun to any given distance, without moving it at such a speed that the Sun's attraction would be unable ever to bring it back again, is 7½ of the period of the planet rotating in a circular orbit at this distance. To get Neptune out to half its present distance [the required launching point, as he had noted earlier] must therefore have taken more than four years, while Jupiter would require about four months. These are roughly the lengths of time which the passing star would have taken to recede to the same distances, and this suggests that the planets may represent masses which started to trail after the star and only gave up the chase after some time. (Russell 1931: 92–3)

Russell was temporarily converted to Jeffreys's collision theory, but he continued to worry about the problem of condensation from high-temperature ejecta.[21] An invitation to give the Page–Barbour Lectures at the University of Virginia in the spring of 1934 provided Russell an opportunity to discuss the subject before a general audience – "I have wanted to get my opinions on this subject out of my system for years," he told S. A. Mitchell (Russell D1933?) –

---

21 Russell (D1932b). In his letters to Jeffreys in 1934 Russell said he preferred the collision theory to the tidal theory but found difficulties in both.

though he apparently did not feel ready to publish a technical critique in an astronomical journal.

In March 1934 Russell wrote to Jeffreys asking for comments on his argument about placing matter into planetary orbits, but stated it in a way that Jeffreys at first misunderstood:

Any matter ejected during a collision must have an initial perihelion distance, not greatly exceeding the solar radius, which means that the angular momentum *per unit mass* will be of the order of one-tenth the value of the Earth, and one-fiftieth that for Neptune. . . . So I have great difficulty in seeing how bodies produced by collision could ever get into orbits like those of the major planets. (Russell D1934a)

Jeffreys, writing his reply while a Cornish chorus sang "The Bonnie Banks of Loch Lomond" all wrong outside his window, thought that the only problem was to get ½₀ or ⅓₀ of the angular momentum of the passing star into the motion of the planets and couldn't see anything obviously impossible about that (D1934). Russell insisted that the problem was more serious:

The embryo planets must also be hauled away from the Sun to at least half their present distances before they are let go, or else they will fly away in hyperbolic orbits; and I don't see how the dickens this can happen. (Russell D1934b)

While thinking about hypothetical stellar collisions, Russell suffered a very real automobile collision that almost killed him before he could get to Virginia to deliver his lectures on the Solar System. The lectures had to be postponed from May to November 1934, and thus the published version was not available until early in 1935. The objections to the encounter theory were primarily the two difficulties discussed in Russell's letters and reviews going back to 1926. The problem of condensation from a high-temperature gas was not considered insuperable, but the argument based on putting planetary material into distant orbits with the appropriate angular momentum for unit mass was presented in detail and seemed to leave no room for rescuing either the Chamberlin–Moulton or the Jeans theory, and "the collision hypothesis fares the worst of all."[22]

Russell's main objection was quickly accepted as fatal to all dualistic cosmogonies; Jeffreys admitted, in his review of Russell's book in *Nature,* that he had no answer to it.[23] The other objection was substantiated four years later by Lyman Spitzer, Jr., a postdoctoral fellow at Harvard who had just received his Ph.D. in astronomy at Princeton and had presumably been encouraged to work on the problem by Russell (Spitzer 1939). Russell wrote to Spitzer that he had long been convinced that condensation of the filament was the most improbable feature of the encounter theories, "and you have gone a long way

22 Russell (1935: 117). A detailed calculation by R. A. Lyttleton (1960) confirmed Russell's conclusion but showed that the tidal theory comes out worse than the collision theory.
23 Jeffreys (1935). Jeans (1938) insisted that the tidal theory is "fundamentally on sound lines" though he had no answer to Russell's objection.

toward proving it" (D1939). Spitzer's calculations showed that a filament of gas at a million degrees would have been almost entirely dissipated into space before it could condense; Russell thought the remainder might conceivably condense to a ring of planetesimals, in which case one might start over again with Chamberlin and Moulton – but no one seemed particularly inclined to follow that route. Instead, cosmogonists flirted with even more complicated schemes, like Russell's "wildish idea . . . that the Sun might once have been a double star, and that the planets were produced by a collision with the companion."[24] The idea of stellar encounter seemed to have been run into the ground.

1.3.5 Theory change: A test case for historical philosophies of science

From the modern viewpoint the period 1895–1935 was one of spectacular progress in physical science. Historians of science have studied in detail the success stories of that period: quantum theory, relativity, nuclear physics, and the expanding universe. Why should we bother with one of the failures – a theory that now seems only a temporary dualistic digression from the main line of development of monistic cosmogonies?

The answer to that question depends on one's beliefs about the purpose of research and writing on the history of science. If one merely wants to display a record of cumulative progress toward the present state of knowledge, then the Chamberlin–Moulton theory deserves no more than a few sentences pointing out its defects and noting its one "positive" contribution, the planetesimal concept. That is in fact the way it is now treated in most texts and review articles. (These accounts have usually been so scanty that a feature of Chamberlin's work that might have been useful to modern cosmogonists, the explanation of how direct rotation would be produced by impact of planetesimals, was forgotten and had to be rediscovered.) One would also mention that the main reason for abandoning the Nebular Hypothesis in 1900 – its failure to explain why the angular momentum of the Sun is so much smaller than that of the planets – is now thought to be fallacious. (The original angular momentum of the Sun could have been transferred to the planets by "magnetic braking," a concept invented after 1935.)

Judging earlier theories in the light of modern knowledge is always precarious, especially so in this case since no theory of the origin of the solar system can now be considered completely satisfactory or generally preferred to its rivals. Thus, the presumption of monistic development, accepted by a majority of cosmogonists from 1946 to 1976, was compromised by evidence for the

---

24 "I had a wildish idea this morning . . ." (Russell D1934a). See also Russell (1935: 135). The idea was independently conceived ("before finishing reading Professor Russell's book") and developed further by R. A. Lyttleton, a young British astronomer who was visiting at Princeton on a Commonwealth Fund Fellowship (Lyttleton 1936).

"supernova trigger" in the following decade, then restored when the trigger theory was abandoned in 1984. While modern cosmogonists favor planetesimal accretion for the terrestrial planets, they still disagree on the mode of formation of the giant planets.

The history of a theory such as the Chamberlin–Moulton cosmogony (and its variant, the Jeans–Jeffreys tidal theory) is valuable for its intrinsic interest: It helps us to understand the creative achievements of earlier scientists without regard to whether their ideas happen to be thought correct at present. Moreover, it may support or undermine certain generalizations about the nature of science as a human activity, and thereby acquire philosophical or sociological interest.

While major 19th-century philosophers such as Comte, Whewell, J. S. Mill, and Schopenhauer displayed considerable interest in the Nebular Hypothesis as an example of scientific theory,[25] 20th-century philosophers of science have paid little attention to cosmogony. For example, Karl Popper (1934/1959, 1962) frequently cites Kant's impact on epistemology and Laplace's ideas about statistical inference, but ignores the fate of their hypotheses about the origin of the Solar System. According to Popper's Principle of Falsifiability, the hypothesis should either have been rejected as soon as retrograde motions of planets and satellites were discovered, or it should have been declared pseudoscientific. But Popper's own confusion about how to apply his doctrine to Darwinian evolution (cf. Popper 1974: 133–43; 1978; 1980) suggests that falsifiability is not an appropriate test for theories about the past.

One widely discussed generalization about science, Thomas S. Kuhn's (1970, 1974) theory of paradigms and how they change in scientific revolutions, might seem to be applicable to the history of cosmogony. As I noted in *Nebulous Earth* 1.4, the Nebular Hypothesis did form the basis for an evolutionary worldview or paradigm in the 19th century. Chamberlin proposed not merely a new theory of the origin of the Solar System, but a new methodology: a "naturalistic" inductive approach utilizing multiple working hypotheses (1.2.3). As a geologist he brought to bear a somewhat different attitude about what kinds of phenomena should be explained by such a theory. Thus, one could say he was trying to establish a new paradigm.

Similarly, in accordance with Kuhn's views it seems obvious that there has not been much cumulative progress in cosmogony. Theorists have lurched back and forth from monistic (Descartes) to dualistic (Buffon) to monistic (Kant, Laplace) to dualistic (Chamberlin and Moulton, Jeans, Jeffreys) to monistic (Weizsäcker, Alfvén, etc.) to dualistic (supernova trigger) and back again to monistic – a pendulum oscillating with ever-decreasing period. Astronomers are no more certain in 1995 than they were in 1895 that the monistic principle is correct, even though knowledge of every aspect of the Solar System has increased by several orders of magnitude.

I do not think that these noncumulative changes in cosmogony are exam-

25 See *Nebulous Earth*, 1.3.2, 1.4.3.

ples of Kuhnian revolutions. The reasons for abandoning the Nebular Hypothesis in 1900 were quite rational and comprehensible to its supporters, most of whom were in fact persuaded to transfer their allegiance to some kind of dualistic theory. Most astronomers did not adopt Chamberlin's naturalistic method or his geological criteria for judging cosmogonic theories; in fact Jeans went in the opposite direction, urging a more abstract deductive approach rather than theorizing tied directly to the latest observations (1919: 18). Yet in spite of these differences in methodology and attitude, the "partial incommensurability" or failure of communication that Kuhn describes as characteristic of relations between followers of different paradigms did not occur, at least not more than one might normally expect when specialists in different fields talk to each other. Indeed Chamberlin himself was acutely aware of precisely this problem and made a special effort to get astronomers, physicists, and mathematicians to check his arguments before publication. (This is an important point established by unpublished letters preserved at Chicago.)

Russell's objections to dualistic theories were likewise fairly persuasive to supporters of those theories; while Moulton and Jeans did not publicly capitulate, Jeans admitted the force of the objections and Jeffreys abandoned his collision theory almost immediately.

One alternative to Kuhn's theory that might seem more compatible with this behavior is the "methodology of scientific research programmes" proposed by Imre Lakatos (1970). Lakatos suggested that scientists work within a research programme (I retain the British spelling here to emphasize the special meaning of this word), which consists of a series of theories, sharing a common "hard core" but adding various different auxiliary hypotheses. The latter serve as a "protective belt" – the hard core is never subject to direct experimental test; if one theory fails to account for observations, a new auxiliary hypothesis may be introduced to save the hard core from refutation. A programme may be characterized as "progressive" (if the hypotheses lead to successful tests) or "degenerating" (if they do not), but it will not, according to Lakatos, be abandoned unless it suffers by comparison with a competing research programme. In this comparison, successful predictions are more important than failures. Moreover, the change from one research programme to another (unlike the change from one paradigm to another in Kuhn's theory) can be rationally justified, at least in retrospect.

In his Lakatosian discussion of the career of continental drift theory, Henry Frankel (1979) identified Chamberlin as the leader of the "contractionist" research programme in geology. According to Frankel, the "hard core" of that programme was the assumption that "the Earth has been contracting periodically since its birth, with the result that the seafloor and continents have interchanged throughout the history of the Earth." Frankel described the planetesimal hypothesis as an "auxiliary hypothesis" that preserved the contractionist hard core from refutation at a time when the discovery of radioactivity threatened the previous assumption that contraction was a result of continued cooling of the Earth since its formation; contraction could now be attributed to the consolidation and rearrangement of a loose aggregate of planetesimals under gravitational forces. Jeffreys was identified as another leader of the same research programme who proposed an alternative auxiliary hypothesis, resurrecting the earlier idea of a cooling molten Earth and showing that it could produce contraction in spite of radioactivity.

Leaving aside the question of whether this description is valid for the history of geophysics we must modify it considerably in order to apply the Lakatos theory to the history of cosmogony. Chamberlin's early papers make it clear that the planetesimal hypothesis was to be considered a "basal hypothesis" that would lead to radical changes in the foundations of geology as well as astronomy. He would not have admitted that it was a "hard core" in the Lakatos sense, to be protected from refutation by a belt of expendable auxiliary hypotheses; his own "method of multiple working hypotheses" forbade such behavior, and indeed he originally introduced the planetesimal hypothesis not as a replacement for the assumption of a molten Earth derived from the Nebular Hypothesis, but only as an alternative working hypothesis to coexist with it.

Nevertheless it is fair to say that Chamberlin did tend to protect the planetesimal hypothesis from refutation by proposing auxiliary hypotheses that could be adjusted to the needs of the moment; and he certainly would not have accepted Jeffreys, who attacked his hard core, as a contributor to his research programme. On the astronomical side he was willing to consider various combinations of large and small stars and nebulae as progenitors of the ancestral spiral nebula, and he eventually abandoned the spiral nebula phase when astronomical evidence made it clear that all observable spirals are orders of magnitude larger than planetary systems.[26] On the geological side he was initially willing to allow various rates of planetesimal accretion so that the heat generated by the process of accretion and rearrangement might or might not cause partial melting; later he decided that the Earth was now a rigid elastic solid and had probably always been so.

This picture of Chamberlin as a Lakatosian scientist is somewhat spoiled by his 1924 announcement, in reply to Reid's criticism, that the hard core of his theory was not the planetesimal hypothesis but the "biparental" origin of the Solar System by stellar encounter. This move seems to me to have been an unwise and quite unnecessary retreat. It took the foundations of the theory dangerously far away from the familiar ground of geology and, as it turned out, made it extremely vulnerable to refutation. Chamberlin and Moulton

26 The deletion of the spiral nebula hypothesis might appear to violate the Lakatos dictum that each subsequent theory in the programme is constructed by *adding* auxiliary hypotheses to the preceding one (Lakatos 1970: 118). However, I do not think one should insist on Lakatos's literal meaning in this case since the purpose of the auxiliary hypotheses is, after all, to protect the hard core from refutation by being subject to refutation themselves. As L. Laudan (1976; 1977: 77) pointed out, it is quite common for new theories to be generated by dropping as well as adding assumptions. (This point was brought to my attention by Michael Gardner.)

thought that Reid's criticism did not refute the planetesimal hypothesis but showed only that it did not give results in agreement with observation when a special model was assumed. Yet they did not do anything to repair the damage. The calculations and arguments later used to revive the planetesimal hypothesis could have been furnished by Moulton and other supporters of the Chamberlin research programme in the 1920s and could have kept it alive in the 1930s – provided they were willing to jettison the binary encounter assumption and accept whatever substitute seemed acceptable on astrophysical grounds.

While Chamberlin and Moulton might have salvaged their programme if they had followed the Lakatos methodology, the historical fact is that they did not. One reason may be that scientists do not in practice maintain the strict distinction between hard core and auxiliary hypotheses; they or their successors become enamored of some of the features of the theory that were earlier supposed to be subsidiary and dispensable. The result may be that the theory is eventually overturned even though no strong evidence has been brought against its original foundations. (The overthrow of the caloric theory of heat is another example of this phenomenon; see Brush 1976: ch. 9.) The situation is further complicated when leadership of the research programme is taken over by scientists who disagree on what parts of it are most essential; as we have seen, it was really the Jeans–Jeffreys theory rather than the planetesimal hypothesis that was eventually rejected by astronomers.

The Lakatosians would presumably argue that this kind of outcome is still in accord with their methodology in the broad sense that it is research programmes rather than theories that are tested and accepted or rejected by scientists. But Lakatos has also stated that a research programme is not judged on the basis of how well it agrees with evidence, but by comparison with a competing programme: *"There is no falsification before the emergence of a better theory"* (Lakatos 1970: 119).

The history of the Chamberlin–Moulton theory provides not one but two counterexamples to this assertion of Lakatos.

First, the Nebular Hypothesis was abandoned because of the criticisms published by Chamberlin and Moulton in 1900, before any alternative theory had been developed to replace it. (Chamberlin repeatedly insisted that he did *not* consider his planetesimal hypothesis an extension of the earlier "meteoritic" theory.) While it is true that most scientists felt the need for *some* explanation of the origin of the Solar System, there was no other fully articulated research programme to which they could transfer their allegiance until 1905. By that time the Nebular Hypothesis was already dead – no one was using it as a basis for research, with the possible exception of W. H. Pickering and F. J. M. Stratton in their work on satellite orbits.[27]

27 See 1.2.4, note 8. Stratton (1906) used the Nebular Hypothesis though he admitted that it was no longer tenable without substantial modification. W. H. Pickering (1917, 1920) continued to support the Nebular Hypothesis supplemented by wandering "planetoids" that could collide with stars.

Second, the encounter theory (including all versions proposed by Chamberlin, Moulton, Jeans, Jeffreys, and others) was abandoned because of the criticisms published by Russell in 1935, before any alternative theory had been developed to replace it. Again, although a few scientists flatly declared that there was now *no* viable alternative[28] (Russell emphasized that the objections to the Nebular Hypothesis were still valid), the Russell–Lyttleton ternary theory soon emerged and won widespread though temporary support. But the highly artificial nature of this theory reflects the fact that it was produced in desperation *after* the refutation of the encounter theory. Other more or less implausible theories quickly followed, but none could command widespread support until Weizsäcker's theory a decade later. Indeed the state of cosmogony between 1935 and 1945 resembles a Kuhnian revolutionary period much more closely than a Lakatosian competition between research programmes.

A satisfactory interpretation of the history of the Chamberlin–Moulton theory ought to take account of the fact that it attempted to bridge two fields of science, astronomy and geology; this would be even more essential in discussing more recent developments where chemistry and nuclear physics are also involved. But a major defect of most philosophical accounts of science is their assumption that theories (or paradigms or research programmes) are generated, tested, and compared within a single field. One exception is logical positivism, which seeks the "unity of science" by *reduction* of theories in one field to those in another more fundamental field. Unfortunately it has been difficult to find legitimate examples of such reduction that fit the logical positivist prescription.[29]

The possibility of reduction presupposes a hierarchy among fields, some being considered more fundamental than others. In the second half of the 19th century Lord Kelvin and his colleagues argued that geology is less fundamental than physics; hence, theories such as the Uniformitarian geology of Charles Lyell must be abandoned if they came into conflict with principles of physics such as the dissipation of energy, and accounts of geological history must be compatible with calculations from heat conduction theory, which indicated that the surface of the Earth must have been hotter than the melting point of rocks a few tens of millions of years ago. The Nebular Hypothesis was thus a means for reducing geological theories (e.g., contraction of the crust resulting from cooling) to an astronomical theory (formation of Sun and planets from a hot nebula). Although alleged geological evidence for contraction and for a former molten state of the Earth was sometimes cited in support of the Nebular Hypothesis, geological evidence that might have contradicted the Nebular Hypothesis was not taken seriously (except by geologists). Partly because of Kelvin's attacks, and partly because of the spectacular advances in other fields during the early decade of the 20th century, geology

28 Jeffreys (1935; 1948: 103; 1952: 290). Armellini (1937). Crommelin (1938). Painter (1940: 62–81). Longwell (1941: 189).
29 Feyerabend (1962, 1965). Brush (1983: ch. 7).

was regarded as a less prestigious, less fundamental science whose theories ought to be reducible to those of physics, chemistry, and astronomy (*Transmuted Past*, 1.3 and 1.4).

One of Chamberlin's major goals after 1895 was to revive and strengthen geology by establishing fundamental principles that were consistent with physics and astronomy but did not require geological evidence to be slighted. Geology was to be an equal partner, not subservient to the passing fancies of those like Kelvin who imposed arbitrary time limits on the age of the Earth based on unrealistic mathematical models. This meant that Chamberlin had to win respect for the "naturalistic method" of geology among physical scientists and mathematicians who preferred quantitative deductive methods. It would be even better if he could show that this method led to new knowledge in astronomy – for example, in the interpretation of spiral nebulae – as well as planetary cosmogony.

What Chamberlin sought was a nonreductive "interfield theory" of the kind discussed by Lindley Darden and Nancy Maull (1977). Two of their conclusions about interfield theories (based primarily on examples from genetics and biochemistry) can be illustrated here.

First, the theory is generated because a problem arises in one field but cannot be solved using the concepts and techniques of that field; however, it is known that another field may be relevant. In this case the problem was the suspected relation between glaciation and the nature of the Earth's early atmosphere, which led Chamberlin to scrutinize the idea that the Earth was initially a hot liquid ball, and eventually to invade the domain of astronomy.

Second, the theory may focus attention on previously neglected items in both fields and predict new items. Moulton's critique of the Nebular Hypothesis, though in part a rehash of earlier criticisms, did make astronomers more aware of the quantitative relations of mass and angular momentum in the Solar System. Chamberlin's suggestion to look for certain kinds of internal motions in spiral nebulae was taken seriously by astronomers.

Darden and Maull also suggested that an interfield theory can generate new lines of research that may in turn lead to another interfield theory. While this did happen in cosmogony, in the sense that modern theories attempt to correlate evidence from several different fields, it is necessary to point out that *inter*field theories tend to be vulnerable to replacement by less comprehensive *intra*field theories. As Frankel (1976) noted in connection with Wegener's theory of continental drift, a theory that ranges over many diverse fields is unlikely to perform as well as a special theory in each individual field, and since most scientists are specialists rather than generalists they judge a theory primarily by how well it accounts for the evidence in their own field. He sees this as the main reason why Wegener's theory was rejected for several decades even though it had greater overall explanatory power than the established theories in geology, geodetics, paleoclimatology, and biology.

Since the Nebular Hypothesis was also an interfield theory, though a reductive one, it was similarly vulnerable to refutation, though as Chamberlin realized it had to be refuted in the domain of astronomy before the scientific community would abandon it. Refutation in the domain of geology was not sufficient or even necessary. But the Chamberlin–Moulton theory had to be defended against attack in both fields; no matter how strong it might be in geology, the fact that it was supposed to explain the origin of the Solar System meant that it could be refuted by Russell's simple arguments in celestial mechanics and by the Russell–Spitzer dissipation argument based on Eddington's theory of the thermal state of the Sun. Moreover, most of the theories that attempted to replace the Chamberlin–Moulton theory (including Jeans's tidal theory) did not take on the obligation of explaining geological as well as astronomical facts; Jeffreys was the only major theorist who was seriously concerned with both fields before the 1950s.

If there had been, in the 1920s and 1930s, a well-defined planetary science community that could have demanded that an interfield theory must only be required to compete with other interfield theories, the Chamberlin–Moulton theory might not have been abandoned so quickly. But Chamberlin's interdisciplinary team at Chicago, which should have been the catalyst for the growth of such a community as well as the source of brainpower for developing his own research programme, did not survive his death. The encounter between geology and astronomy produced a noticeable disturbance because of Chamberlin's incredible energy and perseverance, but the filament of bright young theorists drawn out of the reservoir of scientific talent did not acquire the right amount of personal momentum to go into stable orbits or condense to a family of cooperative followers of a research programme; instead, they dispersed to independent careers or fell back into obscurity.

It is surprising not that Chamberlin failed in his difficult and delicate undertaking, but that he came so close to success. In the words of Henry Norris Russell (1929: 7), the fact that his theory was a subject of lively controversy "in no wise obscures or detracts from the distinction which is due to Professor Chamberlin and his colleagues, for the origination of the most valuable and fertile contribution to the cosmogony of our system which has been made within the memory of living men."

PART 2

Nebular rebirth and stellar death

2.1

Introduction

Attempts to find a plausible naturalistic explanation of the origin of the Solar System began about 350 years ago (Descartes, 1664/1824–1826) but have not yet been quantitatively successful, making this one of the oldest unsolved problems in modern science. Most historical accounts have concentrated on the period from the mid-18th to the mid-20th century.[1] Thus, only those who have actively participated in cosmogonic research or have made a special effort to keep up with the voluminous technical literature are familiar with what has happened in the last 30 years. Popular accounts, even those written by participants, generally fail to do justice to the recent history; review articles are often composed by advocates of a particular theory and don't even mention all of the other theories.

2.1.1 The Solar System in the first space age

Since the period 1956–85 coincides almost exactly with the First Space Age – from *Sputnik* to the *Challenger* disaster – one might expect that new results from lunar and planetary exploration played an important role in the development of ideas about lunar and planetary formation. While this is indeed the case for selenogony, it was not true for the Solar System in general. Theorists occasionally suggested measurements that planetary probes could make to test predictions or provide constraints on models, but there is little evidence that such measurements influenced planetogonic[2] ideas. Ground-based observations were frequently more decisive. Conversely, particular missions or experiments were sometimes justified on the grounds that they would provide planetogonic information when in fact they had little direct relevance to the origin of the Solar System (Witting 1966: ii; 1969: 195).

The space program did have an important indirect effect on our subject by creating a planetary science community and encouraging the development of sophisticated experimental and computational techniques. In the United

1 Whitney (1971). Numbers (1977). Jaki (1978). Merleau-Ponty (1983). Treder (1987). Other works are cited in Brush and Landsberg (1985: 168–83).
2 We use the word for convenience and also to indicate that in discussing the origin of the *Solar System,* one usually includes the origin of the Sun, asteroids, satellites, meteorites, and comets only insofar as they seem relevant to *planet* formation. Like "selenogony" (referring to the origin of the Earth's Moon) and "cosmogony," the word "planetogony" can denote either the origin of the entity or a *theory* of that origin.

States, the National Aeronautics and Space Administration (NASA) funded the research of several theorists and enabled them to work out the consequences of their assumptions on electronic computers. Conferences held to discuss plans and results of the space program provided convenient forums to discuss the origin of the Solar System – which was, after all, prominently mentioned as a major goal of the program in nearly every official announcement about it.

But the space program had a greater impact on ideas about the formation of the Solar System by making possible new kinds of observations of the universe beyond our own system. Infrared, X-ray and gamma-ray telescopes on artificial satellites placed in orbit outside the Earth's atmosphere, combined with high-tech ground-based observations, provided crucial data on the early stages of star formation and on abundances of certain atoms and isotopes considered significant in planetogonic theories. Theories were also influenced by research on Solar System material that is delivered to Earth free of charge: the meteorites. The Allende meteorite, which fell in 1969, contributed significantly more to our understanding of the Solar System than the lunar samples obtained by the *Apollo 11* mission that year.

Has the tremendous quantity of empirical knowledge we now possess about the Solar System answered the question of its origin? Can we explain the existence and properties of the planets any better now than we could 50 or 100 years ago? Is it still true, as Boris Levin complained more than two decades ago (1972), that we have no consistent picture of the formation of the Solar System? Have we simply accumulated a lot of facts that we don't understand?

By "don't understand" I mean that – at least in the late 1980s where my historical account ends – the most elementary question could not be confidently answered: Does a Solar System naturally form by itself from primordial matter, or does it require assistance from a previously existing star or other entity? On the answer to this question depends the frequency with which planetary systems can be expected to form in our galaxy and hence the probability that other intelligent life may exist.

Other "elementary" questions that few theories even tried to answer include: Why are there nine planets? Why does Bode's law or a similar numerical rule give such a good fit to the sizes of most planetary orbits? It seems that a relatively small amount of effort was devoted to the final outcome of the planetary formation process compared with its early stages.[3]

2.1.2 Planet formation as a contemporary process

Although scientists were still profoundly ignorant on these and other aspects of planetogony, they did have a new outlook on the subject, more optimistic

3   McCrea (1974). Levy (1985: 11).

than that prevailing in the 1930s and 1940s. The origin of our own system was no longer seen as a mysterious event that took place in the distant past under conditions vastly different from those prevailing in the universe today, and hence almost inaccessible to objective inquiry. Instead it was believed that the Solar System was formed in an environment similar to that which exists now throughout the galaxy, by processes many of which can be directly observed.

One reason for this view, aside from the already mentioned observations hinting at star and planetary formation, was that the *age* of the Earth and the rest of the Solar System seemed to be established quite accurately and was much less than that of the galaxy or the universe. Up to about 1952 the age of the universe inferred from the Hubble constant for expansion was actually *less* than the best value for the age of the Earth as estimated from radiometric dating (1800 to 2000 million years compared with 3000 to 3500 million years). Even if one ascribed large uncertainties to both estimates, it still appeared that the Solar System was formed during the infancy of the universe and hence under conditions unlike the present ones. The discrepancy was removed when the astronomers revised their distance scale. The accepted value of the age of the Earth, first proposed by Clair Patterson in the mid-1950s, remained at 4500 million years, compared with values ranging from 10,000 to 20,000 million years for the galaxy and the universe.[4] On the basis of these numbers astronomers concluded that the Earth is relatively young[5] and was formed by processes similar to those we might be able to observe in the galaxy today.

Although radiometric dating thus favored what geologists might call a "Uniformitarian" view of the formation of the Solar System, it allowed a "catastrophic" theory of the origin of Earth's Moon to emerge. Before 1969, one could extrapolate the history of the lunar orbit back to an epoch, more recent than the Earth's own formation, when the Moon was captured by or ejected from the Earth; this might even be supposed to have left surviving traces on the Earth's surface. When it was discovered that some lunar rocks are at least as old as the Earth, such theories were excluded. Other evidence from lunar samples seemed to refute *all* pre-1969 theories. Instead, selenogony turned for guidance to a theoretical picture of the Solar System as it might have been just before the present planets emerged from a battlefield of colliding smaller objects and postulated an impact far more spectacular than anything that could plausibly be imagined to occur in the past 4000 million years.

The preplanetary stage became observable in the 1980s: Satellite and ground-based observations detected disks formed from solid bodies surround-

---

4 See *Transmuted Past*, 2.2.7. For estimates of the age of the universe in the 1980s, see Van den Bergh (1981), Edwards (1982), Hodge (1984), Sandage and Tammann (1986), Fowler (1987), Pierce and Tully (1988), Tully (1988).

5 Of course, 4.5 thousand million years is *old* relative to most *earlier* estimates of the age of the Earth; in particular, radiometric dating completely demolishes "young-Earth creationism" (Brush 1982a).

ing nearby stars.[6] Claims to have detected planetary companions of other stars were occasionally made before 1956 but remained of doubtful validity; in the next three decades such claims were increasingly frequent and somewhat more credible.[7] Planetogonists finally had the opportunity to make predictions that might soon be testable on systems other than our own.

6 Aumann et al. (1984). Black et al. (1984). Harper et al. (1984). Smith and Terrile (1984). Appenzeller and Jordan (1987). I remind the reader that this is a history of the subject up to the mid-1980s and does not attempt to cover more recent events.
7 Reuyl and Holmberg (1943). Strand (1943). Van de Kamp (1956, 1986). Gatewood and Eichhorn (1973). Heintz (1976, 1978).

2.2

Methodology

Historians of science generally have little use for philosophical analysis. It would just get in the way of telling a good story. Unfortunately in this case our story lacks a dramatic climax. Despite much talk by the actors themselves about "clues" and "detective work," the mystery has not been solved; the perpetrator has not been unambiguously identified and we aren't even sure we have found the murder weapon. Yet it may be precisely because of the lack of a satisfying conclusion that this period in the history of cosmogony offers an excellent opportunity to see how science makes cumulative progress by small steps and how scientists frustrated by the slow pace of progress, yet certain that they are gradually developing more powerful methods of research, may be forced to articulate their goals and procedures in ways that would be unnecessary if they could point to sensational discoveries. And when scientists speak philosophically it is only fair that we pay attention to what philosophers say about science.[1]

2.2.1 Facts to be explained

Let's begin with the formalities. Every popular novel has to have a bedroom scene. Similarly, every survey article on the origin of the Solar System includes a list of "facts that any satisfactory theory must explain." I have collected in Table 1 all the items included in at least two such lists, and in Table 2 I have noted the priority assigned to these items by the various authors.

Ostensibly the purpose of presenting such a list is to establish criteria for eliminating inadequate theories and for choosing the best from those that are adequate. But that purpose is almost never accomplished since *no* theory is generally believed to give a convincing explanation of all the facts.

One might suspect each author is tempted to stress those facts that his or her own theory can best explain. One or two authors may have succumbed to this temptation, but for the most part those with radically different theories agreed on the list of facts to be explained.

I conclude therefore that the primary function of presenting these lists is merely to inform the reader about the domain of subject matter to be dis-

---

1 A comparable period of slow progress occurred in physics and chemistry in the late 19th century; the philosophical writings of Ernst Mach, Pierre Duhem, Wilhelm Ostwald, and other scientists comprise one legacy of that period.

Table 1. *Facts to be explained by theories*

A. The Sun has a very small part of the angular momentum of the Solar System despite having most of its mass; the giant planets have most of the angular momentum.
B. There are nine planets.
C. The orbital motion of the planets is quite regular: (a) All move in the same direction around the Sun; (b) their orbits are nearly copolanar; (c) their orbits are nearly circular. But the small deviations from regularity may be significant.
D. The set of orbital radii of the planets follows an approximate numerical pattern (Titius–Bode law); the radius of the asteroid belt also fits this pattern.
E. There are two groups of planets: "terrestrial" with masses less than or equal to that of the Earth and high densities; "giant" with much greater masses and lower densities.
F. The giant planets (except Neptune) have many satellites, nearly all with direct orbital motion; the terrestrial planets have few satellites or none.
G. There are distinctive differences in the chemical compositions of the terrestrial and giant planets. Earth and other terrestrial planets have much greater than solar abundances of Li, Be, and B, but much less H and He.
H. Most planets have direct rotation (same sense as orbital motion) but a few are retrograde, and Uranus has its rotation axis nearly perpendicular to its orbital axis.
J. There are numerical regularities in the distances of satellites of giant planets from their primaries.
K. There is a belt of small bodies ("asteroids") between Mars and Jupiter.
L. The Earth has a satellite that is unusually large compared with its primary.
M. Numerous comets and meteors move through the Solar System with somewhat irregularly distributed orbits.

Table 2. *Priorities of facts (Table 1) according to various authors*

	A	B	C	D	E	F	G	H	J	K	L	M
McCrea (1963)	1	2	3	10	5	6		4		8	7	9
Witting (1966, 1969)	1		4	4	8		10					
Harr (1967)			1	2	3	4			4			
Herczeg (1968)	4		1	2	3	8	5	7				
Williams and Cremin (1968)	2	1	3	6	4		5					
Woolfson (1969)	1	2				4	3	6		7	8	
McCrea (1974)		1		5		6	3	4				
Mestel (1974)	2		1	3	5	4			4			
Dormand (1973)	2	1		6		4	3	7		8	8	
Hartmann (1987)	10		1	4	5	9	5	3	9			8
Williams (1979)	1	2	3	5			4					

cussed rather than to provide a serious basis for choosing the best theory. But even this purpose is not well served, for the lists fail to reflect changes in the boundaries of the domain during the period under discussion. In particular, hardly any author listed star formation or isotopic anomalies in meteorites, yet much of the discussion after 1960 centered on those topics. Witting (1966, 1969) is a notable exception; he called for specific measurements directly related to facts of theoretical significance, such as isotopic abundances, and for theoretical and observational research on star formation.

The lists have about as much relevance to the discussion of planetogonic theories as the bedroom scene has to the plot of the typical novel. And indeed they can also mislead the reader by emphasizing creation at the expense of development. How do we know that all the facts listed in Table 1 were established at the *beginning* of the Solar System? Perhaps the regularities of planetary motion are not original but were slowly established over billions of years by dissipative processes and perturbations. "There is no point in building an elaborate theory of the formation of planets to explain some apparently clear-cut present-day property of the system if in fact that property is not really original at all but has come about otherwise" (Lyttleton 1968: 5).

2.2.2 Prediction or explanation?

One step higher on the ladder of methodological analysis is the issue of prediction versus explanation. Should more credit be given to a theory that not only explains known facts about the Solar System, but also allows the theorist to derive facts discovered in the future?

Many scientists have accepted Karl Popper's (1934/1959) doctrine that in order to be truly scientific a theory must lead to predictions that can be refuted by new experiments or observations. This "falsifiability criterion" has recently caused considerable mischief because Popper himself used it to conclude that Darwinian evolutionary theory is not scientific but is merely a "metaphysical research programme"; it is used to explain what has already happened but does not determine what will happen in the future (Popper 1974: 133–43). If applied in Popper's original sense, the criterion would put not only evolutionary biology but much of geology and astronomy outside the boundaries of science, since those disciplines study phenomena that take place over such broad domains of space and time that they cannot be brought into the laboratory for controlled experimentation, and it would seem that predictions could not be checked for thousands or millions of years. The demand that theories make testable predictions has therefore been seen as just another attempt to force all sciences into the mold of physics and chemistry; defenders of the autonomy of other sciences argue that they should be expected to provide only plausible explanations, not predictions.[2] But a recent comprehensive analysis of historical case studies shows that even in the physical sciences, prediction plays a fairly minor role in the assessment of theories (Laudan, Laudan, and Donovan 1988: 18–20, 29; see also Brush 1995).

At the 1984 conference on the origin of the Moon, when one participant argued that we should take seriously only those hypotheses that are testable, another retorted that such behavior would be comparable to that of the drunk who looks for his wallet only under the street lamp.[3] In other words, the set of theories that can be definitively tested by present-day observations is relatively small and may not include the one that is later accepted as correct.[4]

Nevertheless, there have been some striking cases of testable predictions deduced from theories about the past. The test in each case was the discovery of previously unknown evidence about the past rather than the occurrence of a new event, but I see no reason why such a discovery should not count as the test of a prediction. (As W. M. Kaula has remarked to me, it's not the phenomena themselves that are predicted, but the observations thereof.) The classic example is Charles Darwin's prediction that the earliest hominids would be found in the place where the primates most closely related to humans (according to his theory) now live, namely, Africa; the prediction was confirmed more than 50 years after it was published.

A second famous example is the prediction made from George Gamow's "Big Bang" cosmology that space should now be filled with "fossil" background radiation whose frequency distribution is given by Planck's law and corresponds to an absolute temperature of a few degrees Kelvin. The discovery (but not the prediction) was rewarded by a Nobel Prize and is generally considered the best evidence for the theory (Brush 1993).

The third example is more relevant to our present topic and also shows that the *refutation* of a prediction may have a significant effect on the progress of science. Harold Urey, from his hypothesis that the Moon was captured by the Earth after being formed elsewhere in the Solar System, predicted that water and other substances are now present near its surface. Following the analysis of lunar samples, Urey recognized that his predictions about the chemical properties of the Moon had been falsified and he abandoned the theory (4.3.4). In fact most scientists decided, on similar evidence, that *all* pre-*Apollo* theories had been refuted; this cleared the way for a completely different theory to be developed – a theory that would previously have been dismissed as too improbable (the giant-impact theory; see 4.4.5).

An important point, often neglected in philosophical discussion, is that a theory cannot gain credit through successful predictions unless it is already

---

2 Mayr (1965, 1985). Munson (1971). Halstead (1980). Lewin (1982). Laudan (1983: 111). But see Williams (1981) and references cited therein for testable predictions of evolutionary theory.

3 See 4.4.7; cf. Hartmann, Phillips, and Taylor, (1986: vi).

4 "I've got the Kuhn & Popper knee-jerk philosophy of science blues. . . . Any hypothesis can be falsifiable if it's sufficiently dumb." J. Chester Farnsworth, quoted in Horn (1986: 573).

considered acceptable on other grounds. This is illustrated by the case of planetary rings, discussed in Section 3.4.5.

Although the testing of specific predictions is not a common feature of research on the early history of the Solar System, it cannot be completely excluded as a method for evaluating hypotheses. In any case it should be kept in mind that scientists sometimes use the term "prediction" to mean not the forecast of something not yet known, but the deduction of something already known (McCrea 1960: 260–1; Brush 1995).

Explanation is thus a major function of theories. As Lyttleton (1968: 5) pointed out, scientists "cannot really quite relax" until they are assured that "the laws of science are sufficiently comprehensive to allow the solar system to happen" – we demand that the origin of the Solar System be explained without invoking any supernatural events.

2.2.3 Deduction or naturalism?

Indeed the dominant philosophy of most physical scientists is better characterized as the view that theories should be *deductive* than that they should be predictive, the latter being only a special case of the former. If one could derive the major properties of the Solar System from a few simple and plausible postulates, without using adjustable parameters or ad hoc hypotheses, one would have a successful theory even if no *new* facts were predicted.[5]

The most extreme version of deductivism was articulated by James Jeans, who asserted that one should start from a simple initial state, deduce its evolution, and then look around in the real world to see if one could find anything corresponding to the result (Jeans 1919: 18). The opposite extreme is the "naturalistic" or "actualistic" method advocated by T. C. Chamberlin (1.2.2) and Hannes Alfvén: One should start from facts and processes observed in the present-day Solar System and gradually work back to earlier stages, without ever postulating an idealized initial state (Arrhenius 1972/1974; Alfvén 1976/1978a: 26–7). Yet Chamberlin and Alfvén do postulate former states of the Solar System from which they try to infer its evolution following physical laws. In fact, Alfvén's theory has probably generated more testable predictions than any other.

Deduction is the preferred method of theorizing, subject to the important qualification that most theorists no longer aspire to derive the entire process starting from diffuse matter and ending with today's Solar System. Instead they divide the process into distinct parts in which different phenomena are supposed to be important, and are content to tackle one part at a time. Thus, one calculation shows how a dense molecular cloud collapses to a hot protostar; another follows the condensation of a gas-dust nebula to streams of small solid particles in Kepler orbits around a central star; a third shows how plan-

5 But it seems that such a theory would necessarily entail *some* predictions of presently unknown facts; cf. Lyttleton (1961: 54).

etesimals might accumulate to form terrestrial planets. Induction still coexists with deduction, though not too peacefully, especially in meteoritics whose leaders insist that they can confidently infer from the structure and composition of chondrules a former high-temperature stage not recognized in astrophysical models (Boynton 1985: 772; Wasson 1985: 136, 156–7, 184–5; Wood 1985: 701).

2.2.4 Theory change

A more general question of interest to historians and philosophers of science, going beyond issues such as prediction/explanation and deduction/induction, is: How does a scientific community decide to accept or reject a theory? The question arises because detailed study of several specific episodes shows that scientists do not generally follow the rules of "scientific method" as decreed by traditional philosophy of science. They don't abandon a theory just because one of its predictions has been falsified, and they don't change from one theory to another because of the result of a "crucial experiment" (Brush 1974).

A well-known attempt to explain theory change is Thomas S. Kuhn's *Structure of Scientific Revolutions* (1970). Kuhn describes an idealized process in which a "paradigm" comes to dominate a scientific discipline, then is undermined by the discovery of anomalies that it cannot explain, and finally is replaced by a different paradigm that may not be objectively superior in all respects but includes a new way of looking at the world and new criteria for judging theories. Although no actual episode of theory change conforms exactly to Kuhn's description, important features of the idealized Kuhnian revolution can be found in the adoption of Newtonian physics, Darwinian biology, quantum mechanics, and plate tectonics.

Kuhn's theory has limited usefulness in understanding the recent history of planetary cosmogony because there has been no fundamental revolution in planetary science, despite an immense increase in the quantity of empirical data and the emergence of several new theoretical concepts in the past 50 years.[6] There is no single paradigm in the sense that Newtonian or quantum mechanics provided paradigms for physics; instead, researchers in planetary science follow the paradigms they learned in astronomy or geology or chemistry or physics. The phenomenon of "partial incommensurability," which Kuhn invoked to explain the difficulties of communication between followers of the old and the new paradigms within a science, is much more striking in planetary science; it does not disappear through the fading away of an old paradigm, but it may be ameliorated as scientists learn to work within two or

6 Alfvén (1981a: 3) refers to his own theory of plasmas as a paradigm, but according to Kuhn's definition this is not correct unless the scientific community accepts it.

three paradigms at the same time.[7] This outcome would be contrary to Kuhn's theory, which denies the possibility of coexistence of more than one paradigm in a scientific community and generally ignores the interaction of different sciences.

The strongest case for the existence of a single shared paradigm in planetogony was made by George Wetherill (1988), who claimed that the community had now arrived at "a more normal type of science, which is not simply pursued by eccentric, elderly gentlemen who fight with one another's theories," but rather one in which "one can undertake finite tasks and receive rewards in the form of an audience and perhaps even employment." This was because the community had agreed on a "standard model of solar system formation, which of course has many bifurcations and branch points and mainly poses questions rather than gives answers." The existence of such an agreed-upon model gives meaning and value to the solution of special problems such as "whether or not planetesimals would be expected to disrupt as they go by a planet." Yet he considered it inappropriate to call this model a "paradigm" – a term that "has meaning only in retrospect, not in the real time advance of science."

A. G. W. Cameron did accept the term "paradigm"; he asserted that there had been a "transition" in Solar System cosmogony that was "marked by a widespread acceptance among the relevant scientific communities of a central paradigm that would allow individual investigators to address parts of the overall problem." According to this paradigm, star formation is the beginning of a monistic process leading to the Sun, planets, satellites, and comets. Papers that do not assume the central paradigm need not be discussed in a review article (Cameron 1988: 441–2).

Reflecting the acceptance of this terminology, the Astronomical Society of the Pacific scheduled sessions entitled "Constructing a Paradigm for the Formation of Planetary Systems" and "Testing the Paradigm" at its centennial meeting held in Berkeley in June 1989.

I would add that it is only insofar as the community does agree on a standard model or paradigm that one can understand why no credit is given to successful predictions based on *non*standard models.

Yet this limited agreement on a standard model was achieved only quite recently, if at all (3.4.8). To describe the earlier period, the "methodology of

7 "In my opinion our difficulty in understanding the nature of the energetic chondrule- and CAI [calcium-aluminum-rich inclusions]-forming processes stems not from a shortage of data or observational opportunities, but from the fact that the problem lies squarely on the boundary between two dissimilar disciplines, mineralogy and astrophysics. Mineralogists lack the background to understand the forces and processes in the nebular regime where their meteorites were formed; astrophysicists lack the background to evaluate the conflicting interpretations they hear from mineralogists, or to appreciate the reality of the constraints that meteoritic data place on nebular processes. Solution of the chondrite problem will require a degree of collaboration between the two disciplines that has not yet been attempted" (Wood 1985: 701–2).

scientific research programmes" proposed by Imre Lakatos (1970) may be more useful. A "research programme" is not as all-encompassing as a paradigm; it is a series of theories developed for a specific purpose, in competition with other programmes. Lakatos made the important point that scientists do not treat theories as independent entities, each to be tested against evidence and accepted or rejected on its merits; instead, a scientist will continually modify his or her theories in the light of new evidence. A minor part of the theory (an "auxiliary hypothesis") may be dropped and replaced by one that agrees better with observations, but the new theory obtained in this way will share a set of basic assumptions (the "hard core") with the earlier one. Lakatos asserted that scientists do not directly test theories against observations and do not abandon a research programme just because one or several of its predictions have been falsified; instead, they look at the track records of competing programmes and may switch their allegiance to a more successful one. Lakatos claimed to avoid the irrationality of Kuhnian paradigm changes by providing objective criteria for judging the track record of a research programme as "progressive" or "degenerating." He did not claim that scientists actually follow these criteria in the short term, but that the long-term behavior of the community can be explained in this way.

Historians and philosophers of science have published a number of criticisms and modifications of the Lakatos methodology. In an earlier section (1.3.5), I pointed out that Lakatos's claim, that scientists do not abandon a refuted theory unless a better one is available, was contradicted by the general rejection of the tidal hypothesis after 1935 despite the absence of any acceptable alternative. Thus, it should be clear that I do not consider Lakatos's theory an accurate description of theory change in science. Nevertheless, I find it a useful idealized model for discussing the behavior of scientists in choosing and modifying theories, just as Kuhn's theory is a useful idealized model for discussing the behavior of scientists in articulating and switching paradigms. Both theories have stimulated research in the history and philosophy of science by suggesting new hypotheses about the behavior of scientists, even if those hypotheses have not been verified.

2.2.5 Scientometrics

Although my interest is primarily in the intellectual rather than the social or institutional history of science, I think there is some value in compiling data on the support and dissemination of research. In particular, I have made considerable use of the *Science Citation Index* (*SCI*) developed under the direction of Eugene Garfield at the Institute for Scientific Information in Philadelphia. With the help of the *SCI* it is possible to locate most of the articles that referred, during the past three decades, to any specified publication. By actually reading these articles – *not* by just *counting* the citations – one can gain a fairly good idea of how the major planetogonic theories were regarded by the

Table 3. *Publications supported by funding agencies*

	1956–65	1966–75	1976–85	30-year total
NASA	26	71	161	258
NSF	7	30	51	88
AEC/ERDA/DOE	16	12	7	35

Note: NASA (National Aeronautics and Space Administration), NSF (National Science Foundation), AEC/ERDA/DOE (Atomic Energy Commission/Energy Research and Development Administration/Department of Energy).

Table 4. *Institutions of authors of publications*

	1956–65	1966–75	1976–85	30-year total
California Inst. of Tech. and/or Jet Propulsion Lab.	5	14	36	55
University of Chicago and/or Yerkes Observatory	11	17	22	50
Harvard and/or Smithsonian Astrophysical Observatory	6	9	27	42
University of California at San Diego, La Jolla	11	18	13	42
NASA-Ames Research Center, Moffett Field, CA	0	8	24	32
(Schmidt) Institute of Physics of the Earth, Moscow	10	10	7	27
Massachusetts Institute of Technology	2	11	8	21
Goddard Institute for Space Studies, New York	10	7	1	18
Max-Planck-Institutes, Germany	0	2	17	19
University of Arizona, Tucson	2	6	11	19
Royal Institute of Technology, Stockholm, Sweden	3	11	4	18
Rice University, Houston	0	2	15	17
Carnegie Inst. of Wash. and/or Mt. Wilson and Palomar Obs.	3	2	11	16
Yale University	3	8	3	14
University of California at Los Angeles	0	4	9	13
Goddard Space Flight Center, Greenbelt, MD	3	2	7	12
Australian National University, Canberra	4	5	2	11
University of Cambridge	4	6	1	11

scientific community during this period. Unfortunately the *SCI* does not include citations from most of the important conference proceedings and book-length surveys on planetary science; my own research suggests that these publications are absolutely crucial for substantive assessments of theories.[8]

In the course of preparing the bibliography for this chapter (and 4.4), which of course included verifying each citation, I decided that with a small amount of additional effort I could record the institutions at which the authors conducted their research and their sources of financial support. This information is noted only when it is stated in the article itself and only when the article is related in some way to planetogony and was published in the period 1956–85. Tables 3 and 4 present the results of this compilation.

8 The *SCI* does not cover the proceedings of conferences on the origin of the Solar System edited by Jastrow and Cameron (1962/1963), Reeves (1972/1974), Wild (1974), and Brahic (1982); it does not cover the cooperative review volumes entitled *Protostars and Planets* edited by Gehrels (1978) and by Black and Matthews (1985). It does not cover the proceedings of the Soviet–American Conference on Cosmochemistry, edited by Pomeroy and Hubbard (1974/1977). It covers some but not all of the *Proceedings* of the Lunar and Planetary Science Conferences held annually at Houston since 1970; it covers *none* of the volumes of *Lunar and Planetary Science,* containing long summaries of papers presented at those conferences (and in some cases not published in the *Proceedings* or elsewhere, e.g., Cameron and Ward 1976). It apparently does not cover most of the proceedings of the Soviet conferences on cosmogony, *Voprosy Kosmogonii.*

2.3

Nuclear cosmochronology and Hoyle's research programme

2.3.1 Hoyle on the early Sun

During the period when he was working on nucleosynthesis with Fowler and the Burbidges (*Transmuted Past*, 2.3.5), Hoyle was developing his own theory of the origin of the Solar System, qualitatively outlined in a popular book on astronomy (Hoyle 1955: ch. 6) and presented in more detail in a 1960 paper.[1] He suggested that the Sun and planets were formed in "a whole shower of stars" from a cloud enriched by material from a supernova explosion (1955: 83–4). A typical interstellar cloud has a rotational speed of 1 cm/sec, which would be increased to 100 km/sec if it simply contracted to form a star; there must be some process that acts to reduce this speed, presumably Alfvén's magnetic-braking mechanism. Hoyle also invoked magnetic forces to explain how heavy elements (from the supernova) were separated from hydrogen (from the interstellar cloud) after first being mixed with it. Magnetic forces pushed gases away from the Sun but did not move the refractory substances that had condensed out as solids or liquids. This means that the disk of material around the Sun has already become fairly cool – not much more than $1000°K$ – yet his theory also required a sufficient amount of ionized gas to maintain the magnetic coupling between Sun and disk needed to transfer angular momentum.

When J. H. Reynolds (1960) announced his discovery of anomalous ^{129}Xe in the Richardton meteorite, he interpreted it to mean that 350 million years had elapsed between the time of nucleosynthesis and the formation of the meteorite. Fowler and Hoyle (1960: 286), using a more elaborate model of continuous nucleosynthesis over an earlier time period, concluded that the time interval after the last nucleosynthesis was only 200 million years. Reynolds noted the discrepancies between different estimates but considered them of little significance – the important point was that "the possibility that billions of years intervened between the formation of the elements of the solar system and the time its planetary bodies were formed is now conclusively ruled out."[2]

---

1 See his recollections (Hoyle 1986). He had earlier "gotten trapped into a wrong line of attack" (involving multiple stars; see Hoyle 1944, 1945) "because this was the trend of research immediately prior to my own work" (Hoyle 1965: 82).
2 Reynolds (1960: 182); see also Cameron (1962b), Kuroda (1961).

Figure 13. Fred Hoyle at a 1973 conference (American Institute of Physics, Niels Bohr Library, E. E. Salpeter Collection).

But the problem of explaining the production of light elements (deuterium, lithium, beryllium, boron) surfaced again with a new twist. Walter K. Bonsack and Jesse L. Greenstein reported (1960) that four T Tauri stars have about 10 times greater lithium abundance (relative to heavy elements) than the Sun. They suggested that the lithium is formed in the atmospheres of the stars by spallation; this would require a source of high-energy protons. Thomas Gold, who collaborated with Hoyle on a paper discussing the production of solar flares by the sudden release of energy stored in twisted magnetic lines of force (Gold and Hoyle, 1960), suggested that a similar process could account for the high abundance of lithium in the Earth as well as in T Tauri stars (Gold 1960: 275). Gold's picture of the formation of planets was qualitatively similar to Hoyle's – they come from material ejected from the Sun's outer envelope and are then pushed outward as they acquire additional angular momentum from the Sun by magnetic interactions.

2.3.2 The Fowler–Greenstein–Hoyle theory

Hoyle's theory of the origin of the Solar System, the B^2FH project on the origin of the elements, and the new ideas about the formation of lithium were combined into a new theory by W. A. Fowler, J. L. Greenstein, and Fred Hoyle (1961).[3] These authors will be referred to as "FGH." In Hoyle's theory (Gold and Hoyle 1960) angular momentum is transferred from Sun to planets in a process that entails dissipation of magnetic energy through a "sequence of powerful solar flares similar to those which occur on the surface of the present sun." The magnetic energy is converted into kinetic energy of high-speed particles, mostly protons, that travel out along the lines of force to bombard the condensed preplanetary material. The bombardment produced spallation products and neutrons (Fowler, Greenstein, and Hoyle 1961: 395).

In a discussion of the recent suggestions that short-lived isotopes such as ^{129}I, ^{107}Pd, and ^{26}Al were produced just before the formation of the Solar System and incorporated into parent bodies of meteorites before they decayed,[4] FGH doubted that these isotopes could be attributed to "galactic" nucleosynthesis. In what seems to be the first published reference to the idea of a supernova trigger for the origin of the Solar System,[5] they wrote:

It can be argued that a supernova event triggered the condensation of the solar system or that the sun originated in an "association" of stars such as those young stars which seem to be moving radially outward from a common point. Discounting these interesting possibilities, it would seem reasonable to ascribe the relatively late production of the I^{129}, Pd^{107}, and Al^{26} to the synthesis in the solar nebula described in this paper. (Fowler, Greenstein, and Hoyle 1961: 403)

In a longer paper published the next year, FGH pointed out a serious difficulty in their own theory. If Li and B are produced by spallation, theory predicts that roughly equal amounts of ^{6}Li and ^{7}Li will be formed, and likewise roughly equal amounts of ^{10}B and ^{11}B. But the observed abundance ratios are ^{6}Li/^{7}Li = 0.08 and ^{10}B/^{11}B = 0.23. To avoid this difficulty they proposed an auxiliary hypothesis: A large flux of thermal neutrons depletes the ^{6}Li and ^{10}B while augmenting the ^{7}Li through the reactions ^{6}Li$(n, \alpha)^3$H and ^{10}B $(n, \alpha)^7$Li (Fowler, Greenstein, and Hoyle 1962: 150). This process would not be effective in the presence of hydrogen since the thermal neutrons are captured by hydrogen. Hence, they must also postulate that most of the hydrogen present in the original solar nebula has already escaped. But not all of it, since they need a *few* hydrogen atoms to manufacture deuterium from the neutrons by the reaction ^{1}H $(n, \gamma)^2$H. So "the D Li Be B abundances

3   The theory was first announced in the Richtmyer Memorial Lecture in New York City on 2 February 1961 by Fowler, who called it "Deuteronomy" since it proposed to account for the formation of deuterons.

William Alfred Fowler (1911–1995) earned his Ph.D. in physics from the California Institute of Techology in 1936. He was on the faculty at Caltech from 1939 until his retirement in 1982. He received the 1983 Nobel Prize in Physics for his experimental research on nuclear reactions relevant to astrophysics. Bethe (1983) gave an authoritative account of this work. See the autobiographical article (Fowler 1992).

4   Reynolds (1960). Murthy (1960). Fish, Goles, and Anders (1960).

5   The idea of a supernova trigger for *star* formation had been discussed earlier by Öpik (1953, 1955a), Hoyle and Ireland (1960), and Murthy and Urey (1962: 628).

demand a stage for primitive terrestrial material intermediate between the original solar and the present terrestrial composition" (Fowler, Greenstein, and Hoyle 1962: 180).

In describing the physical conditions in the solar nebula, FGH relied on the earlier model of stellar evolution worked out by Henyey, LeLevier, and Levee (1955). This gives a gradual increase in luminosity as the radius shrinks in the early contracting phase of the star, following the relation $LR^{0.78}$ = const. (Schwarzschild 1958). FGH estimated that the temperature at the surface of the solar condensation does not exceed 500°K, and at 1 AU it has dropped to 170°K. Iron and oxidized silicates have already condensed near the primeval sun; ice forms in the region from Venus to Mars and promotes the growth of planetesimals to metric dimensions. In the region of Jupiter and Saturn, solid ammonia plays the role of ice but, in addition, H and He gases remain and can be accreted by gravitational capture.

In contrast to the proposal of Fish, Goles, and Anders (1960) that ^{26}Al was formed by galactic nucleosynthesis and incorporated into asteroid-sized bodies, where it provided a heat source, FGH argued that their theory does not demand such a rapid formation of large bodies (in only a few million years). Instead the ^{26}Al can be produced by "last minute synthesis in the solar nebula," for example, by solar proton bombardment through the reactions ^{26}Mg(p,n)^{26}Al, ^{27}Al(p,pn)^{26}Al, and ^{28}Si(p,^{2}pn)^{26}Al.

2.3.3 Objections to Hoyle's theory

Hoyle presented his theory at a conference at the Goddard Institute in New York City in January 1962 and encountered two objections. First, E. J. Öpik criticized Hoyle's assumption that hydrogen and helium could be selectively lost from the nebula simply because their molecules attain sufficient velocity to escape from the gravitational field of the Sun. According to Öpik, gases may be "blown off" from the nebula, but there is no separation process that would effectively eject hydrogen and helium without getting rid of the other elements as long as they are all in the gas phase (Öpik 1962/1963: 75). I have not found any published reply by Hoyle to this criticism.

The second objection was based on Hayashi's theory (*Transmuted Past*, 2.3.4) that the Sun went through a convective, highly luminous phase early in its evolution. This would prevent water from condensing at distances of the order of 1 AU, as proposed by Hoyle to assist the growth of planetesimals. The existence at this distance of metric-size planetesimals, containing heavy elements that could be split into light elements by proton bombardment as postulated in the FGH theory, was thus called into question (Moss 1968; Tayler 1968).

To counter this threat to his theory, Hoyle undertook a critical examination of Hayashi's theory to see whether his conclusions "are really inescapable or not." This critique was admittedly motivated by a desire to preserve the role of an icy matrix in forming planetesimals in the FGH theory (Faulkner, Griffifths, and Hoyle 1963: 2). On the basis of his own calculations Hoyle had to admit that a fully convective, highly luminous stage must occur, although the theoretical temperature and luminosity could be somewhat reduced by postulating the presence of additional free electrons produced by high-energy particles, or by inserting a magnetic field. The Hayashi effect could explain so many astrophysical facts as to make it

virtually certain that very deep convection did occur in the Sun as it approached the main sequence . . . if it were not for the necessity for the condensation of water within the elementary planetesimals, we should feel that the observational evidence distinctly favoured the completely convective models of Hayashi. But on the other hand the requirement for the condensation of water appears so strong that an episode of low luminosity for the primitive Sun seems essential. (Faulkner, Griffiths, and Hoyle 1963: 9)

Hoyle eventually decided to abandon the hypothesis that planetesimals are formed with the help of ice in the region of the terrestrial planets. He reformulated his theory, keeping the assumptions (1) that the contracting Sun becomes rotationally unstable and ejects a disk of material, and (2) that angular momentum is transferred to this disk by magnetic coupling, but with the new assumption that (3) these processes occurred during Hayashi's "overluminous" stage. Temperatures throughout the nebula would be much higher than in his previous theory; as a result, only metallic Fe, MgO, and SiO_2 would condense in the region of terrestrial planets. Water would condense only at the distance of Uranus or Neptune. To aggregate the planetesimals without ice he had to rely on the stickiness of hot metallic iron – but this problem was dismissed in one sentence. Without primordial water, Hoyle's Earth now had to acquire water (and other volatiles) by capturing gases ejected later from the Sun. The major advantage of the theory was claimed to be its explanation of "why the inner planets are largely made up of iron and magnesian silicate" (Hoyle and Wickramasinghe 1968: 415). It also implied that the Earth was formed hot ($>1000°K$) rather than cold: not necessarily molten at first but with the possibility of large-scale convection and energy release associated with the flow of iron toward the center.

2.3.4 Decline of the Fowler–Greenstein–Hoyle theory

The FGH theory was facing other difficulties in persuading the experts to accept its explanation for the origin of light elements. Cameron (1962b) dismissed it as "unlikely" and inconsistent with chemical and isotopic data on meteorites (1963b; 1965: 570). V. Rama Murthy and Harold Urey (1962) preferred to believe that these elements were synthesized before the formation of the Solar System. McCrea complained that it had too many adjustable parameters (1963: 286). Within two or three years Fowler and Hoyle were beginning to have serious doubts about their own theory. A paper by Burnett, Fowler, and Hoyle (1965) noted that meteoritic and terrestrial material had the same isotopic composition for Li, Gd and K; if the meteorites had been formed farther from the Sun, for example, in the asteroid belt, they would have experienced a different particle flux and (according to the FGH theory) should have different isotopic compositions. (This was one of the tests suggested by McCrea [1963], and Cameron used this result as a major objection to the FGH theory [1965: 570–1].)

By 1970 the FGH theory was dead. Experimental and theoretical research by R. Bernas, H. Reeves, and their colleagues at Orsay showed that it would be difficult to explain the observed light element abundance ratios by spallation reactions.[6] Reeves, Fowler, and Hoyle (1970) proposed an alternative model for the production of light elements: bombardment of interstellar matter by cosmic rays. Thus D, He, Li, Be and B were already present in the solar nebula and their origin has nothing to do with the origin of the Solar System. This conclusion seemed to be generally accepted by the time of the Nice conference in 1972 (Reeves, 1972/1974: 51–2).

Although the FGH theory was abandoned, the idea of accounting for anomalous abundances of certain elements or isotopes by irradiation in the early Solar System was occasionally revived on an ad hoc basis (Clayton, Dwek, and Woosley 1977: 300).

2.3.5 Decline and transmutation of Hoyle's theory

The demise of the FGH theory did not force Hoyle to abandon his broader theory of the origin of the Solar System. But for several years he published nothing on the subject except for a brief note on volcanism, and his theory gradually went out of favor.[7] He continued to worry about the source of volatile substances in the terrestrial planets, which should consist only of refractory materials according to his theory. In 1972 he proposed that the volatiles were ejected as ices from Uranus and Neptune and eventually found their way to the Earth and other terrestrial planets (Hoyle 1972: 333–4). His 1979 book, *The Cosmogony of the Solar System,* elaborated this idea and connected it with the hypothesis that life originated on these particles, now identified as comets.

Dallaporta and Secco (1973, 1975) worked out Hoyle's theory quantita-

6   Bernas et al. (1967). Bernas et al. (1968). Reeves (1968). For other evidence against the FGH theory see Burnett, Lippolt, and Wasserburg (1966), Murthy and Sandoval (1965). The abundance and possible sources of boron were discussed by Cameron, Colgate, and Grossman (1973).

7   After the first *Apollo* moon landing Hoyle suggested that igneous lunar and terrestrial rocks could have been formed in the hot phase of the solar nebula rather than by volcanism after the formation of those bodies (Hoyle 1969: 401). Stubbs (1970) reported on a discussion of Hoyle's theory at a meeting of the International Astronomical Union.

tively; they concluded that the magnetic field of the proto-Sun was too weak for his basic mechanism to be effective. Nevertheless in the late 1970s there seemed to be some evidence from meteorite studies for a strong magnetic field in the early Solar System[8] and indications of such a field in the lunar crust.[9] In the 1980s, several observations were reported of magnetic fields around young (T Tauri–type) stars.[10]

Although Hoyle's theory was frequently mentioned as a leading contender in the 1960s,[11] it had dropped out of sight by 1985. Specific criticisms of his theory were occasionally published: A solar nebula whose mass was only $0.01M_\odot$ would not provide enough material in the region of the outer planets for them to grow to their present sizes during the lifetime of the Solar System (Safronov in Hoyle 1971: 202); his ideas about how a cloud would collapse to form a solar nebula are not in agreement with modern ideas on this process (Cameron in Hoyle 1971: 200; Whipple 1979); his mechanism for pushing rocky material away from the sun will not work (Whipple 1964, 1979); his mechanism for pulling volatiles back to the terrestrial planets is "hard to believe" (Page 1980: 325); and his account of the action of magnetic fields is faulty (Okamoto 1969: 48; Woolfson 1971: 268; Alfvén and Arrhenius 1976b: 259). Urey judged Hoyle's entire theory to be "quite artificial" (1963a: 154). But these objections are not much stronger than those made against other theories, especially in view of the acknowledged fact that Hoyle's overall picture is in several respects qualitatively similar to generally accepted ideas (Whipple 1979; Friedlander, 1985: 293).

I attribute the failure of Hoyle's theory to retain its popularity to the fact that it didn't look like a "progressive research programme" in the sense defined by Lakatos. Being associated with the ill-fated FGH theory was certainly no help, though that need not have been a fatal defect. But, as noted, Hoyle changed one of his original auxiliary hypotheses in order to conform to the Hayashi theory of the early superluminous Sun, after having previously stated that his explanation of the aggregation of planetesimals was in conflict with Hayashi's theory. This move would be called a "degenerating problemshift" in the Lakatos scheme unless Hoyle had succeeded in using his modified theory to make a successful new prediction. But in fact Hoyle never claimed to make any predictions at all, and he made no effort at all to use new data from the space program or from meteorite analysis to develop his theory. Nor did he take advantage of the later demise of Hayashi's superluminous Sun to revive his original theory. In his 1979 book he did not refer to any recent

---

8 Banerjee and Hargraves (1971, 1972). Brecher (1971, 1973). Brecher and Arrhenius (1974). Lanoix, Strangway, and Pearce (1978). Sonett (1978). Nagata (1979). Sugiura, Lanoix, and Strangway (1979).
9 Sonett, Colburn, and Schwartz (1975). Banerjee and Mellema (1976).
10 Gershberg (1982). Appenzeller and Dearborn (1984). Gnedin and Red'kina (1984). Lago (1984). Bouvier and Bertout (1986). Gnedein, Pogodin, and Red'kina (1986). Uchida (1986).
11 Wood (1962). Cameron (1963b: 23). McMahon (1965: 228). Herczeg (1968: 187, 188). Woolfson (1969: 157). Dormand (1973). Dorschner (1974).

observations except for the study of lunar samples, and even there he was not able to come to any definite conclusion about the origin of the Moon. The reader of that book might well conclude that most of it could have been written 30 years earlier. The failure of Hoyle's programme seems to find a reasonable explanation in terms of the Lakatos methodology, or indeed under most orthodox accounts of scientific method.   .

2.4

Cameron's programme

During the 1960s the two most popular theories (at least in the English-speaking world) were those of Hoyle and A. G. W. Cameron.[1] Cameron, like Hoyle, was much occupied with the problem of element synthesis in stars in the late 1950s. A nuclear astrophysicist who had been working at Atomic Energy of Canada Ltd. (Ontario), Cameron was visiting Caltech in 1960 when Reynolds published his results on xenon isotopic anomalies. This discovery, suggesting that ^{129}I had been present in some meteorites at the time of their formation, "transformed the direction of much of my research," Cameron recalled in 1985. "I was already interested in the meteorites as an interesting problem in connection with the abundance of the elements, and now it was clear that there were clues in the meteorites about the time scale on which they had been formed. Therefore, I spent the rest of my time at Caltech thinking about problems of star formation and the origin of the solar system and their related time scales. . . . Eventually all of this thinking was published in one lengthy paper" (Cameron 1986b).

2.4.1 Collapsing clouds

In December 1960 Cameron sent Harold Urey "the first draft of a paper I have written on the formation of the sun and planets. This work represents a kind of vacation from nucleogenesis that I have been taking since last spring. . . . In the course of this cosmogony I do not find any place for your primary, secondary, and tertiary lunar-size objects in the asteroid belt" (Cameron D1960).

Cameron also rejected Kuiper's gaseous protoplanets on the grounds that they would be subject to thermal disruption by the Jeans criterion unless they were much larger than the effective thickness of the nebula (Cameron 1960/1962: 41; 1973c: 385).

These views on the formation of planets were not firmly held and proved to be subject to modification on short notice. The hard core of Cameron's programme was his assumption that the early stage of collapse of the solar nebula must be consistent with a plausible model for formation of a solar-mass star from the interstellar medium under presently observable conditions,

---

1 Wood (1962). Whipple (1964). McMahon (1965: 228). Herczeg (1967/1969). Kaula (1968: 429). Dorschner (1974).

Figure 14. A. G. W. Cameron (A. G. W. Cameron).

and with the time scale determined from isotopic anomalies interpreted on the basis of a reasonable theory of nucleogenesis (Cameron 1960/1962; 1962b).

Although Cameron started with the same kind of assumptions as Hoyle – they agreed on what phenomena had to be explained and what type of explanation would be acceptable – their theories were quite different. Cameron started with a very massive solar nebula derived from a cloud compressed by external pressure, and with angular momentum corresponding to its rotation around the center of the galaxy, whereas Hoyle postulated only enough mass to restore solar composition to the planets and enough angular momentum to make the early Sun rotationally unstable (Cameron 1963b: 23–5).

Cameron could criticize Hoyle for failing to explain how the extra angular momentum that the solar nebula inherited from its galactic rotation was lost (Cameron 1963b: 23). In an earlier decade Hoyle could have defended his minimum-mass, minimum-angular-momentum nebula as a naturalistic assumption based on the well-known present state of the system as opposed to deduction from a hypothetical initial state. But Cameron also claimed the benefits of naturalism because his initial state was consistent with known conditions in the present galaxy. Cameron had enlarged the domain of the problem from the Solar System to the galaxy, though it remained to be seen whether he could cover planetary formation, which had been central to the original domain.

Cameron attributed isotopic anomalies to "a period of nucleosynthesis in the galaxy which enriches the interstellar medium with fresh radioactivities a relatively short time before the formation of the solar system" (Cameron 1963d), in place of the FGH hypothesis that these radioactivities were produced by spallation in the solar nebula.[2]

Cameron agreed with Hayashi that the early Sun went through a convective, highly luminous phase (Cameron 1962b; Ezer and Cameron 1962, 1963); as noted earlier, Hoyle resisted this idea for several years since it conflicted with the hypothesis that planetesimals could be formed with ice in the inner Solar System. Cameron proposed that the planets were formed in the primitive solar nebula *before* the Sun (1964/1966: 238; 1965: 572; 1974: 68); residual gas later streamed inward to form the Sun.[3] Nascent planets were then exposed to a high-temperature environment when the Sun went through its Hayashi phase.

Another difference between Cameron's and Hoyle's basic assumptions is that Cameron did not accept the Alfvén hypothesis, adopted by Hoyle, that transfer of angular momentum from the Sun to the solar nebula takes place primarily by magnetic braking in an ionized gas. Instead, Cameron attributed this transfer to turbulent viscosity (1962b, 1969a, 1969b).

As D. ter Haar and Cameron pointed out in their historical review (1963; 34–5), *none* of the existing theories is satisfactory, because they lack a quantitative basis; their authors suggest processes that might lead to our Solar System but fail to demonstrate by rigorous calculation that they would actually do so. Cameron attempted to satisfy this demand by deducing the quantitative consequences of a series of models for the collapse of a gas cloud, using electronic computers to solve the hydrodynamic equations. The problem seemed similar to the collapse of larger gaseous spheres to form galaxies, discussed by Leon Mestel (D1962), who found that two kinds of disks can be formed by rotational instability: one with nearly uniform mass distribution, the other "axially condensed." In the present case, the uniform disk presumably will eventually form a binary star system whereas the axially condensed

---

2   Cameron (1963d) accepted Kuroda's (1961) suggestion that spontaneous fission of the extinct radionuclide ^{244}Pu accounts for the Xe isotope abundances in the Earth's atmosphere. R. O. Pepin concluded that this hypothesis could account for the observed anomalies at least as well as the FGH hypothesis (Pepin 1964: 209).

3   Cameron suggested that the planets could have acquired their atmospheres in this way but immediately pointed out that this suggestion seems to be refuted by the xenon isotopic abundances, which indicate that the atmosphere was acquired gradually from the Sun throughout terrestrial history. He regarded this as an objection to his theory of the formation of the Solar System (Cameron 1965: 580). To avoid the objection he added yet another auxiliary hypothesis: The Earth lost its primordial atmosphere when it became rotationally unstable and spun off the Moon. But this process (G. H. Darwin's selenogony, recently revived by D. U. Wise) would work only if the Earth were already greatly deformed and rotating close to the limit of stability; that might be so if it had been formed from two bodies of comparable size in the primitive solar nebula (1965: 582). This idea reappeared with some modifications in the Cameron–Ward (1976) selenogony.

one may form a single star surrounded by planets (Cameron 1962a: 341; 1963b: 26–7).

But the actual calculations showed that no central stellar body is formed in hydrostatic equilibrium with the disk (Cameron 1963d; 1965: 571). Instead, ringlike condensations were formed; Cameron conjectured that these might break up into separate disks that could be precursors of the giant planets (1969a). The "Sun must be formed as a result of gaseous dissipation processes" (Cameron, 1974: 58). Other theorists also found that the collapse of a rotating cloud gives a ring with no central condensation, and concluded that the process would result in a binary or multiple star system.[4]

2.4.2 The viscous accretion disk

Another result of the computer calculations was that dissipative processes such as angular momentum transfer take place much more rapidly – only a few thousand years – than previously assumed (Cameron and Pine 1973).

Thus, the principal goal of the research shifted to the calculation of sequences of models of the primitive solar nebula in which dissipation occurred during the formation. This goal was achieved during the winter 1975–76. Contributing in a very material way to the achievement of this goal was the development of the theory of the viscous accretion disk. (Cameron 1976/1978: 55)

The concept of a viscous accretion disk, on which Cameron now based his theory, had been developed by D. Lynden-Bell and J. E. Pringle (1974)[5] to account for the high and rapidly varying radiation from T Tauri stars. They concluded that

whatever the dissipation mechanism, the basic form of evolution is the expansion of the outermost parts to carry all the angular momentum together with the collection of an ever increasing fraction of the mass towards the centre. This process is much slower in systems of larger scale and this fact encourages us to see analogies between the present state of the Galaxy and a very much earlier stage of the solar system that was a spinning disc of gas and dust. (1974: 604)

4 Larson (1972a, 1974). Black and Bodenheimer (1976). See Bodenheimer and Black (1978) for a brief history of these calculations.
5 Historical note by Cameron (1976/1978: 55): "The basic elements of such a theory were published many years ago by Lüst (1952), but this important pioneering paper did not become well known. The theory was twice again independently worked out, by Lynden-Bell and Pringle, but these authors discovered each others' work and recently published a joint paper outlining the theory."
   According to Lynden-Bell and Pringle (1974: 604) the work originated in the former's 1960 thesis: "We worked out the basic similarity solutions for the evolution of time-dependent Newtonian discs under the action of viscosity but failed to find any solutions that evolved under their own self-gravity. However, in many of the more recent applications the self-gravity is negligible so these solutions are now of greater interest."

In the following decade accretion disks were invoked to explain a number of exotic astronomical phenomena such as quasars, cataclysmic variables, and powerful X-ray sources in binary systems (Lin and Papaloizu 1985: 985). The existence of a well-developed mathematical theory for this model seems to compensate for the failure to show that it evolves from a specific previous state. Thus, the domain of Solar System cosmogony has shrunk a little bit from the ambitious galactic scale previously envisioned. Accretion disk theory also covers up ignorance of physical processes by lumping the effects of several unspecified sources of turbulence into a single parameter, thus sacrificing or postponing a more fundamental understanding of how the solar nebula works (Cabot et al. 1987: 451).

Cameron also inferred from his new model that the solar nebula will be unstable against the formation of rings that may condense to giant gaseous protoplanets in the outer Solar System (Cameron, 1976/1978: 64; 1978b: 5; DeCampli and Cameron 1979). Cameron continued to defend this mode of formation for the outer planets despite criticism and arguments for the alternative planetesimal model (Lin and Papaloizou 1980; Mizuno 1980; Gautier and Owen 1985; Podolak and Reynolds 1985). He also claimed that the terrestrial planets could be formed from giant gaseous protoplanets; after a core of refractory material condensed at the center of the protoplanet, the gaseous envelope would be evaporated as the temperature rises in the inner regions of the dissipating solar nebula (Cameron, DeCampli, and Bodenheimer 1982; Cameron 1985a: 1096).

The accretion disk model led Cameron to estimate substantially lower maximum temperatures for the inner part of the Solar System than in his earlier models. This estimate helped to undermine the condensation-sequence models, which had relied on the assumption that all material in the region of terrestrial planets had once been completely vaporized (3.3.3).

But in 1983, calculations on star formation adapted to data for T Tauri stars again led to a high-luminosity phase of the early sun (Mercer-Smith, Cameron, and Epstein, 1984). Temperatures in the nebula during later stages of accretion would

exceed the condensation temperature of iron to surprisingly large radii . . . small bodies will be totally evaporated to a distance beyond that of the formation of Mars. Bodies of planetary size, such as remnant cores of condensed matter left over from the evaporation of giant gaseous protoplanets, may survive this period. (Cameron, 1984a: 119)

In particular, the temperature at the distance of Mercury could be in the range 2500° to 3500°K; Cameron therefore proposed that Mercury was first formed as a much more massive planet and then lost most of its rocky mantle by vaporization, leaving behind a high-density core (Cameron 1984a, 1985c).

Cameron's frequent reversals on such matters as the temperature of the solar nebula, the supernova trigger (2.5), and the origin of the Moon (4.4) have sometimes puzzled other scientists. In his autobiographical notes he acknowl-edges that "many of my friends are never sure what my current thoughts on a subject are" since he is "ready at a moment's notice to abandon a favorite hypothesis when presented with a good reason." The "good reason" may be a new calculation as often as a new observation. Moreover, Cameron reserves the right to reinterpret evidence in a way consistent with his own views (Cameron 1986b).

2.5

Isotopic anomalies and the supernova trigger

In the mid-1970s Cameron revived the theory briefly alluded to earlier: the "supernova trigger." Supernova explosions had been invoked as a cause of star formation since Öpik's 1953 suggestion, and the idea that isotopic anomalies in the Solar System could be explained by injection of nuclei recently synthesized in such an explosion – an explosion that also caused the collapse of the presolar nebula – had been discussed by Cameron and others in the 1960s. But these ideas remained speculative and had to compete with other speculations such as that by Fowler, Greenstein, and Hoyle (1961, 1962), which postulated the formation of light nuclei in the early Solar System by spallation. The supernova might also be replaced, as a source of compression to initiate collapse of the presolar nebula, by the density waves thought to account for the spiral-arm structure of galaxies.[1]

Three events in 1969 led to a revival of the supernova trigger theory. First, the Fowler–Greenstein–Hoyle theory was abandoned on the grounds that the most probable mechanism for the production of light elements was the bombardment of interstellar matter by galactic cosmic rays (Reeves, Fowler, and Hoyle 1970). Second, the fall of the huge Allende meteorite in Mexico made material for the study of isotopic anomalies suddenly much more easily available.[2] Third, the *Apollo* lunar landing project involved the development of extremely sensitive instruments for analyzing the isotopic composition of samples and stimulated the interest of the scientific community in "hands-on" Solar System research.

2.5.1 Magnesium, xenon, oxygen

Of the many isotopic anomalies, the most intriguing was the possible excess of ^{26}Mg, considered as the decay product of ^{26}Al. J. R. Simanton and his colleagues had discovered in 1954 that the latter nuclide has a previously unknown ground state (below the known 6-second positron-emitting state) that decays by positron emission to ^{26}Mg with a half-life of less than a million

---

1 Lin and Shu (1964). Lin (1971: 97). Wetherill (1975b: 298).
2 See Begemann (1980), L. Grossman (1980, 1981). Pillinger (1984) gives a survey of isotopic meteorite research during the past 50 years.

years (later found to be about 720,000 years; Simanton et al. 1954). Harold Urey (1955a) proposed that ^{26}Al in the early Solar System could have been a source of heat to melt meteorites but then rejected this mechanism because it would have melted the Moon. It was revived and developed by the Anders group at Chicago.[3] Because of its short life the ^{26}Al must have been produced fairly recently, perhaps by proton irradiation of magnesium in the reaction ^{26}Mg(p,n)^{26}Al (Fowler, Greenstein, and Hoyle 1962: 192; Reeves and Audouze 1968). Everyone assumed, with Cameron (1962b), that the time interval between initial collapse of the presolar nebula and formation of meteorites must have been much more than a million years.

In 1970 W. B. Clarke and his colleagues reported a 4 to 6 percent excess of ^{26}Mg in the meteorites Bruderheim and Khor Temiki (Clarke et al. 1970). But David Schramm, Fouad Tera, and Gerald Wasserburg (1970) could find no anomalies in several samples including the ones analyzed by Clarke's group.[4]

Schramm (1971) stated that there is no evidence for the presence of ^{26}Al at the time of final solidification of the meteorites, although it could have been a significant heat source before solidification. He discussed the possible synthesis of ^{26}Al by silicon or carbon burning in supernovae but remarked that "time scales in the early solar system make it more likely that ^{26}Al, if present in planets, was synthesized by a proton irradiation in the early solar system" (1971: 249).

Two Australian scientists, C. M. Gray and W. Compston, reported finding excess ^{26}Mg in the Allende meteorite in 1974. In agreement with Schramm (1971) they concluded that the parent ^{26}Al was made within the Solar System by proton bombardment of light elements. But their results were regarded as inconclusive by American scientists (e.g., Lee, Papanastassiou, and Wasserburg 1976), as were the preliminary results of Lee and Papanastassiou (1974).[5]

---

3 Fish, Goles, and Anders (1960); see also Murthy and Urey (1962). Anders wrote, in a 1989 letter to me: "Urey had completely disowned this idea by 1956, and strongly resented our attempts to revive it. Our paper submitted to *Astrophysical Journal* in 1959 was rejected, and an eminent astrophysicist wrote: 'The time scale necessary for the operation of short-lived radio-activities ($\leq 10^8$ years) is just impossible. . . . The current trend in astronomical ages makes your time-scale ridiculous.' Our revised paper met with a friendlier reception when resubmitted in 1960, presumably because Reynolds had meanwhile found evidence for extinct 16–Myr I^{129}."

4 In order to distinguish isotopic anomalies due to material coming from distinct nucleosynthetic processes from those due to later fractionation processes in the Solar System, one has to look for effects that are nonlinear in the mass-number differences. Thus, a fractionation process that tends to separate an isotope of mass number i from one of mass number j will generally produce an effect proportional to $(i - j)$. The abundance ratio of the two isotopes would be

$$R_{ij} = R_{ij}^{0}(1 + k(i - j))$$

where R_{ij}^{0} is the normal ratio and k is the fractionation factor.

5 The view that Gray and Compston's work was inconclusive "is uncharitable and reflects the chauvinism of the U.S. scientists" who expressed that opinion, according to Ringwood. Ringwood states that "the quality and reliability of Compston's mass-spectrometry is internationally accepted. There is no doubt that they resolved and measured a real effect and recognised it for what it was. They should therefore receive unqualified credit as discoverers of the ^{26}Mg excess" (letter to S.G.B., 24 August 1988).

Of the numerous other isotopic anomalies discovered and discussed in the 1970s, the most important were those in the noble gases and oxygen. The xenon anomalies had become more complicated since Reynolds's work in the early 1960s. In 1969, three papers independently proposed that a "strange" xenon component, discovered by Reynolds and Turner (1964), came from fission of a superheavy element with atomic number about 114 (Anders and Heymann 1969; Dakowski 1969; Srinivasan et al. 1969). The Anders group continued to support this hypothesis for more than a decade against proposed supernova and other explanations, finally abandoning it in 1983 (R. S. Lewis et al., 1983).

In 1972 O. K. Manuel, E. W. Hennecke, and D. D. Sabu reported that carbonaceous chondrites contain two isotopically distinct components of trapped xenon that cannot be explained by nuclear or fractionation processes, and suggested that isotopes 131 through 136 might have been produced by a high flux of thermal neutrons on ^{235}U. This flux could be due to an early deuterium-burning stage in the outer region of the Sun or to the irradiation of planetary material before accretion as proposed by Fowler, Greenstein, and Hoyle.[6] At the same time David C. Black (1972) found in the carbonaceous chondrite Ivuna a component of neon that he called "E" and suggested an extra-Solar System origin for it.[7]

In 1973 Robert N. Clayton, Lawrence Grossman, and Toshiko Mayeda found that the oxygen in certain minerals in carbonaceous chondrites is depleted in isotopes 17 and 18.[8] They attributed this to admixture of a component of almost pure ^{16}O that "may predate the solar system and may represent interstellar dust with a separate history of nucleosynthesis." Since ^{16}O is produced by alpha-process reactions in stars, one might expect that ^{16}O-rich samples would also have an excess of isotopes with other integral numbers of alpha particles such as ^{24}Mg and ^{28}Si (compared with other stable isotopes of those elements).

Donald Clayton (1975b: 768) argued that the ^{16}O anomaly found by Robert Clayton, Grossman, and Mayeda (1973) proves that it is possible to form grains containing material produced by nucleosynthesis before it is diluted by interstellar matter, and proposed that the xenon anomalies were also due to

6 They do *not* mention a supernova here though they later were credited with this idea by Cameron and Truran (1977).
7 Schramm (1978) said this was the first anomaly for which the most reasonable explanation seemed to be to postulate primitive material of different isotopic composition. Donald Clayton (1979a: 162) said that Black's suggestion about the origin of the "Ne–E" component was "a far-reaching conclusion, perhaps the first of its kind based on good data soundly analyzed rather than on pure speculation. Nevertheless, the argument had little impact on astrophysics or on solar system science at the time. It was really after the ^{16}0 anomaly that Black's discovery was well remembered, and is now regarded as being of fundamental importance."
8 Schramm (1978: 387) said this work "initiated much of the present activity" concerning isotopic anomalies. Robert Norman Clayton (b. 1930) was born in Canada and educated at Queen's University, Ontario. He completed his graduate work with a Ph.D. at Caltech in 1955 and has been on the faculty of the University of Chicago since 1958. For further details see Wood (1982b). On Lawrence Grossman see 3.3, note 10.

grains formed near exploding stars.[9] E. E. Salpeter (1974) had argued that a supernova would be surrounded by a cold dense gaseous shell where grains could form. During the next few years Donald Clayton elaborated the hypothesis of "presolar grains" as an alternative to the conventional assumption that meteorites condensed only after the formation of the Solar System and thus reflected the isotopic composition prevailing at that time (3.3.3).

The first generally accepted proof of the presence of ^{26}Al in the early Solar System came late in 1975 when Gerald Wasserburg's group at Caltech announced their discovery of a large anomaly in the isotopic composition of magnesium in a chondrule from the Allende meteorite.[10] The ^{26}Mg excess was nonlinear so it could not be attributed to fractionation effects. According to the report by Typhoon Lee, D. A. Papanastassiou, and Wasserburg (1976), ^{26}Mg is enriched by about 1.3 percent, and "there is a strong correlation *in this chondrule* between the ^{26}Mg excess and the Al/Mg ratio so that the most plausible cause of the anomaly is the *in situ* decay of now-extinct ^{26}Al."[11]

Previously, isotopic anomalies had been attributed to production of ^{26}Al by irradiation in the early Solar System, but the Caltech group doubted this explanation on the grounds that such irradiation would have produced other anomalies that are *not* observed, for example, ^{53}Cr from decay of ^{53}Mn (Lee, Papanastassiou, and Wasserburg 1976: 112). Another paper left open this possibility, however, stating that ^{26}Al could be attributed either to the "injection of freshly synthesized nucleosynthetic material into the solar system immediately before condensation and planet formation, or local production within the solar system by intense activity of the early Sun" (Lee, Papanastassiou, and Wasserburg 1977: L107; see also Lee 1978: 226).

Donald Clayton, Eliahu Dwek, and S. E. Woosley (1977) criticized the irradiation model in more detail, showing that a proton fluence large enough

9 "It is, I think, one of my most creative ideas," though strongly resisted by other scientists (Clayton 1975a: 64). The idea that interstellar grains had survived the formation of the Solar System and might be present in meteorites had also been suggested by Anders (1964) and Cameron (1973b). The resistance of the community is indicated not only by Clayton's own testimony, but by the delay in publication of his papers, e.g., a paper submitted in April 1979 but not published until December 1981 after arguments with "four epochs of anonymous referees" (D. D. Clayton 1981a: 374, 386).
    Donald Delbert Clayton (b. 1935) received his Ph.D. at Caltech in physics, in 1962. He taught space science and physics at Rice University starting in 1963; he is now on the faculty at Clemson University.
10 Gerald Joseph Wasserburg (b. 1927) received a Ph.D. in geology at the University of Chicago in 1954, while maintaining a commitment to the physicochemical approach of Harold Urey. He has been on the faculty at the California Institute of Technology since 1955. For his role in science policy making see Chapman (1982: 204).
11 Lee, Papanastassiou, and Wasserburg (1976: 109). The result had been announced at the winter meeting of the American Physical Society in Pasadena, 29 December 1975 (Lee, Papanastassiou, and Wasserburg 1975); their paper, submitted to *Geophysical Research Letters* in October 1975, appeared in the January 1976 issue of that journal, but its first two paragraphs were missing so it had to be republished in the next issue.
    Typhoon Lee received his Ph.D. in astronomy at the University of Texas in 1977. Dimitri. A. Papanastassiou was born in Athens in 1942. He came to the United States and obtained his Ph.D. from the California Institute of Technology in 1970. He has been a research fellow and research associate since 1970, specializing in mass spectrometry.

to produce the inferred quantity of ^{26}Al must have created anomalies in ^{36}Ar, ^{80}Kr, and other isotopes. Failure to observe those anomalies, together with discoveries of other anomalies that could not be produced by irradiation such as ^{16}O and ^{202}Hg, weakened the credibility of the irradiation model.[12]

2.5.2 Revival of the supernova trigger

Several scientists at the spring 1976 meeting of the American Geophysical Union discussed the possibility that a supernova explosion shortly before the formation of the Solar System could be responsible for the recently discovered isotopic anomalies (Cameron and Truran 1977: 447). The first published discussion of this possibility based on what is now considered reliable evidence was that of Lewis, Srinivasan, and Anders (1975), in a paper on Xe isotopes in the Allende meteorite, but these authors concluded that a supernova origin of the anomalies was unlikely. Soon afterward Sabu and Manuel (1976) proposed that the data of Lewis, Srinivasan, and Anders (1975) indicated that a supernova did explode in the vicinity of the present Solar System (see also Manuel and Sabu 1975/1976). Unlike other theories, the Sabu–Manuel hypothesis assumed that the Sun already existed before the explosion and had formed a binary system with the star that was to explode. But their interpretation of the xenon data has generally been considered unacceptable by other scientists (Anders, letter to S.G.B., 1989).

In July 1976 Cameron and J. W. Truran submitted to *Icarus* their paper "The Supernova Trigger for the Formation of the Solar System." The basic idea was the same one proposed by Cameron 16 years earlier: "The supernova responsible for injecting short-lived radioactivities into nearby interstellar clouds may also have been responsible for triggering the collapse of those clouds to form stars and accompanying planetary systems" (Cameron and Truran 1977: 448). But now there was much better evidence for those short-lived radioactivities, and the best alternative explanation (that they had been produced by irradiation in the early Solar System) had been discredited. The hypothesis that supernova explosions can form new stars was gaining increased support from astronomical observations; shortly after Cameron and Truran began to circulate their paper, William Herbst and George Assousa (1977) reported new observations of supernova-induced star formation in Canis Major.

12 This argument was reinforced by Schramm (1977, 1978) and Reeves (1978b). Lee (1978) showed that an irradiation model could explain the ^{16}O and ^{26}Al anomalies quantitatively and is consistent with the nearly normal Mg, Ca, and Ba observed in most samples; hence, "We may not need a supernova to explode immediately before the formation of the solar system" (1978: 226). However, he admitted that the model could not explain the large ^{48}Ca excess recently discovered by Lee, Papanastassiou, and Wasserburg (1978) and other anomalies in heavy elements that seemed to require large neutron fluxes that would be hard to achieve in the early Solar System. Thus, "At least some isotopic anomalies are to be explained by an extra-solar source" (Lee 1979: 1605).

The supernova trigger theory quickly became enormously popular, receiving wide publicity in both the technical journals and the press.[13] The Cameron–Truran hypothesis was attractive because it promised to explain many diverse phenomena by a single event. But it promised more than it could deliver, as Cameron himself soon realized. In a report on the 8th Lunar Science Conference in March 1977, Cameron was quoted as saying his work was "trying to make a synthesis of a lot of different ideas into a single picture, and perhaps it was too ambitious an attempt." He thought perhaps he and Truran had gone too far in attempting to explain the heavy-element anomalies, though their description of light elements was satisfactory (Spruch 1977: 19). Three years later Cameron retreated further by admitting that a class of anomalies known as "FUN" (for fractionated unknown nuclear) could be better explained by changes in the proportions of products from different nucleosynthetic processes, as Donald Clayton had long argued, than by postulating a single nearby synthesis site (Consolmagno and Cameron 1980).

At the same conference in 1977, Wasserburg told a reporter that "the discovery of new isotopic effects, which are related to nuclear, chemical, and kinetic effects, is taking place very nearly on a weekly basis. Therefore, anyone trying to play God is in a crap game with very rapidly changing rules" (Spruch 1977: 19). Nevertheless, Wasserburg's group found independent support for the recent injection of nucleosynthetic material when they discovered unexpectedly large amounts of ^{107}Ag, presumably produced by decay of ^{107}Pd, in the Santa Clara meteorite (Kelly and Wasserburg 1978). At the end of 1978 Wasserburg was quoted by a Washington *Post* reporter as saying, "Something went off with a helluva bang just before the solar system was born" (O'Toole 1979).[14]

While the Cameron–Truran trigger theory was clearly the most widely accepted in the late 1970s, several other theories invoked supernovae in somewhat different ways to explain the formation of the Solar System

1. Wilbur Brown proposed a model for formation of the Solar System from massive fragments of the shell ejected by a supernova (1970, 1971a, 1971b,

---

13 The *SCI* (which I have augmented by scanning a few additional publications) lists the following numbers of citations for the years 1977–86: 1977: 11; 1978: 28; 1979: 27; 1980: 24; 1981: 18; 1982: 6; 1983: 12; 1984: 10; 1985: 10; 1986: 7. Articles in newspapers and general or popular scientific magazines include Edmunds (1977), Spruch (1977), Sullivan (1977a, 1977b), Falk and Schramm (1979), O'Toole (1979).

14 In an article in *Sky and Telescope* (July 1979), Sydney Falk and David Schramm mentioned the discovery of excess ^{107}Ag and concluded, "Whether or not a supernova directly caused the collapse of the solar nebula, the injection of supernova grains from a nearby event over 4 ½ billion years ago seems to be the only explanation for the existence of the short-lived species aluminum-26 and palladium-107 at the time the solar system was formed" (1979: 22). Kelly and Wasserburg, in a letter to the editor of *Sky and Telescope,* pointed out that the Santa Clara meteorite probably contained material that had accreted, melted, differentiated metal from silicate, and then solidified again, whereas the Allende meteorite had probably never been molten since its formation. Thus, the discovery of evidence for the extinct radioactivity ^{107}Pd in Santa Clara was *stronger* evidence for a recent nucleosynthesis "(perhaps in a supernova)" than the discovery of evidence for ^{26}Al in Allende (Kelly and Wasserburg 1980: 15).

1974). Herbst and Assousa (1978: 369) called this the first suggestion that a supernova was involved in the origin of the Solar System, though Brown himself cited earlier papers by Cameron (1962b) and Hoyle (1945). A modified version of this model was presented under the name "snow plow model" by Herbst and Rajan (1980), on the basis of work by Roger Chevalier (1974; Chevalier and Theys 1975) showing that the expanding shell of gas from a supernova may be compressed to a high enough density to initiate star formation. The advantage of this model is "that there is no need for supernova ejecta to penetrate and mix with a pre-existing cloud" (Herbst and Rajan 1980: 42), a difficulty with the trigger model raised by the calculations of Steven Margolis (1979). D. A. Wark (1979) also used this model to discuss the condensation of the presolar nebula.

2. Donald Clayton's model proposed that presolar grains could be formed as condensations in supernova ejecta ("SUNOCONS"), but since the grains were solidified before the formation of the Solar System, their isotopic anomalies do not indicate a single recent nucleosynthetic event but could come from several earlier supernovae (D. D. Clayton 1977, 1978a, 1981a, 1982, 1986).

3. Hubert Reeves proposed a "Bing Bang" model in which "the sun was most likely born amidst a fireworks of supernovae" (1978b: 400). Like Donald Clayton, Reeves proposed that ejecta from many supernovae have been incorporated into our Solar System even though no single supernova played the role of "trigger." His emphasis is more on the astronomical side, whereas Clayton's is more on the chemical; Reeves proposes that the Sun was born in an OB association such as Orion, where a large molecular cloud was compressed by passage through a galactic spiral arm and stars of various masses began to form. The heaviest stars evolve gradually to the supernova stage, and the remnants of their explosions sweep across the region where other stars like the Sun are still forming.

4. Thomas Gold's flypaper model proposes that an already-collapsing cloud catches the ejecta from a supernova. Gold has not published the details of this model (there is only a brief reference to it by Donald Clayton 1977: 267), and it is not clear how much it differs from the others.

Perhaps the most bizarre idea is the proposal of Manuel and Sabu (1975/ 1976) that the supernova was actually concentric with our Sun, which formed on its remnant core, while the planets condensed from the debris of outer layers (see also Manuel 1981). One consequence of this model is that the Sun's interior should contain a significant amount of iron (Manuel and Hwaung 1983), a conclusion reached for completely different reasons by Carl Rouse (1983, 1985), though it contradicts the view generally accepted since 1929 that the Sun is mostly hydrogen and helium.

It might appear that acceptance of any but the last of these hypotheses would imply rejection of the monistic principle in favor of a model requiring at least one other star to assist the formation of planets orbiting our Sun. In the classical dualistic theories, that would entail a very low probability for planetary systems elsewhere in the galaxy and thus an even lower probability for the existence of intelligent extraterrestrial life. But the new theories are not so pessimistic: A supernova can trigger the collapse of many clouds without having to be very close to any one of them, so the older estimates of the chance or stellar encounters do not apply. More generally, the clouds that collapse to form stars and planets are strongly affected by the presence of nearby clouds and other stars, so the monistic–dualistic dichotomy may simply be obsolete.

2.5.3 Rejection of the trigger hypothesis

Up to 1984 it was widely believed that one or more supernova explosions were directly involved in the formation of the Solar System, although there were several competing models based on this idea (see end of 2.5.2). A crucial assumption of most of these models was that the ^{26}Al that decayed to produce the ^{26}Mg found in Allende must have been synthesized in a supernova. Truran and Cameron (1978) attempted to support this assumption by a new and more comprehensive analysis of the process by which ^{26}Al is synthesized. They suggested that ^{14}N, produced in the usual carbon cycle, is converted to ^{18}O by absorption of an alpha particle and subsequent positron emission. Two more alpha particles and ejection of a neutron give ^{25}Mg. The dominant mechanism that produces ^{26}Al is absorption of a proton by ^{25}Mg, but it is destroyed by the reaction ^{26}Al $+ n \rightarrow {}^{26}$Mg $+ p$. Thus, the amount of ^{26}Al produced depends on the concentration of free neutrons and protons (as well as on the cross sections of these reactions). Heavy elements may be competing to absorb the neutrons and thus decrease the rate of destruction of ^{26}Al. Taking account of all these factors, Truran and Cameron estimated that the ratio ^{26}Al/^{27}Al should be between 4×10^{-4} and 2×10^{-3}. They concluded that "the supernova event forming ^{26}Al occurred between 2 and 3.7 million years prior to condensation of solar nebula material" – the implication being that their calculation supports the trigger theory.

Other groups confirmed that ^{26}Al can be produced in supernovae. W. D. Arnett and J. P. Wefel (1978) calculated the production of ^{26}Al in the carbon shell of a massive (12M$_\odot$) star and found a ^{26}Al/^{27}Al ratio of about 10^{-3}, within the range estimated by Truran and Cameron, but based primarily on quasihydrostatic rather than explosive burning. S. E. Woosley and Thomas A. Weaver found a similar result (1980) by considering explosive neon burning rather than carbon burning. In either case the products would eventually be ejected in a supernova explosion. E. Vangioni-Flam, J. Audouze, and J.-P. Chièze (1980) calculated much higher ratios of ^{26}Al/^{27}Al, up to 0.1, but only under conditions of very high temperatures and densities.[15]

But evidence soon emerged that ^{26}Al can be produced more abundantly in

---

15 These calculations depend on assumptions about the equilibrium between the ground state and the short-lived isomeric state of ^{26}Al; see Ward and Fowler (1980).

other processes. Henry Nørgaard argued (1980) that ^{26}Al can be produced from ^{25}Mg in the outer envelope of red giant stars, giving a ^{26}Al/^{27}Al ratio ranging from 0.5 to 1.0 in some cases. Since "such stars are known to be losing mass at a considerable rate and . . . there is strong observational indication of the presence of grains in the outer atmosphere of these stars," Nørgaard suggested that they could have contributed to the ^{26}Mg excess in Allende. He noted the discovery by B. Srinivasan and E. Anders (1978) in the Murchison meteorite of isotopic anomalies in the noble gases of the kind expected to result from nuclear processes in red giants; this was additional evidence that dust grains from red giants had been injected into the Solar System.

Wolfgang Hillebrandt and Friedrich-Karl Thielemann (1982), following up earlier work by M. Arnould, H. Nørgaard, F.-K. Thielemann, and W. Hillebrandt (1980), proposed that nucleosynthesis in novae could be a significant source of both Ne–E and ^{26}Al. They obtained ^{26}Al/^{27}Al production ratios of about 1, although they considered the total production rate of ^{26}Al quite uncertain because several of the relevant proton capture rates were not accurately known.

In 1981 Simon P. Worden, Timothy Schneeberger, Jeffrey Kuhn, and John Africano questioned the need for a supernova to produce ^{26}Al on the basis of their analysis of flare activity on T Tauri stars. They suggested that "the expected proton flux from these events may explain early Solar System abundance anomalies without recourse to nearby supernovae" (p. 520). "While estimates based on T Tauri energetics cannot refute the supernova theory [as an explanation of isotopic anomalies], we find the consistency of the irradiation models with the flux estimates considerably more satisfactory than appealing to the special circumstances of a supernova to explain the abundance anomalies" (p. 526). They conclude: "The total proton flux expected from the flares is consistent with the irradiation model for solar isotopic abundance anomalies, thus precluding the necessity for a nearby supernova" (Worden et al. 1981: 527; see also Feigelson 1982).

Another piece of research in nuclear physics strengthened the hypothesis that ^{26}Al can be produced more abundantly in sites other than supernovae. Arthur Champagne, Albert Howard, and Peter Parker (1983a, 1983b, 1983c) found a low-lying resonance in the reaction ^{25}Mg $+ p \rightarrow$ ^{26}Al $+ \gamma$, indicating that its rate would be 10 orders of magnitude greater at low stellar temperatures than previously estimated. They noted that "a supernova explosion is still a most efficient dispersal mechanism but may not be the primary production route. The actual source of ^{26}Al in the early solar system is therefore still open to question" (Champagne, Howard, and Parker 1983c: 689).[16]

These calculations acquired new significance when the gamma-ray telescope on the High Energy Astronomical Observatory satellite (HEAO-3) re-

16 Fowler (1984) said his view that ^{26}Al could not be synthesized in supernovae at high temperatures because the large cross section for ^{26}Al(n,p)^{26}Mg was confirmed by measurements of Skelton, Kavanagh, and Sargood (1983) on ^{26}Mg(p,n)^{26}Al.

vealed relatively large amounts of ^{26}Al throughout the galaxy.[17] Donald Clayton (1984) argued that these amounts could not have been synthesized by supernova explosions if current calculations of the production ratio are correct. "The observed ^{26}Al is more likely due to about 10^8 dispersed novae, or to a single old (10^4 − 10^6 yr) supernova remnant that today surrounds the solar system. If the ^{26}Al is dispersed, the high interstellar ratio today . . . calls into question the requirement that a supernova trigger for formation of the Solar System was the cause of a concentration 3-times larger then" (1984: 144). He stated that novae are better candidates and that the value of ^{26}Al concentration inferred from Allende "was simply the average interstellar value at that time, negating the need for a "supernova injection" of ^{26}Al into the forming solar system" (1984: 145).

During the preceding decade Clayton had been undermining the supernova trigger hypothesis from another direction by showing that heavy-element anomalies could be more plausibly interpreted in terms of presolar grains (2.5.2). He proposed that the barium and neodymium isotopic anomalies found by McCulloch and Wasserburg (1978a) should be interpreted as extinct radioactivities resulting from radioactive decay within interstellar grains, and that the Ca−Al-rich inclusions in Allende are not condensates from a hot gaseous solar nebula but admixtures of precondensed matter formed by heating, with some separation of r-process and s-process products taking place during accumulation processes (D. D. Clayton 1978b; cf. McCulloch and Wasserburg 1978b). Clayton stated that the remeasurement of the neodymium cross sections by Mathews and Käppeler (1984) "totally vindicated my approach and my suggestions" (letter to S.G.B., 1988). Similarly he argued that isotopic anomalies of strontium (Papanastassiou et al. 1978) and samarium (Lugmair, Marti, and Scheinin 1978) could best be explained by gas/dust fractionation in the protosolar accumulation rather than by supernova injection (D. D. Clayton 1978c, 1979b).

Early in 1984 Cameron (1984b) announced that the reasons that led him and Truran to propose the supernova trigger no longer seemed compelling. Many of the isotopic anomalies could be explained without postulating injection from a nearby supernova (Consolmagno and Cameron 1980); processes in red giants and novae could account for more copious production of ^{26}Al than supernovae; and the HEAO-3 observations proved that ^{26}Al was indeed copiously produced. Others might continue to support the trigger hypothesis,[18] but for Cameron it was time to put it back on the shelf.[19]

---

17 Mahoney et al. (1982, 1984). ^{26}Al was expected to provide one of the sharpest lines in the diffuse galactic background; see Ramaty and Lingenfelter (1977), Lingenfelter and Ramaty (1978), Ramaty (1978).
18 McSween (1984). Couper and Henbest (1985: 126). Goldsmith (1985: 368). Lee (1986).
19 He revived it 11 years later (Cameron et al. 1995).

PART 3

Planetogony and plasma

3.1

Safronov's programme

By 1960 the hypothesis that planets formed by gravitational collapse of massive gaseous protoplanets, advocated primarily by G. P. Kuiper, had been abandoned by most planetogonists (Urey 1956a; Ruskol 1958/1964; Cameron 1960/1962: 41). The most popular alternative was accretion of solid particles, with or without the presence of gas during the later stages of planetary formation.

3.1.1 Accretion of particles in the protoplanetary cloud

Accretion theories originated in the 19th century when they were associated with the idea that the Earth and other planets were built up from meteoritic material. T. C. Chamberlin (1903a, 1905) revived the idea under the name "planetesimal hypothesis," giving it both astronomical and geological respectability (1.2). On the astronomical side he removed the objection that planets formed by accretion would have retrograde rotation, by showing that the coalescence of planetesimals in intersecting elliptical orbits would somewhat favor the formation of objects with prograde rotation. His conclusion was confirmed long afterward by modern planetary theorists.[1] On the geological side Chamberlin worked out the properties of an Earth assembled slowly enough to remain cold and solid throughout its early history. His theory avoided the consequences (especially the excessively small age) of the assumption that it had condensed from a hot fluid ball.

Although several scientists such as Urey and Ringwood discussed the chemical aspects of the formation of terrestrial planets (3.3), there were few attempts before 1970 to develop quantitative physical models of the accretion process itself. This seems odd in view of the fact that powerful theoretical methods for treating very similar processes were widely known in the physical sciences. The kinetic theory of gases, formulated by James Clerk Maxwell and Ludwig Boltzmann, had been actively developed for a hundred years; it provided systematic techniques for computing the properties of systems of colliding particles and could be modified to take account of inelasticity of collisions, combination and fragmentation of particles, their nonspherical shape, spatial inhomogeneities, external fields, and so on (Brush 1972, 1976;

1 Artem'ev and Radzievsky (1965). Giuli (1968a,b). Harris (1977). See 1.3.2, note 20, for further discussion.

Figure 15. V. S. Safronov (V. S. Safronov).

Hirschfelder, Curtiss, and Bird 1954). Physical chemists had worked out approximate theories to describe coagulation and chemical reactions in fluid media. Astrophysicists were familiar with the application of stochastic models to systems of interacting stars (Chandrasekhar 1943). And, when analytic techniques could not adequately handle more complicated "realistic" models, electronic computers were available to grind out numerical solutions. It appears to me that most of the theoretical research on planetary accretion done in the 1970s and 1980s – with the possible exception of some projects requiring very fast, large-memory computers – could have been done at least 10 or 15 years earlier if anyone had been interested.

In fact the only person who seems to have been seriously interested in pursuing this kind of research during the 1960s was Viktor S. Safronov at the O. Yu. Schmidt Institute of Earth Physics in Moscow. Following the ideas of Otto Schmidt and other Soviet cosmogonists, Safronov worked out in considerable detail the dynamical and thermal aspects of a model of colliding, accreting, and fragmenting solid particles.[2]

2 Schmidt (1944, 1949, 1958). B. Levin (1948, 1953, 1956). Gurevitch and Lebedinsky (1950). For general surveys see Safronov (1983/1995), Safronov and Vityazev (1983/1985). The capture process on which Schmidt originally based his theory is discussed by Lyttleton (1968: 28–32). Witting (1966: 83) wrote that "Schmidt's theory appears to be on solid ground as far as the boundary conditions [facts to be explained] are concerned; none are violated, and the theory is able to explain many of the dynamical boundary conditions well and completely." English translations of many of the major Soviet articles on planetogony may be found in A. Levin and Brush (1995).

The earliest comparable work in the West is that of Stephen H. Dole (1970), but his calculation was much less ambitious than those of Safronov. For biographical information on Safronov see 1.1, note 19.

Although a few of his papers appeared in English translation shortly after publication (Safronov 1959, 1962a, 1965/1966, 1966/1967), Safronov's achievements were not generally recognized in the West until 1972 when an English-language version of his 1969 book, *Evolution of the Protoplanetary Cloud,* became available (Safronov 1969/1972). Since then the Safronov model or one of its variants has been the most popular explanation for the formation of the terrestrial planets. It has also played a major role in the leading theories of the origin of the giant planets and their satellites, as well as asteroids, comets, and meteorites.

Safronov urged a division of labor in cosmogony: The problem of the origin of the protoplanetary cloud (PPC) could be treated separately from the problem of its evolution into planets, and that problem in turn was distinct from the history of planets after their formation. He preferred the hypothesis of common formation of the Sun and PPC over Schmidt's assumption that the PPC was formed elsewhere and later captured by the Sun, but considered himself a proponent of Schmidt's ideas since his model pertained only to the second stage. Thus, Safronov's theory did not compete with those of Hoyle and Cameron in trying to explain the formation of the Sun. He did dispute Cameron's assumption that the PPC is very massive (2 to 4 solar masses), preferring a low-mass PPC (about 0.05 solar mass). He also rejected the assumption of Weizsäcker, Cameron, Hoyle, and others that turbulence plays an important part in the evolution of the cloud.

Starting with a relatively low-mass, gas-dust cloud in which any primeval disordered motions have been damped out, Safronov assumed that dust particles would settle to the central plane and grow to centimeter size. As suggested by Edgeworth (1949) and by Gurevich and Lebedinsky (1950), the dust layer breaks up into several condensations by local gravitational instability. These condensations then combine and contract.

Coagulation theory goes back to the work of Marion von Smoluchowski on Brownian movement at the beginning of the 20th century, as presented in Chandrasekhar's influential review article (1943). In Safronov's first model fragmentation by collisions is ignored; the coagulation coefficient is assumed to be proportional to the sum of the masses of two colliding bodies. The number of particles with mass m is found to vary approximately as $m^{-\frac{2}{3}}$ for long times, except for large m where an exponential damping factor becomes important. Fragmentation does play a role, especially when the relative velocity of two colliding particles is high. But if the relative velocity is very small, the particles will tend to move in similar orbits and collide so rarely that growth cannot occur. Safronov argued that, as the particles grow, encounters that don't lead to collisions will increase the relative velocities. The relative velocities most favorable for growth are those somewhat less than the escape velocity, which of course depends on the mass of the particles. The average relative velocity tends to increase as the particles grow so that it remains in the range favorable for further growth (Wetherill 1980a: 5; Fisher 1987: 224-6).

Safronov also concluded that when one body in a region happens to become significantly larger than the others, it will start to grow even faster because its effective cross section for accretion of other bodies is enhanced by gravitation. In this way a single planet can emerge in each "feeding zone" within the PPC and then sweep up the rest of the material in that zone.

Safronov (1959) noted the importance of high-speed impacts of a few large bodies in the formation of the Earth, a feature he attributed to B. Yu. Levin. He estimated that the formation of the Earth was essentially completed in 10^8 years, and that in spite of the large impacts, the initial temperature inside was only a few hundred degrees. Using an equation derived by Lyubimova (1955) he found that heating by contraction would raise the central temperature to about 1000°K at the end of the formation process; radioactive heating would later raise this to several thousand degrees. Thus, the 19th-century scenario – cooling from an initial temperature of several thousand degrees – was completely reversed. Here Safronov's model was in agreement with Western studies of the thermal history of the Earth (e.g., Urey 1951).

Using a theoretical relation between the impacts of small bodies on the accreting planets and the resulting inclination of their axes of rotation, Safronov estimated from the observed inclinations that the largest bodies striking the Earth during its formation had masses about 1/1000th that of the present Earth (Safronov 1965/1966; 1969/1972: 134). Thus, the large tilt of the Uranian axis was ascribed to impact of a body having 1/20 the mass of that planet.

If the initial temperature of the Earth was only a few hundred degrees, one might think that planets further from the Sun started out much colder – perhaps cold enough to freeze hydrogen and helium from the PPC. But Safronov (1962a) argued that the gas-dust layer is so thin that the Sun's radiation goes not only through it but along its surface so that it can be scattered into it through a boundary layer. This effect would keep the temperature from falling below 30°K at the distance of Jupiter and 15°K at the distance of Saturn. Thus, these planets could not condense hydrogen directly but could only accrete it gravitationally after reaching a sufficiently large mass at a later stage of their growth.

A major drawback of Safronov's theory was that the estimated time for formation of the outer planets, using the equations derived for the terrestrial planets, was about 10^{11} years. In addition to the obvious disadvantage of requiring a time longer than the present age of the Solar System (4.5×10^9 years) to form these planets, it is inconvenient not to have a fairly massive proto-Jupiter present while Mars is being formed, if one wants to attribute the small size of Mars (relative to Earth) to interference from its giant neighbor.

To alleviate this difficulty Safronov assumed that the outer regions of the PPC originally contained a much larger amount of material, much of which was ejected by gravitational encounters with the growing embryos of massive planets. This hypothesis would accelerate the early stages of the accretion process, while gravitational trapping of gas would accelerate the later stages (1969/1972: ch. 12; 1970/1972). But the ad hoc or qualitative nature of these hypotheses damaged the credibility of the theory.

The extremely low initial temperature of the Earth also created a problem if one wanted to explain the segregation of iron into the core. Safronov was temporarily attracted by the idea that the Earth's core is not iron but silicate, chemically similar to the mantle but converted to a metallic fluid by high pressure. This was the hypothesis of V. N. Lodochnikov (1939) and W. H. Ramsey (1948, 1949), widely discussed in the 1950s (*Nebulous Earth*, 2.4). As pointed out by Levin (1962) it had cosmogonic advantages that Safronov recognized (1969/1972: 152). But the postulated silicate phase transition proved elusive, and it was shown both experimentally and theoretically that silicate compounds did not have high enough density at core pressures. So Safronov was forced to accept either the traditional iron core or a compromise iron oxide core, with a correspondingly higher internal temperature (1972: 445–46).

Safronov's programme lacked the glamor of more ambitious schemes that promised to explain the formation of the Sun as well as the planets from a simple initial state, and it encountered difficulties in explaining the properties of the present Solar System. Yet he was successful in building up a body of basic theory that turned out to be useful as a starting point for other cosmogonists.

3.1.2 The Americanization of Safronov's programme

The *SCI* gives a rough measure of the visibility of selected publications in the Western scientific community. Of course, one cannot get any information about the nature of the reception or influence of those publications from citation counts alone, and citations not listed in the *SCI* may turn out to be more important than those that are (2.2.5, note 1). Bearing in mind these caveats, I still think it is significant that the total number of citations (excluding those by Safronov and other Soviet scientists) of all of Safronov's publications from 1961 through 1971 was only 25. For comparison, one paper by Cameron (1962b) was cited 101 times in this period (excluding self-citations). Starting in 1972, the year when Safronov's *Evolution of the Protoplanetary Cloud* was first available in English translation, and going through 1982, that book was cited 107 times by non-Soviet scientists, and Safronov's earlier papers were cited 31 times. So his visibility in the West was more than fivefold as great in the second 11-year period, primarily because of the English translation of his book.

Looking at the papers that cited Safronov in the 1970s, one finds that almost all of them contain favorable remarks even when disagreeing on specific technical points. Here are some examples of the Western response to Safronov's work.

1. Peter Goldreich and William W. Ward, in a note added in proof at the end of an influential paper on the formation of planetesimals by gravitational instability in a dust disk without the need to invoke "stickiness," admitted

that Safronov had given a similar discussion, which they had apparently read only after finishing their own work (Goldreich and Ward 1973: 1061). Safronov subsequently received partial credit for what nevertheless was most often called "Goldreich–Ward instability."

2. S. J. Weidenschilling (1975) supported Safronov's suggestion that matter ejected from Jupiter's zone could deplete the zones of Mars and the asteroids.

3. R. J. Dodd and W. Napier (1974) reported that numerical simulations based on Safronov's model confirmed his conclusion that a dominant nucleus arises that quickly incorporates lesser objects; the simulation gives correct values for the rotation rates of terrestrial planets but not for Jupiter and Saturn.

4. Joseph Burns (1975) suggested that the angular momentum of Mars could be attributed to the impacts of the last few bodies falling on it as in Safronov's theory.

5. S. F. Singer (1977) agreed with Safronov that the observed obliquities of the planets can be explained by late impacts.

6. In an elaborate calculation of the thermal evolution of the Earth and the Moon based on Safronov's model, Kaula (1977b) found that accreting planetesimals would add enough heat to the Earth to bring about core segregation if not vaporization; he also inferred from his results that an impact origin of the Moon (4.4) is more likely than binary accretion.

7. P. Farinella and P. Paolicchi (1977) found from their theory results on the mass distribution consistent with those of Safronov.

8. J. N. Goswami and D. Lal (1979) stated that their observations of particle tracks in chondrites provide evidence for Safronov's accretion model and against Cameron's (1978a, 1978b) gas collapse model.

9. W. K. Hartmann and D. R. Davis (1975) acknowledged that they had been "influenced by some of the early Soviet accretion theories, published in the 1950s and 60s," in developing their ideas about lunar origin, although they had not studied Safronov's (1969/1972) book in detail.

Many other scientists simply quoted and used Safronov's results without bothering to discuss their validity.

In 1976, George Wetherill at the Carnegie Institution of Washington announced the first results of his calculations on a modified version of Safronov's theory.[3] Wetherill's work was motivated in part by photographs of Mercury's surface taken by the *Mariner 10* spacecraft on 29 March and 21 September 1974, analyzed by Bruce Murray's group (1974/1977, 1975a). It appeared that Mercury, like the Moon, had suffered a "late heavy bombardment" after its

---

3 George West Wetherill (b. 1925) received a Ph.D. in physics at the University of Chicago in 1953 and then obtained a research position at the Department of Terrestrial Magnetism (DTM) of the Carnegie Institution of Washington (CIW). From 1960 to 1975 he was professor of geophysics and geology of the University of California, Los Angeles. In 1975 he was appointed director of DTM, CIW.

Figure 16. George W. Wetherill (George W. Wetherill).

formation (Wetherill 1975a). Hence it was likely that there was a high flux of asteroid- or Moon-sized bodies throughout the inner Solar System, 4000 to 4500 m.y. ago.

Wetherill's research, unlike Safronov's, made extensive use of computer simulation. Although he confirmed many of Safronov's results, he found one important difference. When the Earth is half-formed, its feeding zone merges with that of Venus. The resulting perturbations produce higher relative velocities and thus reduce the cross section for capture of planetesimals by massive bodies. This will prevent runaway growth of the largest embryo in each zone. The second-largest body in the Earth's zone may then have a mass as large as $\frac{1}{20}$th of the Earth's rather than only $\frac{1}{1000}$th.

Such large bodies, though having only a transient existence in the final stage of accretion, will produce substantial heating by their impacts on the terrestrial planets and the Moon (Kaula 1979). Since Safronov accepted the conclusion that the Earth has been heated by large impacts during its formation (Safronov and Kozlovskaya, 1977; Safronov 1978, 1981), Wetherill could say (1981) that every current theory predicts high initial temperatures for the formation of planets.[4]

4 Murray, Malin, and Greeley (1981: 9) stated that this was one of the two major new ideas in planetogony that "have gained increasing acceptance since the space age began," the other being

A group at Tucson announced another numerical simulation project based on a modification of Safronov's theory (R. Greenberg et al. 1978, 1978b). They supported the idea that large bodies were prevalent in the early Solar System by showing that planetesimals as large as those generated in Wetherill's scheme could have been generated without invoking perturbations by proto-Venus (Greenberg 1979). Further numerical results (Greenberg 1980) generally supported Safronov's analytic work but contradicted his conclusion that relative velocities of planetesimals would tend to be comparable with the escape velocity of the dominant body. More of the total mass of the system was found to be in smaller planetesimals, which would collide mostly with each other and therefore tend to have smaller velocities; hence, when they did collide with a larger body they would be more likely to accrete and promote its runaway growth (cf. B. Levin 1978a, 1978b). One consequence of this result was that Uranus and Neptune could grow "in a reasonably short time, well below the actual age of the system, without the need for ad hoc assumptions about excess mass or artificially-low relative velocities among the icy planetesimals" (Greenberg et al. 1984).

A paper published by Wetherill in the final year covered by this chapter suggests that a modified Safronov model may be able to explain the existence of four terrestrial planets starting from 500 bodies each of mass 2.5×10^{25} kg (one-third lunar mass) (Wetherill 1985). But this result is clearly stochastic and depends on the existence of large impacts. Several runs gave three or four planets but none reproduced precisely the observed distribution of masses and distances. So the best theory of the formation of terrestrial planets was not quite capable of explaining the simplest properties of those planets as known 200 years ago.

"heterogeneous accretion." Ringwood pointed out that he had persistently advocated it during the 1960s when the cold origin was a generally accepted dogma (letter to S.G.B., 24 August 1988).

3.2

The giant planets

3.2.1 Gaseous condensation models

During the 1950s it was generally held that the interiors of the Jovian planets are mostly or entirely solid, consisting mostly of hydrogen with smaller amounts of helium and heavier elements.[1] These heavier elements ("ices" of H_2O, CH_4, NH_3 and dust) may have separated into a small central core (Öpik 1962a; Peebles 1964). It seemed likely that they had formed by gravitational instability in the gaseous primordial solar nebula, although Öpik (1962a: 255) argued that accretion from a cloud of solid particles was also a plausible origin.

In 1968 William Hubbard, on the basis of Frank Low's (1966) estimate of the excess thermal radiation of Jupiter, concluded that it must have an internal energy source. Estimating its central temperature as about 10^5 K, he inferred that a rigid atomic or molecular lattice could not exist; the interior must behave like a convecting fluid.[2] He developed thermal models of both Jupiter and Saturn on this basis (Hubbard 1968, 1969).

In the 1970s two competing models for the origin of the giant planets were developed (Williams 1979). One treated them as miniature protostars: contracting gaseous subcondensations from the primordial nebula that reached a maximum temperature that was not high enough to initiate thermonuclear reactions (so they couldn't become real stars) and then cooled down. A rock/ ice core could form later, by precipitation of dust inside the collapsed cloud or by capturing particles from outside.[3] Safronov (1974: 100) criticized this model, arguing that gaseous condensation could not explain the formation of Jupiter, for example, because one would need 60 times its mass to be initially present in its zone in order to produce gravitational instability. It would then be difficult to explain why only $\frac{1}{60}$ of the original gas ends up in the planet.

The other model, proposed by Cameron (1973d) and worked out in detail by Perri and Cameron (1974) and by Podolak and Cameron (1974a, 1974b),

---

1 Ramsey (1951). DeMarcus (1958). Wildt (1958: 244; 1961: 197–202). Öpik (1962a: 248). Öpik's paper was written in part as a critique of the only serious alternative, Alfvén's hypothesis (1954), which implied that these planets consist primarily of C, N, and O.

2 Öpik had also concluded, on the basis of older measurements, that Jupiter radiates more energy than it receives from the Sun. For additional measurements of the temperatures and estimated energy fluxes of Jupiter and Saturn see Aumann, Gillespie, and Low (1969), Ingersoll et al. (1980), Hanel et al. (1981, 1983).

3 Bodenheimer (1974, 1976, 1977). Graboske et al. (1975). Pollack, Burns, and Tauber (1979). Pollack et al. (1977). A. S. Grossman et al. (1980).

might be considered an application of Safronov's programme, although it was introduced for other reasons. They postulated that solid material would first accrete up to a critical size that would then cause the surrounding gas to become unstable and collapse onto it.

But, as noted in 2.4.2, when Cameron reformulated his nebular models on the basis of accretion disk theory, he concluded that all the planets were formed from giant gaseous protoplanets. According to Cameron (letter to S.G.B., 6 September 1988), the motivation for introducing this hypothesis was that Safronov-type theories predicted that several thousand million years were needed for the formation of Neptune:

I considered that to be entirely unacceptable and adequate grounds for rejecting the theory entirely (although, of course, not the mechanisms which obviously played some role in planetary accumulation). What appeared to be necessary was an alternative theory that could get Neptune together in a reasonable time, and if it could do this, should it not also be a faster way of assembling the other planets as well? It seemed likely that such an alternative theory would probably involve gravity, which is capable of acting quickly. And thus I was led to investigate the mechanism of gravitational instability in the gas of the nebula (very early when there was little mass around anyway).

DeCampli and Cameron (1979) applied this hypothesis[4] to the formation of giant planets. They noted that it had earlier been rejected because it was believed that rocky cores exist in the giant planets; "It is argued that micronsized grains could not settle out of a gaseous object with mass ~ 1 M_J and density $\geq 10^{-12}$ g cm^{-3}. Second, it is argued that such distended objects would be tidally destroyed by the proto-Sun. These arguments have suggested that the rocky core had to come first, followed by the accretion of solar composition" (DeCampli and Cameron 1979: 368). On the contrary, DeCampli and Cameron claimed that "rapid grain *growth* can take place inside protoplanets, so that the infall of grains to the center is very rapid compared to the Kelvin–Helmholtz time of the protoplanet prior to dissociation of molecular hydrogen." Moreover, in the Cameron accretion disk model, "distended gaseous protoplanets are not unconditionally destroyed by solar tidal forces." The De-Campli–Cameron model was defended by Cameron in later papers (1979b; 1985a: 1096; Cameron, DeCampli, and Bodenheimer 1982) and further developed by Bodenheimer et al. (1980).

After 1985, Cameron recalled, he "abandoned this idea and its consequences . . . when some reasonable ways of forming Neptune quickly were suggested. The most important suggestion was due to Lissauer, who postulated a much more massive solar nebula from Jupiter on out. However, in the meantime I had examined the case for Mercury on both planetary accumulation pictures." In work with Fegley, Benz, and Slattery, Cameron found that

4 McCrea (1960, 1963). Dormand and Woolfson (1971, 1974, 1977, 1989). Donnison and Williams (1974, 1978, 1985). Schofield and Woolfson (1982a, 1982b).

"Mercury in the GGP [giant gaseous protoplanet] scenario barely squeaked by as possible but improbable, whereas Mercury in the Wetherill scenario turned out to be very plausible. All of these things coming together convinced me Wetherill was right" (letter to S.G.B., 6 September 1988).

3.2.2 The Kyoto programme and the nucleation model

Going back to the situation in the 1970s, we see that there were now two distinct theories for the formation of the giant planets; neither seemed to have been developed far enough to make specific predictions about observational data. In order to find a crucial test, planetary scientists turned to a theory developed by Chushiro Hayashi and his colleagues at Kyoto University.[5] In particular Hiroshi Mizuno, a member of the Kyoto group, worked out a quantitative application of the theory to the formation of Jupiter and Saturn. Mizuno's model (1980) can be plausibly interpreted as an extension of the Safronov model, and its success is regarded as a victory of the planetesimal accretion theory over the gaseous condensation theory.[6]

Hayashi's group published a series of papers on protostars and the solar nebula during the 1960s and 1970s. The papers by Hayashi, Adachi, and Nakazawa (1976) and by Hayashi, Nakazawa, and Adachi (1977) took up the growth of protoplanets from 10^{25} g to the mass of the Earth (10^{28} g) or Jupiter's core (10^{29} g). They concluded that the capture of planetesimals could be accelerated by gas drag, so that the Earth would be formed in 10^7 years and Jupiter's core in 10^8 years.

Mizuno, Nakazawa, and Hayashi (1978) then investigated the instability of a gaseous envelope surrounding a planetary core. Perri and Cameron (1974) had found that if the core mass of proto-Jupiter is greater than about 70 Earth masses, the envelope can no longer be in hydrostatic equilibrium and collapses onto the core. They assumed that the envelope is adiabatic, but Mizuno et al. argued that its outer layer should be isothermal because of its low opacity. In this case the critical mass is reduced to only 15 Earth masses for proto-Jupiter and 6 for proto-Saturn. A similar calculation was reported by Harris (1978a). These values are roughly consistent with those of Slattery (1977), who found that both Jupiter and Saturn have cores with about 15 Earth masses.

In a more elaborate calculation using a three-layer envelope (isothermal, radiative, and convective), Mizuno (1980) found that the critical core mass is

5 Chushiro Hayashi (b. 1920) received his D.Sc. at Kyoto University in 1954 and has been on the faculty there since 1954 (appointed professor of physics in 1957).
6 Mizuno's work was known to only a small group of experts. The popular book by David Fisher, which is generally quite accurate on many aspects of the recent history of planetary cosmogony, gives a rather misleading account of this episode and fails to mention either Hayashi or Mizuno (Fisher 1987: 162–3). The *SCI* would be of no help unless one already knew that Mizuno's paper was important and read all the papers that cited it, rather than just counting them, since the citation rate is still lower than that of the abandoned supernova trigger theory of Cameron and Truran (1977).

nearly independent of the protoplanet's distance from the Sun. With a reasonable value for the grain opacity in the envelope, this critical mass comes out to about 10 Earth masses, an acceptable value for all four giant planets.

Mizuno's remarkable result for the critical core mass and the Mizuno–Nakazawa–Hayashi scenario for formation of giant planets were quickly acclaimed by experts on planetary structure and evolution.[7] Stevenson (1982b) and Pollack (1984) noted that the Kyoto "nucleation" model still left unanswered some important theoretical questions, such as the conditions that actually determine collapse. Stevenson (1982b) and Bodenheimer (1982) also suggested that the nucleation model has not yet solved the difficulty of excessively long accretion times for giant planets characteristic of Safronov's theory. But Stevenson (1982b) also concluded that *no* current model of giant planet formation satisfies all the observational constraints.[8]

3.2.3 A crucial test?

The empirical observation that was claimed to provide a crucial test between the Kyoto nucleation model and the gaseous condensation model was the infrared spectrometer (IRIS) measurement of the carbon/hydrogen ratios in giant planet atmospheres by the *Voyager* spacecraft. D. Gautier and T. Owen argued in several papers that the observed enhancement of carbon in the four giant planets relative to its solar abundance can be explained by the nucleation model: methane ice as well as grains of refractory materials and ices of H_2, NH_3, and so on accreted to form cores. The accretion process heated the core, releasing methane, which then enriched the surrounding gaseous envelope. The giant gaseous protoplanet model of Bodenheimer (1974) and Cameron (1978, 1978b), on the other hand, implies, according to Gautier and Owen, that the composition remained solar and homogeneous during collapse. So the observed carbon enhancement favors the nucleation model over the GGP model.[9]

Pollack (1985) disagreed, arguing that the enhanced C/H ratio can be explained on either model. He stated that the nucleation model still had difficulty forming giant planets quickly enough, especially if one needs Jupiter before Mars; on the other hand the GGP model has trouble forming cores for Jupiter and Saturn.

Others argued that the crucial test should be the ice/rock ratio in Uranus

---

7 Hubbard and MacFarlane (1980). Hubbard (1981). Lunine and Stevenson (1982). Safronov and Ruskol (1982). Stevenson (1982a). Smoluchowski (1983b). Weidenschilling (1983). Weidenschilling and Davis (1985).
8 See Boss (1988a, 1989) for the status of this problem in the late 1980s.
9 Gautier et al. (1982). Gautier and Owen (1983, 1985). See also Hubbard (1981), Torbett, Greenberg, and Smoluchowski (1982), Weidenschilling (1983: 209), Baines, Schempp, and Smith (1984), Courtin et al. (1984).

and Neptune; but the results of this test were not yet conclusive at the end of 1985.[10]

Cameron (1985a) apparently saw no conflict between Mizuno's model and his own. He suggested that giant protoplanet cores may be formed in the inner parts of the nebula, then moved by tidal interactions to the outer part, where they capture the surrounding gas to form massive envelopes by Mizuno's process.

10 Podolak and Reynolds (1984). Pollack (1984). Fisher (1987: 163).

3·3

Chemical cosmogony:
The terrestrial planets

In the preceding section we saw that a possible crucial test between two theories turned out to involve chemical composition, although chemical considerations were not central to either theory. We now turn to a group of theories that depend on chemistry in a much more direct way: those designed to explain and predict the chemical composition of the *terrestrial* planets. Although much of this literature is devoted to technical details, an important general question emerges: Did Solar System material pass through a stage when the temperature was high enough to vaporize and mix it, so that all evidence pertaining to its possible previous existence in a condensed state was lost? Was the Solar System "born again" with no "memory" of a previous incarnation, or can we identify the place where its atoms were synthesized and learn how we are descended from the rest of the universe? Thus, we arrive by a different route at the same problem addressed by the supernova trigger hypothesis.

3.3.1 Urey and the formation of terrestrial planets

Chemical cosmogony, or more generally cosmochemistry, acquired its importance in the U.S. scientific community after World War II primarily because of the efforts of Harold Urey (1893–1981). As winner of the 1934 Nobel Prize (for his discovery of deuterium) and expert on nuclear chemistry, Urey had the requisite prestige and energy to lead the younger generation of physicists and chemists into planetary science, a subject scorned earlier in the century as not worthy of the best minds.[1]

Urey became interested in the formation of the Earth when he agreed to give a course entitled "Chemistry in Nature" with Harrison Brown at the University of Chicago in 1948 or 1949.[2] To prepare his first lecture on the heat balance of the Earth he read Louis B. Slichter's 1941 article and was surprised to learn that the temperature of the Earth must actually be rising

---

1 On Urey's influence in attracting people into planetary science see Ringwood (1979: v), Taylor (D 1980: 2–3), Sagan (1981), Ezell and Ezell (1984: 17). For the low status of planetary science before the 1960s and its subsequent revival see *Transmuted Past*, 1.4, Sagan in Shklovskii and Sagan (1966), Whitaker (1985), Tatarewicz (1986, 1990). A biographical note on Urey is in 1.1.4, note 5.
2 Urey (1952: ix); on Brown see K. R. Smith, Fesharaki, and Holdren (1986).

Figure 17. Harold C. Urey (American Institute of Physics, Niels Bohr Library).

rather than falling. He wrote: "This led on to consideration of the curious fractionation of elements which must have occurred during the formation of the earth. One fascinating subject after another came to my attention, and for two years I have thought about questions related to the origin of the earth for an appreciable portion of my waking hours" (1952: ix).

In a long article (1951) and a comprehensive book (1952b), Urey presented his views on the planets and the Moon from a physicochemical perspective. He assumed as a matter of course that the original nebula "was once completely gaseous and at very high temperatures" (1951: 237) and undertook to determine the sequence in which different chemical compounds would condense as the nebula cooled (1951/1952). He had initially supposed that the Earth accumulated at about 900°C as a "concession to traditional high-temperature assumptions relative to the earth's origin," but quickly revised this estimate downward on the basis of chemical reasoning (Urey 1953: 290). He suggested that the accumulation of the Earth must have started at temperatures below 100°C. Much higher temperatures, such as those assumed in Eucken's (1944) theory, would be incompatible with the presence of iron sulfide and silicates mixed with the metallic iron phase in meteorites, since iron sulfide is unstable in the presence of cosmic proportions of hydrogen and iron above 600°K. Silicon dioxide and silicates are unstable at higher temperatures, yet both are present in meteorites. Although he expected that gravitational contraction of the growing Earth would have generated higher temperatures, these could not have been greater than about 1200°K without contradicting geological evidence (Urey 1951). Consideration of the abundances of volatile elements at the Earth's surface made it "overwhelmingly obvious" that the high-temperature origin hypothesis is invalid (Urey 1953: 286).

Urey thus assumed that the terrestrial planets accumulated at low temperatures from small solid planetesimals; they initially consisted of a grossly homogeneous mixture of silicate and iron phases. The iron would initially be in an oxidized condition in the presence of cosmic proportions of water vapor; it was therefore necessary to postulate a later high-temperature stage during which the iron was reduced and partially fractionated from the silicates. At that time iron was thought to be much less abundant in the Sun (even after removing hydrogen and helium) than in the Earth, so it was necessary to find some process that could concentrate iron in the terrestrial planets. Urey went to considerable lengths in devising schemes to fractionate iron from silicates in a manner consistent with other processes needed to allow the planets to retain volatile compounds (Ringwood 1966a: 46).

Having initially adopted Kuiper's giant protoplanet theory, Urey soon began to have doubts about that theory. It was difficult to understand how silicates could have evaporated from them to the extent necessary to explain the composition of the terrestrial planets while still retaining some water, nitrogen, and carbon (Urey 1954a).

Two years later Urey abandoned the hypothesis that protoplanets (in the sense of large masses of gas and dust of solar composition) had been involved in the formation of the terrestrial planets. Instead he postulated that two sets of objects of asteroidal and lunar size, called "primary" and "secondary" objects, were accumulated and destroyed during the history of the Solar System. The primary objects were suddenly heated to the melting point of silicates and iron, perhaps by explosions involving free radicals triggered by solar-particle radiation. After cooling for a few million years these primary objects "were broken into fragments of less than centimeter and millimeter sizes. The secondary objects accumulated from these . . . and they were at least of asteroidal size. These objects were broken up . . . and the fragments are the meteorites" (Urey 1956a: 623). The reason for constructing this scheme was to explain the presence of diamonds (presumably formed only at very high pressures) in meteorites.

At the first Symposium on the Exploration of Space (April 1959), Urey suggested that "the moon may be one of these primary objects, as I realized after devising what seemed to me a reasonable model for the *grandparents* of meteorites" (1959: 1727). As the *New York Times* headlined one of his speeches two years later, "Urey holds moon predated earth" and is one of the few relics of an early stage of the Solar System (Sullivan 1961). Urey could therefore prescribe a set of chemical and physical observations to be made from the Moon's surface to give information not only about meteorites, but also the formation of the planets.

Figure 18. A. E. Ringwood (A. E. Ringwood).

3.3.2 Ringwood's programme

As Urey turned his attention increasingly to the Moon, other scientists took up the challenge of reconstructing the chemical history of the early Solar System. One was A. E. Ringwood,[3] an Australian geochemist and cosmogonist. Ringwood proposed to interpret the densities of the Earth, Venus, and Mars as representing different redox states of primordial condensed material of chondritic or solar composition. Previous interpretations of the densities of these planets had assumed that iron/silicate ratios were a free parameter and Urey had invoked complex processes in the solar nebula that fractionated iron from silicates prior to accretion. Ringwood did not consider such processes

3 Alfred Edward Ringwood (1930–1994) was educated at Melbourne University, earning his Ph.D. there in 1956. He was a research fellow and professor at the Australian National University in Canberra from 1959. See the obituary by Brett (1994).

necessary because he rejected the supposed fact (subsequently disproved on other grounds) that meteorites are greatly enriched in iron compared with solar composition (see 3.3.3).

Like Urey, Ringwood (1960) assumed that the Earth formed by accretion of planetesimals in a cold gas-dust nebula, and that meteorites can provide clues to the nature of the primeval material. Carbonaceous compounds would initially be mixed with nonvolatile oxides, silicates, and ices. The heat generated by accretion would raise the temperature high enough to allow carbon to reduce iron oxide to metallic iron; the Earth would melt enough to allow the denser iron to sink to the center. At the same time H_2O and CO_2 produced by the reduction reactions would provide a dense atmosphere. This atmosphere would absorb solar radiation and further raise the temperature. But then one has to explain what happened to the atmosphere.

Ringwood suggested that the loss of the primeval atmosphere resulted from the catastrophic formation of the core, which suddenly reduced the total moment of inertia and generated additional heat. If the Earth's rotation was already fairly rapid, this additional rotational speed (to conserve angular momentum) could make it unstable against fission. Part of the mantle flies off and goes into orbit around the Earth, becoming the Moon; the primeval atmosphere is stripped off at the same time and dissipates into space.

In the earlier fission hypothesis for the origin of the Moon (4.2.3) there was a difficulty in making the proto-Earth unstable because the present Earth–Moon system does not have great enough angular momentum; Ringwood's theory allowed the missing angular momentum to be carried off from the system by the escaping atmosphere.

Carbonaceous chondrites, according to Ringwood (1962), are similar in chemical composition to the Earth. Unlike other meteorites, they have the same abundances of nonvolatile elements as the Sun (with the possible exception of iron and copper), and those abundances are consistent with those calculated from nucleosynthetic models. They contain some iron that, after reduction and heating, could have constituted the core. But they contain large amounts of volatile substances, suggesting that they have not undergone the kind of thermal evolution that other meteorites have experienced. Moreover, the fact that iron and nickel are found to be completely oxidized in these chondrites indicates that they have always been cold.

Ringwood's hypothesis that the primordial Earth-substance resembled carbonaceous chondrites, composed of low-temperature minerals, was threatened by the discovery that high-temperature minerals have been replaced by low-temperature minerals in carbonaceous chondrites, hence the high-temperature minerals were earlier (DuFresne and Anders 1961, 1962; Sztrokay, Tolnay, and Foldvari-Vogl 1961). Ringwood (1963) retreated somewhat from the position that carbonaceous chondrites are primordial, conceding that they must have been radioactively heated for a short time (not more than 10^8 years), as suggested by Fish, Goles, and Anders (1960), but insisted that they are still "the nearest approach which we possess to primordial material."

In order to construct an Earth model from carbonaceous chondrites, Ringwood found that not only iron and nickel but another metallic component must be transferred from the mantle to the core. Since SiO_2 is the common oxide most easily reduced to metal after the oxides of iron and nickel, he proposed that the core contains some silicon (Ringwood 1964/1966: 296). Since silicon is less dense than iron, this hypothesis was qualitatively consistent with the shock wave compression experiments indicating that pure iron is too dense to be the sole constituent of the core (Al'tshuler et al. 1958).

Another geochemical influence was that the core was not in chemical equilibrium with the mantle when it formed; otherwise the iron would have removed essentially all of the Ni, Co, Cu, Au and Pt from the mantle as it separated. This was another piece of evidence favoring a cold rather than hot origin for the Earth (Ringwood 1964/1966). But Ringwood also suggested, contrary to Urey, that the Earth was extensively or completely melted at some time in its early history (1966a: 71–2).

Ringwood considered his own scheme for the evolution of the Earth to be much simpler than Urey's; the latter postulated a complex multistage process involving high-temperature processing of the material (e.g., in lunar-sized bodies) before it was assembled into the Earth, whereas Ringwood's did the job in a single step.

In keeping with the desired simplicity of his theory, Ringwood then abandoned his earlier hypothesis that the primeval material had been subjected to radioactive heating in the nebula, and with it the assumption that this material is similar to carbonaceous chondrites. Instead he postulated a higher proportion of hydrogen in the primordial material and gave a more important role to a primeval atmosphere, consisting primarily of H_2, CO, and H_2O, in reducing iron oxides.

3.3.3 High-temperature condensation

In the early 1960s several events encouraged cosmogonists to include a high-temperature stage in their scenarios for the formation of terrestrial planets. Astrophysical models proposed by Hoyle (1960), Hayashi (1961), Ezer and Cameron (1963, 1965) and others implied a superluminous phase for the early Sun, perhaps the same phenomenon as the copious mass ejection observed in T Tauri stars (Herbig 1962). Paul W. Gast (1960) found that alkali metals are depleted in the Earth's upper mantle as compared with chondrites, suggesting that some volatilization had occurred during the Earth's formation. John Wood (1958, 1962) proposed that chondrules are direct condensates from the solar nebula; they could have formed near the Sun's surface, then been pushed out by the process described in Hoyle's (1960) theory. The T Tauri stellar wind might provide the brief high-pressure surge needed to allow them to condense (Wood 1963: 165). Thus, chondrules are surviving planetesimals of the type from which the terrestrial planets condensed (Wood, 1963).

Edward Anders became a leading advocate of the hot-origin hypothesis.[4] He accepted Wood's proposal for the origin of chondrules (Anders 1962/1963) and considered this an argument in favor of an early high-temperature phase for the solar nebula. He pointed out that after Urey had proposed his cold-origin theory, new evidence indicated that many volatile elements are depleted in chondrites, implying a high-temperature process. But "no model involving a common, unitary history of chondritic matter can account for this abundance pattern. One is driven to the assumption that chondritic matter is a mixture of at least two kinds of material of widely different chemical histories" (Anders 1964: 5–6). One kind has been significantly more depleted than the other and was therefore separated at much higher temperatures.

Hans Suess (1962/1963) recalled that direct condensation of chondrules from a gas phase had been popular 30 or 40 years earlier, but that Urey had persuaded him to abandon it in the 1950s. But now, with new evidence and the recognition of different kinds of chondrules, the idea could be revived. Contrary to Burbidge, Burbidge, Fowler, and Hoyle (1957) (B^2FH), who assumed that Solar System material is a mixture of atoms from several sources, Suess (1964, 1965) argued that the solar nebula was quite homogeneous. "Among the very few assumptions which . . . can be considered well justified and firmly established, is the notion that the planetary objects . . . were formed from a well-mixed primordial nebula of chemically and isotopically uniform composition. At some time between the time of the formation of the elements and the beginning of condensation of the less volatile material, this nebula must have been in the state of a homogeneous gas mass of a temperature so high that no solids were present. Otherwise, variations in the isotopic composition of many elements would have to be anticipated" (Suess 1965: 217).[5]

A pioneering calculation of the molecular equilibria and condensation in a

4 Edward Anders was born in Latvia in 1926. He has taught chemistry at the University of Chicago since 1955.

5 In his 1987 book Suess insisted that the isotopic anomalies discovered in the 1970s constitute only a "minute fraction" of Solar System material, while the rest "shows no measurable indications of incomplete mixing of genetic components. This can only be explained by assuming that both the R- and the S-components (as defined by B^2FH, see Transmuted Past, 2.3.5) were gaseous when the mixing occurred. At some time, a practically homogeneous gas mass must have existed with a temperature sufficiently high (higher than ca. 2000°K) and a total gas pressure sufficiently low that no condensed matter was present." But he admitted that chondrules cannot be explained by direct condensation from such a gas (Suess 1987: 91–3).

Hans Eduard Suess (1909–1993) was born in Vienna and received his Ph.D. in chemistry from the University of Vienna in 1935. He was appointed research associate at the University of Hamburg in 1937 and worked with Hans Jensen on the theory of nuclear stability. (Jensen won the Nobel Prize for his "shell model" of the nucleus that developed out of that research.) In 1949 Suess went to the University of Chicago to work with Harold Urey and Willard Libby. After serving as a chemist with the U.S. Geological Survey from 1951 to 1955, he was appointed professor of chemistry at the University of California, San Diego. He retired in 1977 (Pellas 1977).

solar nebula was carried out by Harry C. Lord (1965), with support and encouragement from Urey (cf. Urey 1951/1952).[6] Previous calculations had been limited to only a few major species or assumed conditions more appropriate to stellar envelopes. Lord considered 150 species in a gas with cosmic elemental abundances, at temperatures of 2000°K and 1700°K and total pressures of 1 atm and 5×10^{-4} atm. He found that Al_2O_3, W, ZrO_2, and $MgAl_2O_4$ are condensed at 2000°K and 1 atm pressure; nothing condenses at 2000°K and the lower pressure. At 1700°K many molecules condense, including oxides of Ti, V, Ca, Mg and Zr. (Silicates were not included because of the inadequacy of thermodynamic data for them.)

John Larimer, in Anders's group at the University of Chicago, generalized Lord's calculations to determine the temperatures at which a number of elements and compounds would condense, using pressures indicated by Cameron's (1962b, 1963d) models of the solar nebula.[7] He attempted to trace the entire cooling history of a gas of cosmic composition in order to account for the fractionation patterns observed in meteorites (Larimer 1967; Larimer and Anders 1967, 1970). In particular, Larimer used the same kind of data that Urey had earlier used to infer low-temperature formation to support high-temperature formation. The elements Pb, Li, In and Tl, which are strongly depleted in chondrites, are among the last to condense.

Anders (1968) argued that evidence on the depletion of volatile elements, obtained by the precise techniques of neutron activation analysis, made it necessary to reverse Urey's conclusion that the Earth and meteorites had accreted at temperatures of about 300°K. Elements that are depleted by factors of 10 to 100 in ordinary chondrites, such as Hg, Tl, Pb and Bi, often occur in nearly their "cosmic" abundances in carbonaceous and enstatite chondrites (Reed, Kigoshi, and Turkevich 1960). Anders concluded that the Earth and ordinary chondrites accreted at about 600°K.

The Anders group argued that their high-temperature condensation hypothesis was justified by Cameron's models, which made it "virtually certain that the nebula passed through a stage of catastrophic collapse when temperatures rose to $>> 2000°K$, causing complete vaporization of any preexisting solids" (Larimer and Anders 1970: 367). Temperatures would be as high as 500°K, as far out as the asteroid belt (Anders 1968: 296). Moreover, Suess (1965) stated:

It is almost an axiom that the solar nebula was well mixed in an isotopic and elemental sense. Certainly no *isotopic* differences have yet been found that might be attributed to incomplete mixing of material with different nucleosynthetic histories . . . we are

6 Harry Chester Lord III received his Ph.D. in chemistry from the University of California, San Diego, in 1967, with a dissertation entitled "High Temperature Equilibria and Condensation, Solar Wind Interactions."

7 John William Larimer (b. 1939) was educated at Lehigh University, receiving a Ph.D. in geology in 1966. He was on the faculty of the University of Chicago from 1964 to 1977, and since then has been professor of geology at Arizona State University.

probably justified in assuming that the solar nebula once had completely uniform elemental composition. (Anders 1971: 2)

The attractive idea that meteorites are direct condensates from the primordial solar nebula was apparently refuted by the fact that the abundance of iron in the solar atmosphere is 5 to 10 times smaller than in meteorites.[8] Several more or less plausible mechanisms to separate iron from silicates in the solar nebula had been proposed.[9] Urey (1967) had concluded that probably no meteorite is accurately representative of the composition of the solar nebula. But in 1969 Garz and Kock (1969) found a systematic error in earlier determinations of the oscillator strengths of iron lines. As a result, the solar abundance had to be increased by an order of magnitude; the corrected value was now in good agreement with meteoritic abundances (Garz et al. 1969; Pagel 1973: 5; Ross and Aller 1976).[10]

At the same time new evidence emerged for the hypothesis that some meteorites are early high-temperature condensates from the solar nebula. Shortly after the fall of the Allende meteorite in February 1969, Ursula Marvin, John Wood, and J. S. Dickey (1970) pointed out that its Ca–Al-rich phases have the composition to be expected for early condensates according to Lord's (1965) calculations. This interpretation was supported by Lawrence Grossman (1973) and his colleagues.[11]

John Lewis (1972b) extended the Larimer–Anders approach by using Cameron's (1969a) temperature–density–pressure profiles for the early solar nebula and the recent upward revision of the solar iron abundance.[12] He showed that the model could explain the density trends in the inner Solar System without invoking any special mechanism for iron/silicate fractionation. The high density of Mercury follows from its condensation at temperatures so high that

8 Goldberg, Müller, and Aller (1960). Aller, O'Mara, and Little (1964). Urey (1964/1966). Goles (1969). Stuart Pottasch (1963) found a higher value but his results were ignored, according to Ringwood (1974). Ringwood (1966b: 123–8) argued that uncertainties and discrepancies in abundance determinations were so large that there was no justification for the conclusion that iron is significantly less abundant in the sun than in meteorites; see later comments by Goles (1969: 127) and Ringwood and Anderson (1977).
9 S. R. Taylor (1965). Banerjee (1967). Harris Tozer (1967). Tozer (1968). See also Weidenschilling (1978).
10 Later data on the composition of the Sun's corona and photosphere indicated that the iron abundance should be raised another 40 percent (Breneman and Stone 1985), suggesting that metorites are unrepresentative of the solar nebula because they contain *too little* iron. Don L. Anderson (1989) discussed the implications of this result for models of the Earth. But there were still discrepancies in solar iron abundances inferred from different spectral lines (Blackwell, Bodenheimer, and Pollack 1984) and a need for more accurate atomic data (Grevesse, 1984).
11 Grossman (1973, 1975). Grossman and Clark (1973). Grossman and Larimer (1974). Grossman and Olsen (1974). Olsen and Grossman (1974). Ganapathy and Grossman (1976). Grossman and Steele (1976). Lawrence Grossman was born in Toronto in 1946 and received his B.Sc. at McMaster University in 1968. He earned his Ph.D. in geochemistry at Yale University in 1972 and has been at the University of Chicago since then.
12 John Simpson Lewis (b. 1941) received his Ph.D. at the University of California, San Diego, in 1968. He was on the faculty of the Massachusetts Institute of Technology from 1968 to 1982 and has been professor of planetary science at the University of Arizona since 1982.

$MgSiO_3$ is only partially retained but Fe metal is condensed (see also the discussion of this point by Grossman 1972); the densities of the other terrestrial planets are accounted for by "different degrees of retention of S, O and H as FeS, FeO and hydrous silicates produced in chemical equilibrium between condensates and solar-composition gases" (Lewis 1972b: 286). Lewis predicted that Earth's outer core is an Fe–FeS melt; Venus has essentially no sulfur but a massive core of Fe–Ni alloy; Mars has virtually no free iron but may have a core of FeS. Only Earth has heavy alkali metals in its core, giving it a large internal heat source (from the decay of ^{40}K) and a resulting magnetic field.

3.3.4 Inhomogeneous accumulation

Another version of the initially uniform, high-temperature hypothesis was proposed in 1969 by K. K. Turekian and S. P. Clark.[13] Rather than assuming that the Earth was initially homogeneous and later evolved into its core–mantle–crust structure by a segregation process, they proposed that the present stratification directly reflects the sequence of condensation: Iron condensed first and formed the core, then silicates condensed around it to form the mantle, and finally the volatile elements and gases were collected. Outer layers of the Earth would be more oxidized than inner ones because of the changing nature of the nebular gas during cooling, as hydrogen was expelled from the Solar System. This scenario would avoid the problem of getting rid of immense quantities of CO and CO_2 resulting from the reduction of iron oxides by carbon, as in Ringwood's model.

The Turekian–Clark model, known as the "inhomogeneous accumulation" or "heterogeneous accretion" hypothesis, was based like the Larimer–Anders model on a condensation sequence starting with a low-pressure gas at 2000°K, but differed from it in one significant feature. Accretion was assumed to be rapid compared with cooling of the gas. As each element or compound condensed, it was assumed to be sequestered inside a solid body so that it could no longer react chemically with the remaining nebular gas. (This is an old idea, going back to Ampère 1833.) The late-condensing material that forms the crust and upper mantle has never been in contact with the core. This explains, for example, the puzzle pointed out by Ringwood: The nickel content of basalts is much higher than would be expected if they had ever reached chemical equilibrium with an iron–nickel alloy; the absence of chemical equilibrium between core and mantle is hard to explain on Ring-

13 Turekian and Clark (1969, 1975). Clark, Turekian, and Grossman (1972). Karl Karekin Turekian (b. 1927) received his Ph.D. from Columbia University in 1955. He has been on the faculty of Yale University since 1956 and is currently professor of geology and geophysics.
   Sydney P. Clark, Jr. (b. 1929) received his Ph.D. in geology from Harvard University in 1955. He held positions as research geophysicist at Harvard and the Carnegie Institution of Washington from 1955 to 1962 and has been professor of geophysics of Yale University since 1962.

wood's hypothesis but easier in the inhomogeneous accumulation theory (Clark, Turekian, and Grossman 1972). But the same feature prevents the inhomogeneous theory from explaining the presence in meteorites and the Earth of those minerals that were apparently formed by chemical reactions between gases and previously condensed compounds, such as troilite (FeS) (J. S. Lewis 1974a; Wood et al. 1981). It is also inconsistent with hypotheses that assume the Earth's core must contain sulfur or silicon in addition to iron (Goettel 1976; Wood et al. 1981).

During the 1970s there was considerable discussion of the merits of homo-geneous versus inhomogeneous condensation.[14] Some authors questioned whether the assumption of thermodynamic equilibrium could legitimately be used to describe the condensation process, in view of the presence of nuclea-tion barriers[15] and the likelihood that solid particles are likely to be much cooler than a surrounding gas.[16] But new developments in astrophysics threat-ened to make *all* these theories obsolete, by undermining the basic assump-tion that the terrestrial planets and meteorites were formed from material that had been completely vaporized when the Solar System was formed.

3.3.5 Rejection of high-temperature condensation

The first challenge to the "hot-origin" postulate came from calculations of Richard B. Larson on the dynamics of a collapsing protostar. He found that "a star of one solar mass first appears on the Hayashi track with a much smaller radius and luminosity than the very large values which have com-monly been assumed," and for more massive stars there is no Hayashi phase at all (Larson 1969: 287; 1972b; 1988). Thus, the Sun may have been formed without reaching very high temperatures until after the planets had been ac-cumulated. During the last decade the role of a hypothetically hyperactive early Sun in planetogony has been considerably diminished (Kaula 1986). A. G. W. Cameron, whose earlier models had provided much of the justification for high-temperature condensation models, announced in 1973 that his latest calculations with M. R. Pine indicated that "the temperature will not rise high enough to evaporate completely the interstellar grains, contained within the gas, beyond about one or two astronomical units" (Cameron 1973b: 545). Thus, it is possible that some of the meteorites in our museums are interstellar grains that survived the formation of the Solar System without being vaporized.

Cameron then abandoned the Cameron–Pine model (2.4.2) and adopted a new model in which the Sun is formed not at the beginning of the accretion

---

14 J. S. Lewis (1973a, 1973b, 1981). Goettel (1976). Kerridge (1977). Walker (1977: 184). J. S. Lewis, Stephen, and Noyes (1979). Ringwood (1979). Wood (1979). Wood et al. (1981: 647). Murray, Malin, and Greeley (1981: 9–10). Kuskov and Khitarov (1982). J. V. Smith (1982). Henderson-Sellers (1983).
15 Blander and Katz (1967). Blander and Abdel-Gawad (1969). Grossman and Larimer (1974: 91). Donn (1975). Goettel and Barshay (1976/1978).
16 Arrhenius and De (1973). Tozer (1976/1978).

period but throughout that period; the temperature in the region of planet formation would be only "a few hundred degrees" (Cameron 1975a: 37). Subsequently he stated this conclusion more sharply:

At no time, anywhere in the solar nebula, anywhere outwards from the orbit of formation of Mercury, is the temperature in the unperturbed solar nebula ever high enough to evaporate completely the solid materials contained in interstellar grains. For some time a number of people have argued that the entire solar nebula started out at a high temperature and cooled while solids underwent a sequence of condensation processes. In fact, there was no available energy source for any such high temperatures to have been initially present. (Cameron 1978c: 63)

The reason why the temperatures are low is that

during the collapse of the interstellar cloud fragment, the energy released by the compression of the gas is readily radiated away, and most of the collapse of the gas cloud occurs with interior temperatures that are likely to be close to 10 K. When the material falls onto and merges with the primitive solar accretion disk, there is plenty of time for the infall energy of accretion to be radiated away into space. . . . The temperature in the disk can . . . be increased only if the disk contains much more mass or if the viscous dissipation per unit mass is increased.

But the highest reasonable values of those parameters have already been assumed, and indeed the estimated temperatures are more likely to be too high than too low (Cameron 1978a: 469; see also 1979a: 998).

A conflict thus developed between astrophysics and meteoritics. In the words or meteoriticist John Wasson:

At the present time most numerical models of cloud collapse yield the result that temperatures were never above about 1000K $\geq$ 1 AU from the axis of the forming solar system. In contrast, most meteorite researchers hold that higher temperatures were necessary to account for a variety of elementary fractionations found between groups of meteorites, between members of a single group, and between components of a single meteorite. (Wasson 1978: 489)

Wasson argued that simple aggregation of interstellar grains could not have produced the observed range of properties of chondrites. He concluded that meteoritic evidence required maximum nebular temperatures greater than 1500°K in the region from 1 to 3 AU and insisted that "satisfactory astrophysical models for the formation of the solar system must be able to generate" such high temperatures (Wasson 1978: 501).[17]

Wasson continued to defend the high-temperature hypothesis throughout

17 John Taylor Wasson (b. 1934) received a Ph.D. in nuclear chemistry at the Massachusetts Institute of Technology in 1958. He has been on the faculty of the University of California, Los Angeles, since 1964.

the period covered by this chapter, suggesting that astrophysicists should be willing to modify their models in order to agree with meteoritic evidence rather than expecting meteoriticists to look for ways to produce high-temperature assemblages in a low-temperature nebula (1985: 156, 184). J. R. Arnold (1980) also maintained that Solar System material was completely mixed at high temperatures, despite the view of astrophysicists.

John Wood, who worked in the same institution as Cameron, pointed out several times that meteoriticists were basing their theories on models that Cameron himself proposed but has now rejected (Wood 1979: 161; Wood and Motylewski 1979: 913–14).[18] Yet there was still strong evidence from the Ca–Al-rich inclusions in Allende and from other meteorites that material was condensed from hot gases in the early Solar System (Wood and Motylewski 1979: 914; Wood and Morfill 1988: 342). To resolve the conflict he suggested that chondrites were produced in regions of localized transitory heating of the nebula (Wood 1979: 165). For example, the infalling interstellar material might have been heated on passing through a standing shock wave as it entered the nebula (Wood 1982a).

In the late 1970s and early 1980s, most meteorite researchers concluded that meteorites did *not* provide strong evidence that the solar nebula was hot throughout. Insofar as meteorites appeared to have been formed at high temperatures, other explanations such as local heating events might be found (J. V. Smith 1979: 11). Anders wrote, "Researchers realized that most if not all high-T features of meteorites required only *local*, small-scale heating events, not necessarily a large-scale hot nebula" (letter to S.G.B., 1989). Direct condensation did not seem to provide a satisfactory account for refractory inclusions in Allende or for chondrules in general.[19] According to R. N. Clayton, T. K. Mayeda and C. A. Molini-Velsko (1985: 765), existing data on Ca–Al-rich inclusions *are totally incompatible with a simple history of a single stage of condensation during monotonic cooling from an initially hot gas, the first-order framework on which many cosmochemical models have been built.*"

The discovery of isotopic anomalies in meteorites (2.5.1) also encouraged scientists to abandon the hot-nebula hypothesis, since that hypothesis as formulated earlier by Suess (1965: 217) and Anders (1971: 4) implied that the nebula material was well mixed. The easiest way to account for the anomalies was to assume that presolar grains had survived without being vaporized (Smith 1979; Wood 1981).

Donald D. Clayton was one of the strongest critics of the hot-nebula hypothesis and an advocate of the view that surviving presolar grains carry a

---

18 John Armstead Wood (b. 1932) received a Ph.D. in geology at the Massachusetts Institute of Technology in 1958. Aside from a three-year research appointment at the University of Chicago (1962–5) he has been at the Harvard-Smithsonian Astrophysical Observatory since 1960.

19 Boynton (1975). Kurat, Hoinkes, and Fredriksson (1975). Kerridge (1977: 48; 1979). Herndon (1978). J. S. Lewis, Barshay, and Noyes (1979). Gooding et al. (1980). Leitch and Smith (1981). MacPherson and Grossman (1981). Wood (1981). MacPherson et al. (1984a, 1984b). McSween (1987: 58).

"cosmic chemical memory" (1981b, 1982) that may provide the key to the origin of the Solar System. He argued in a series of papers that the concept of "high-temperature thermal condensation in the early solar system," which meteoriticists had come to accept as an established fact, should be completely abandoned (1978a: 110; see also 1979a, 1980a, 1980b). He complained that his own hypotheses had been treated "rudely" by the cosmochemical community (1979a: 168) but by the early 1980s was able to claim widespread support for his views.

The measurement of rare gas isotopes in the atmosphere of Venus by the American *Pioneer Venus* probe and the Soviet *Venera* lander mission, in December 1978, yielded additional evidence against high-temperature condensation. It was found that the abundances of ^{36}Ar and ^{38}Ar are relatively about 100 times greater on Venus than on Earth (Blamont 1980/1982; Hoffman et al. 1980). This is just the opposite of what would be expected from the equilibrium condensation theory if Venus had been formed at a higher temperature than Earth, unless one invoked additional hypotheses such as a strong pressure gradient in the nebula or some process to incorporate volatiles into the planetesimals accreted by planets late in their formation (Pollack and Black 1979), or implantation of argon isotopes by the solar wind (Wetherill 1980b, 1981b). Attempts to explain the pattern of rare gas abundances in terrestrial planet atmospheres continued in the 1980s, but with little credence being given to high-temperature condensation theories (Donahue and Pollack 1983).

Although much of this evidence is equally damaging to all high-temperature condensation theories, it is easier to show that it is inconsistent with the Lewis theory because that theory made more specific statements than others about the properties of the terrestrial planets. Lewis himself placed considerable emphasis on testing theories against observational data. He wrote (1973c: 34):

Remarkably, it seems that theoreticians have devoted very little of their time to comparing the results of their modeling to the present observational data on the solar system. Faced with the exciting possibilities inherent in designing one's own solar system, many authors have lacked the self-discipline to see to it that their creative art emulates nature. (There is often the heady implication that theory is its own excuse, and nature can fend for itself.) The theorist's art might very well apply to some undiscovered solar system, but what we are really interested in right now is *ours*. There is an enormous wealth of data on the Earth, the Moon, meteorites, the terrestrial planets, and, in the near future, the outer planets and their satellites, all of which will require assimilation. There will be ever-increasing pressure on the theoreticians to make their models bear some resemblance to the available observational data.

One could argue that even though the original nebular model on which Lewis based his calculations was withdrawn by its inventor, Cameron, and replaced by another one that excluded the possibility of temperatures high enough to vaporize interstellar dust in the region of the terrestrial planets, the Lewis condensation model still should be tested against observational data. In view of its initial success in explaining the densities of the terrestrial planets in terms of the specific substances expected to condense at the temperatures and pressures corresponding to the positions of those planets in the solar nebula, it deserves to be judged by its success in predicting other planetary properties.[20] (Recall that van der Waals's equation of state gives a reasonable first approximation to the properties of gases and liquids, over a much greater region of densities and temperatures than those in which the molecular assumptions from which it was derived are valid; a theory can be useful even if it is based on erroneous principles.)

In this connection it is interesting to look at one particular problem: the difference between Venus and the Earth, with respect to sulfur content. Before he developed his equilibrium condensation theory, Lewis (1968) had predicted that sulfur compounds should be found in the Venusian atmosphere, although he estimated that the dominant cloud-forming species are compounds of mercury (Lewis 1969). But his condensation calculations led him to state that Venus has virtually no sulfur (1972b: 288), and this together with the absence of FeO is what makes its density slightly lower than that of the Earth. The validity of this explanation of the Venus–Earth density difference was disputed by Ringwood and Anderson (1977) (see also Zharkov 1983: 140) and defended by Goettel, Shields, and Decker (1981) (see also Phillips and Malin 1983). But the discovery of large quantities of sulfur compounds in the atmosphere of Venus,[21] shortly after Lewis has stated that Venus has no sulfur, looked like a direct refutation of his theory.[22] Lewis could still point to his earlier statements about sulfur compounds in the Venusian atmosphere as having been confirmed, yet, as he pointed out, these statements were based on an assumption of Earthlike composition that was superseded by his later model (Lewis and Kreimendahl 1980).

In their 1984 monograph, Lewis and Prinn continued to use the equilibrium condensation model. They claimed that the peak temperatures and pressure profiles calculated from Cameron's more recent (1978b) model in the region of condensation of preplanetary solids are similar to those in his earlier model; they admitted that "it is by no means sure or even likely that solid solar system materials were fully vaporized in the nebular phase" but suggested that their "condensation temperatures" could be interpreted as "the *highest* temperature at which gases and solids were intimately mixed" (Lewis and Prinn 1984: 9, 59, 67). These statements acknowledge the fact that the planetary science community has largely abandoned the basic principles on

20 Fanale (1976). Pollack et al. (1976). Solomon (1976). Pollack (1979). Lange and Ahrens (1982).
21 Sill (1972). A. T. Young (1973), crediting L. D. G. Young. Cruikshank (1983: 5).
22 Ringwood and Anderson (1977: 249). McGetchin and Smyth (1978: 514). Lewis (1974b) suggested that the sulfur could have been added later by comets and meteorites; Ringwood and Anderson called this "ad hoc."

which the Lewis model was originally based; perhaps the best justification for retaining it is to have a *simple* model that explains *some* features of the planetary system, as a basis for comparison with more realistic models (cf. McSween 1989: 151). And of course, one could always hope that astrophysical fancy would once again favor high-temperature models (Cameron 1984a, 1985c; Boss 1988b).

3·4

Alfvén's electromagnetic programme

Although the cosmogony of Hannes Alfvén was first introduced several years before the theories of Hoyle, Cameron, Ringwood, and Safronov, I discuss it last for two reasons: First, it has not been as widely accepted as an explanation for the formation of the planets; second, it pays more attention to the smaller members of the Solar System – satellites, rings, and asteroids.

3.4.1 Methodology

The development of Alfvén's theory is fairly well described by the Lakatos "methodology of scientific research programmes" mentioned in 2.2.4. There is a well-defined "hard core" – the postulate that electromagnetic phenomena in plasmas are of primary importance in cosmogony (and in space science in general). The hard core is surrounded by auxiliary hypotheses: the strong magnetic field of the early Sun, clouds of different composition falling toward the Sun, the critical velocity effect, accretion of planetesimals, jet streams, and partial corotation. Several of these auxiliary hypotheses are separately testable in the laboratory and/or in space. The programme cannot be tested as a whole, but individual scientists may judge for themselves whether its track record is progressive or degenerating.

Alfvén's cosmogony is also of philosophical interest because he explicitly advocates a methodology of "actualism" (2.2.3). This methodology seems to have two distinct components: First, one should start from the present state of the system and try to infer what might have happened at successively earlier times, rather than postulate a particular initial state and see whether the present state can be deduced from it; second, one should try to explain the development of a system in terms of physicochemical processes that can be observed in operation at present.[1] Alfvén has not always observed the first rule but he has done quite well in following the second, so that other scientists have credited him with discovering specific new phenomena even if they don't accept his general cosmogony.

---

1 "We should not look for *the* cosmogonic theory . . . which solves the whole problem at once, but for a number of theories clarifying the great multitude of detailed questions of which the total cosmogonic problem consists" (Alfvén 1967b: 223).

Figure 19. Hannes Alfvén (American Institute of Physics, Niels Bohr Library, Weber Collection).

3.4.2 Magnetic braking

The first and perhaps the most important example of Alfvén's actualistic style is magnetic braking. He showed that an ionized gas surrounding a rotating magnetized sphere will trap magnetic field lines, acquire rotation, and thereby slow down the rotation of the sphere (Alfvén 1942a). Ferraro (1937) had obtained this result earlier but did not suggest its possible use in cosmogony. Alfvén (1942b, 1943a, 1946) proposed that the early Sun had a strong magnetic field, and that its radiation ionized a cloud of dust and gas, which then trapped the magnetic field lines and acquired most of the Sun's original angular momentum. Magnetic braking would resolve one of the major difficulties in the original nebular hypothesis, which implied that the Sun should be spinning very rapidly after the planet-forming rings had been spun off from the cooling, contracting nebula.

As noted in 2.3.1, Hoyle's theory involved a form of magnetic braking, and other scientists developed different versions of this idea.[2] But the postulate that the early Sun's magnetic field was strong enough to make this process an

2 Lüst and Schluter (1955). Schatzman (1962). Mestel (1968). Mogro-Campero (1975). On the differences between Hoyle's and Alfven's approaches see Hoyle (1986: 35–6).

important factor in redistributing angular momentum was rejected in the 1970s because there seemed to be no independent evidence for it.[3] Such evidence may yet come from studies of remnant magnetism in meteorites or from observations of T Tauri stars (see references cited in 2.3.5). There is also considerable doubt that the nebula was sufficiently ionized to make magnetic braking effective as compared with other mechanisms (Tscharnuter 1984).

While there is no direct evidence that the Sun lost its angular momentum by magnetic braking, measurements of stellar rotation from the Doppler shifts of spectral lines showed that stars in later stages of evolution generally rotate more slowly than those in earlier stages. There seemed to be some fairly universal process by which a star loses most of its angular momentum at a particular stage of its evolution (Struve 1950; Kraft 1967). So one can no longer use the slow rotation of the Sun as a conclusive argument against the nebular hypothesis.

Alfvén did not advocate a nebular hypothesis in the sense that both the Sun and planets evolved from a single cloud. Instead, he proposed that a previously formed Sun encountered several clouds of neutral gas that were ionized and stopped at different distances from the Sun and eventually condensed into planets. In later publications (e.g., Alfvén 1960b) these clouds were identified as fragments left at the periphery of a cloud from which the Sun formed.

3.4.3 Critical ionization velocity

To explain why clouds of specified chemical composition stopped at particular distances from the Sun, Alfvén introduced his "critical velocity hypothesis": A cloud of neutral gas will start being ionized when it encounters plasma with relative velocity v such that its kinetic energy becomes equal to the ionization energy,

$$\tfrac{1}{2}mv^2 = eV_{ion},$$

where V_{ion} is the ionization potential of the atoms. The velocity cannot exceed this critical value until the ionization is almost complete (Alfvén 1954, 1960a). The ionized cloud is then stopped by the braking action of the Sun's magnetic field.

Cloud A, consisting mostly of H, He, C and some metals, has the highest ionization potential, so it falls furthest into the gravitational field of the Sun, to a distance determined by equating the potential energy $GM_\odot m/R$ to its kinetic energy $\tfrac{1}{2}mv^2$. The cloud then cools and refractory elements condense, later forming planetesimals. Clouds B, C, and D are stopped at greater distances from the Sun; because of the lower temperatures at those distances,

---

3 Cameron and Pollack (1976: 63). J. W. Freeman (1976/1978). Prentice (1978: 364). Dai and Hu (1980).

hydrogen and other gases can condense from clouds C and D in the region of the outer planets.

Although Alfvén was not able to give a satisfactory explanation of the atomic mechanism by which ionization occurred at the critical velocity, he was able to find extensive experimental evidence for the validity of his formula (Brush 1990). Even if one rejects the idea that it has anything to do with the formation of planets, one has to accept the critical velocity phenomenon as a substantial contribution to plasma physics, inspired by a cosmogonic problem.

The critical velocity effect could be applied to the formation of planetary rings. According to Alfvén (1960a, 1960b), Saturn has rings because its cloud is braked at a distance 7.9 times its radius, inside the Roche limit, whereas Jupiter's is braked outside the Roche limit at 22 radii and therefore forms satellites instead. Uranus supposedly would have no rings because the critical velocity is not reached before the cloud hits the planet; the Uranian satellites were formed from the next (D) cloud, which has different chemical composition.

Alfvén's theory implied that the giant planets would consist mostly of elements in the carbon–nitrogen–oxygen group. This contradicted the views accepted in the 1950s that those planets are composed mainly of hydrogen and helium (3.2.1). Some scientists considered the failure to account for the chemical composition of planets a fatal objection to Alfvén's theory.[4]

3.4.4 The $\frac{2}{3}$ effect

But Alfvén was more interested in explaining the physical rather than chemical properties of the Solar System. In 1967 he revived another cosmogonic plasma hypothesis from his early work, giving it the name "partial corotation." Whereas other discussions of magnetic braking assumed a tendency for the entire plasma to rotate at the same angular velocity as the magnetized central body, implying infinite electrical conductivity, recent space physics observations (Persson 1963, 1966) indicated the presence of electric fields parallel to the magnetic field. In this case the transfer of angular momentum to the plasma will be restricted and only partial corotation achieved. If the plasma later condenses to grains that move in Kepler orbits, with critical velocities equal to those of the plasma element, then their orbital eccentricity will be $\frac{1}{3}$. They cross the equatorial plane at the circle $r_n = 2r_0/3$ where r_0 = original distance of plasma element from the sun. If there is a small body ("embryo") moving in a circular orbit in the equatorial plane, it will absorb the grains so they eventually move in circles at a distance $\frac{2}{3}$ of that at which they condensed (Alfvén 1942b: 24–5). Similarly a solid body moving in an orbit within the original plasma will cast a "cosmogonic shadow" – a gap in the

4 B. Levin (1962). Urey (1963a). Goettel and Barshay (1976/1978). Tai and Chen (1977). I. P. Williams (1979).

distribution of particles formed by condensation at ⅔ of the radius of its orbit (Alfvén 1967a).[5]

Alfvén proposed that the inner boundary of the asteroid belt is the shadow of Jupiter and that the Cassini division in Saturn's ring system is the cosmogonic shadow of the satellite Mimas, whose orbit is 3/2 times as large (Alfvén 1942b: 25; 1967a; 1968b; Alfvén and Arrhenius 1973: 164–7). The A ring extends from this division out to the Roche limit; it has only medium intensity because grains had to fall through Mimas's orbit in order to form it, and some were captured by Mimas. The B ring is formed by grains originally between the Roche limit and Mimas. The C ring is weak because it comes from a region partially swept by grains of the A ring.

According to most theorists, the structure of Saturn's rings is primarily determined by resonances with satellite orbits. Alfvén argued that this explanation was not quantitatively sufficient. He proposed to use the new satellite "Janus," whose discovery was reported by Dollfus (1967, 1968), as a crucial test of the resonance and corotation hypotheses. According to Dollfus, the Janus orbit is at a distance of 1.6×10^{10} cm from the center of Saturn, so its cosmogonic shadow should be at ⅔ of this or 1.07×10^{10} cm. The resonance gap would be at $(\frac{1}{2})^{\frac{2}{3}} = 0.63$ of its orbit, or 1.01×10^{10} cm. Alfvén argued that his own hypothesis was supported in this case because there is a minimum in the luminosity curve at 1.06×10^{10} cm but none at 1.01×10^{10} cm (Alfvén 1968b; Alfvén and Arrhenius 1973: 167).

Unfortunately for the impact of this test, later observations from the *Voyager 1* and *2* space probes failed to confirm a Saturnian satellite at 1.6×10^{10} cm (Stone and Miner 1981, 1982); the name Janus was subsequently assigned to one of the two co-orbiting satellites at 1.51×10^{10} cm (Alfvén 1983; Lissauer and Cuzzi 1985).

Alfvén noted a luminosity minimum at 1.11×10^{10} cm and argued that it is also of cosmogonic origin; he predicted the existence of a previously undiscovered Saturnian satellite at 2.80 Saturn radii, one magnitude fainter than Janus (Alfvén 1968b). But no satellite has yet been discovered at this distance (Stone and Miner 1981, 1982; Lissauer and Cuzzi 1985), and Alfvén did not mention this prediction in his more recent papers on Saturn's rings although he claimed new confirmations of his theory from the *Voyager* results (Alfvén 1981b, 1983; Alfvén, Axnäs, and Lindqvist 1986).

Alfvén's interpretation of the structure of Saturn's rings received little support from scientists outside of his own group. Pollack (1975), one of the few who even mentioned Alfvén's theory in print, rejected it in favor of the resonance theory; Franklin and Colombo (1970: 338), in their paper on resonance theory, dismissed Alfvén's hypothesis as "speculative." Later review articles ignored Alfvén entirely (Cuzzi 1978; Pollack 1978; Pollack and Cuzzi 1981; Lissauer and Cuzzi 1985; Cuzzi et al. 1984); and an eyewitness report on the

---

5 Vytenis Vasyliunas (1987) disputed the theoretical reasoning leading to the ⅔ factor. (I thank David Stern for this reference.)

discussions of scientists working on the *Voyager* project failed to mention Alfvén's name (Cooper 1983).

3.4.5 The rings of Uranus

In December 1972 Bibhas R. De, a student of Alfvén, submitted a paper entitled "On the Possibility of the Existence of a Ring of Uranus" for publication in *Icarus*. He inferred from the critical velocity hypothesis that Earth, Jupiter, and Uranus should originally have had ring systems. The Earth's original rings were probably swept up or dissipated by the Moon after its capture; in the case of Jupiter, the satellite Amalthea, near the Roche limit, probably formed from matter that would have otherwise become a ring, although "it is still possible that some particles remain in orbit around Jupiter within its Roche limit." Uranus, whose satellites seem to have a regular pattern and were therefore probably not captured, has its innermost satellite well beyond the Roche limit, so a ring system should have survived within that limit (De 1972/1978: 341).

The paper was rejected by *Icarus* when first submitted, although one referee later stated that he had recommended its publication (Öpik 1977). After the rings of Uranus were discovered in March 1977 (Elliot, Dunham, and Mink 1977; Millis, Wasserman, and Birch 1977; Miner 1990: ch. 4), De tried again to get it published. In a letter to Carl Sagan, then editor of *Icarus,* De noted that the paper had originally been rejected because it was based on "Alfvén–Arrhenius numerology." But, De argued,

my prediction was based on . . . an astrophysical model – the Alfvén–Arrhenius model. All astrophysical predictions are necessarily based on a model, and a successful prediction in part vindicates the model. And surely a paper that scientifically predicted the existence of a ring around a specific planet has to be significant vis-a-vis your journal. Your rejection of the paper reflected to me not the spirit of Icarus, but rather that of cautious Daedalus – and also a lot of scientific parochialism on the part of your referees. Having been a graduate student at the time, I did not have the courage to contradict the rueful remarks of your referees. (De, letter to C. Sagan, 26 March 1977, quoted by permission)

While his paper was being reconsidered by *Icarus*, De attempted to secure some public credit for his prediction by contacting Brian G. Marsden, director of the Bureau for Astronomical Telegrams at the Smithsonian Astrophysical Observatory in Cambridge. But Marsden discounted the scientific value of De's 1972 paper:

I do not think that a general remark of this type, backed up with some theoretical ideas though it may have been, can really be classed as a *prediction*. If you could have specified the distance from Uranus more precisely, or if you could have said that the rings would consist of something like five extremely narrow structures, it would have been a different matter; but I am sure that it must have occurred to other astronomers that there was no real reason to believe that Saturn was unique in having rings, and that Uranus was an excellent second candidate. Such thoughts would have been completely independent of the Alfvén–Arrhenius ideas. After all, it is a straightforward observation that Saturn has a well-developed regular satellite system in rings. The only other planets known to have well-developed regular satellite systems are Jupiter and Uranus. If the rings are made of ice (say), Uranus obviously becomes a better candidate than Jupiter, and in any case, direct detection of a Jovian ring would probably have been much easier than a Uranian ring. As a matter of fact, I believe that A. G. W. Cameron made a "prediction" on much these grounds, but he couldn't predict the detailed structure of the Uranian rings either.[6]

It may be true that the Uranian rings *could* have been qualitatively anticipated from other cosmogonic theories, and it is certainly true that astronomers from William Herschel onward thought they had glimpsed them, but I am not aware of any specific published prediction based on an accepted theory. A leading expert on planetary science remarked, just before the discovery of the rings, that "the reason why Saturn alone has rings may be explainable by a 'condensation' theory according to which only the early Saturnian environment had the right nebular density and temperature for the nucleation and growth of ring particles from the gas phase" (Stevenson 1978: 404). Sagan, on the other hand, recalled that he was "really puzzled" at that time as to why Saturn was the only planet to have rings (1988: 13; 1975: 29). Elliot, one of the 1977 discoverers, was quoted in the press as saying that the discovery "caught everyone by surprise" (Sharma 1977).

De protested that his prediction "was based on a fundamental phenomenon of plasma physics known as the critical velocity effect," which was predicted by Alfvén in the 1950s and subsequently confirmed by laboratory experiments. In view of other successful predictions from Alfvén's theory, this one should not be dismissed as the lucky result of a " 'shot-gun approach' of making a large number of predictions a few of which may come true. . . . If indeed Cameron and anybody else had made a similar documented prediction on the basis of specific physical arguments . . . they should go on record as predictors of the rings as well" (letter to B. G. Marsden, 9 May 1977, quoted by permission).

Three new *Icarus* referees remained unsympathetic to De's work (anonymous reports quoted in letter from Sagan to De, 31 May 1977). All complained that De had not said anything specific about the structure of the

---

6 Marsden, letter to B. R. De, 3 May 1977, quoted by permission. Marsden (letter to S.G.B., 1988) pointed out that Cameron (1975b: 283), while not actually stating that Uranus *now* has rings, suggested that it may have had them previously, since ice, the major component of Saturn's rings, is stable at the distance of Uranus. He said the Uranian rings would have been lost because of perturbations by the satellites. A more definite (but still qualitative) prediction that *all* planets have rings, because of plasma corotation effects, was made by Gold (1964: 193).

Uranus rings and had not based his prediction on a quantitative deduction from accepted physical ideas. One compared it to Velikovsky's claims "to have predicted all sorts of things, but among most astronomers his predictions do not command respect, since they are not based upon logical claims of reasoning." But another warned:

There is, however, an obvious political problem with rejecting it again since the title is so pertinent and the paper's conclusion so correct. To make matters worse, the Alfvén–Arrhenius ideas have received nearly the scorn reserved for Velikovsky and so the public might wonder whether we are open to new ideas.

In response, De insisted that his prediction had been based on a model that was as well established as any other cosmogonic model (in view of the experimental evidence for the critical velocity effect) and that to demand a more specific description of the structure of the rings as part of the prediction would be expecting more than any other theory of the formation of the Solar System had achieved (letter to C. Sagan, 7 June 1977).

Sagan justified his rejection of the resubmitted paper on the grounds that having sent it to a large number of referees, he could not find one who advocated its publication:

The essential problem is, as you know, the feeling of all the referees that we are engaged in a fallacy sometimes called the enumeration of favorable circumstances – that is, that erroneous theories, if there are enough of them and if they make a sufficiently large number of predictions, must on occasion make a subsequently validated prediction. (Letter to B. R. De, 12 July 1977, quoted by permission)

Here Sagan ignored De's claim that Alfvén's theory had not made any *incorrect* predictions and had made several correct ones:

In addition, however, the essential argument of your paper has already appeared in the Alfvén–Arrhenius volume published as a NASA special report. Either of these reasons alone and certainly the two of them together constitute in my opinion grounds for rejection of the resubmitted paper.

(The NASA report was published after 1972 [Alfvén and Arrhenius 1976] so this could not have been a legitimate reason for the original rejection of De's paper.)

De published the paper a few months later in an issue of *Moon and Planets* commemorating Alfvén's 70th birthday. One scientist said that the successful prediction enhances Alfvén's credibility as a ring analyst and makes his general theory more plausible (McLaughlin 1980), but others share the view of the *Icarus* referees that the prediction has no scientific value because it is based on an incorrect theory. "To be right for the wrong reason does not hold much weight in scientific circles" (Elliot and Kerr 1984: 74). This view prevails even

though in this case the "wrong reason" includes the critical velocity phenomenon, which has been experimentally confirmed (Newell 1985: 99; Brush
1990). Most publications on the rings of Uranus (e.g., Elliot and Nicholson
1984) do not mention De or Alfvén at all.[7]

3.4.6 The rings of Jupiter

As Sagan suggested in a 1988 letter to me, "A useful historical calibration of
this question [about the value of De's prediction] would be to consider the
prediction of V. S. Vsekhsvyatsky that Jupiter has a ring." That prediction had
been made in the early 1960s, based on a volcanic eruption theory, but it was
ignored even though it was published in the principal Soviet astronomical
journals and available in English translation in major Western libraries
(Vsekhsvyatskii 1962). Sagan recalled that the theory

involved huge volcanoes on the Galilean satellites of Jupiter spewing out debris. Since
one of the most surprising discoveries of Voyager 1 is that the innermost Galilean
satellite Io does, in fact, have six or eight large active volcanoes, Vsekhsvyatskii, naturally enough, wanted credit for that as well. But what was Vsekhvyatsky's argument
for the volcanoes? It was that this is the only explanation he could imagine for the fact
that many short-period comets have aphelia in the vicinity of Jupiter's orbit. The
volcanoes were needed to spit out comets. (The argument wholly neglects the celestial mechanical explanation, which goes back at least to the time of Laplace, that
long-period comets have their orbits regularized by gravitational interactions with
Jupiter.) Here, as with Velikovsky, we have an example of correct − even surprising −
"predictions" based on wholly erroneous reasoning. Every now and then a wrong
argument will reach a right conclusion. As I said at the time, I think this is an
example of the statistical fallacy of the enumeration of favorable circumstances.
(Sagan, letter to S.G.B., quoted by permission)

Other American scientists seem to have dismissed the priority claim as just
one more Soviet attempt to claim credit for discoveries made elsewhere; only
the popular magazine *Sky and Telescope* reminded its readers that
Vsekhsvyatskii had "espoused" the idea of a Jovian ring for many years
(Beatty, 1979: 426; Shabad 1979).

A later prediction of Jovian rings did win a small measure of recognition for
its authors. In 1976 Mario Acuña and Norman Ness, in an attempt to explain
an anomalous dip in radiation intensity in data obtained by the *Pioneer 11*
spacecraft, proposed as the third, least likely, hypothesis, the "remote" possibility that a previously unobserved satellite or ring at 1.83 Jovian radii might
exist (Acuña and Ness 1976: 2921). This "exciting hypothesis" was supported

---

7 An exception is Petelski et al. (1980), who do give credit. The current view (early 1990s),
  that the rings must have been formed recently because they are rapidly dissipated by the
  atmosphere of the planet, can now be used as another reason to ignore the Alfvén–De theory,
  which assumes a cosmogonic origin, according to a referee of one of my papers (Brush 1992).

rather more enthusiastically by Fillius (1976: 925). But, according to Burns, Showalter, and Morfill (1984: 201), "since other explanations for the reduced flux levels were possible, little heed was paid to the suggestion."

Acuña and Ness followed up their proposal by asking Bradford Smith, head of the *Voyager* imaging team, to look for a ring or satellite at that distance during the planned *Voyager* encounter with Jupiter (Elliot and Kerr 1984: 97–8). At the urging of Tobias Owen, the search was made and led to the discovery of a Jovian ring by *Voyager 1* in October 1977. Candice Hansen, a *Voyager* technician, was the first to see the image (Elliot and Kerr 1984: 100–101).

Smith's initial announcement overlooked the Acuña–Ness prediction, stating that "the discovery of the ring was unexpected in that the current theory that treats the long-term stability of planetary rings would not predict the existence of such a ring around Jupiter" (Elliot and Kerr 1984: 103). But the published paper did give credit to the Acuña–Ness forecast:

The camera had been oriented to record the central region of a possible ring system with the same scale as Saturn's; it was thus entirely fortuitous that it imaged the outer edge of the Jovian rings. However, Acuña and Ness had predicted a ring or an undiscovered satellite from an analysis of charged particle data from Pioneer 11, a prediction in remarkably close agreement with this discovery. (Smith et al. 1979: 955)

Yet in a public lecture in 1982 Smith again asserted Jupiter's ring was "unexpected" – "Theory had argued against Jupiter's ability to retain a planetary ring" (Smith 1983: 87).

The Acuña–Ness proposal was more acceptable to the planetary science community than De's prediction of Uranian rings, because it had a clear physical basis and included a specific quantitative statement about the probable location of the ring; since critical ionization velocity was not theoretically understood it could not be a legitimate basis for prediction even though it was an established phenomenon, and in any case the prediction that another planet has rings was of little value without a definite description of the nature and location of those rings.[8]

But most references to Acuña and Ness do not call their proposal a "prediction" – rather it was a "suggestion" or even an "unrecognized detection" of the ring (Smith et al. 1979: 955; Jewitt 1982: 44). Indeed, it was not a real "prediction" in Popper's sense. It was not a deduction from a theory that could have lost credit if no ring had been found. Instead, it was only a "suggestion" of one possible way to explain some puzzling data – and not the most plausible one even for its authors. Conversely, the discovery of the Jovian ring did not enhance the credibility of any theory.

8 There is some doubt as to whether the observed ring can actually account for the absorption of charged particles that led to the Acuña–Ness proposal (Burns, Showalter, and Cuzzi 1980: 344). There is a possible analogy with the prediction of Neptune and Pluto on the basis of perturbations of known planets, leading to the discovery of those planets; the forces exerted by those planets then turned out to be inadequate to explain the perturbations that led to their detection (Hoyt 1980).

3.4.7 Jet streams

Planetary scientists respected Alfvén's contributions to plasma physics and cosmic electrodynamics (for which he received the 1970 Nobel Prize in Physics); few of them felt qualified to criticize his ideas on magnetic braking, critical velocity, and partial corotation in plasmas. But his next hypothesis involved only classical mechanics and generated a large literature, both pro and con. He proposed that inelastic collisions of solid particles moving in Kepler orbits will tend to focus them into "jet streams."

The jet stream idea was first applied to the Hirayama asteroids; Alfvén (1968a) proposed that the "Flora" family "contains three groups of bodies travelling in almost identical orbits, thus constituting three jet streams." This would provide an alternative to the earlier view that these families resulted from exploded planets, and would make the asteroids an intermediate stage in the formation of planets (Alfvén and Arrhenius 1970). Alfvén and Arrhenius (1970) therefore urged NASA to undertake a mission to an asteroid to determine its chemical composition.

In his 1968 paper Alfvén mentioned only briefly the more general significance of the jet stream phenomenon in cosmogony. He cited Kiang's (1966) suggestion that the phenomenon may be due to a viscosity effect and remarked: "It can be shown that viscosity interaction may produce focussing of small bodies into jet streams. However, it is beyond the scope of the present paper to develop a theoretical explanation" (Alfvén 1968a: 102).

Detailed calculations of the dynamics of systems of particles indicated that Alfvén's jet stream effect – a sort of "negative diffusion" – could indeed occur, but only if the collisions were sufficiently inelastic.[9] Alfvén's interpretation of the Hirayama family was supported by several scientists,[10] but others argued that asteroid data cannot be explained by the jet stream hypothesis[11] or opposed the hypothesis for other reasons.[12] Alfvén and others argued that the *narrowness* of planetary rings can be explained as a jet stream effect.[13]

3.4.8 The evaluation of theories

Although scientists often reject specific hypotheses, they rarely publish comprehensive critiques of entire research programmes. The historian must search

---

9 Baxter and Thompson (1971, 1973). Trulsen (1971, 1971/1972, 1972a, 1972b). White (1972: 304).
10 Danielsson (1969). Chapman, McCord, and Johnson (1973). Ip and Mendis (1974: 240). Ip (1975, 1976a, 1976b). Hämeen-Antilla (1977: 437). Shukhman (1984).
11 Napier and Dodd (1974). Gradie and Zellner (1977). Degewij, Gradie, and Zellner (1978: 648).
12 Arnold (1969). Whipple (1974: 86–7). Brahic (1975). Tai and Chen (1977). Kaula (1977a). Pratap (1977: 448). Henon (1978). Gradie, Chapman, and Williams (1979). Safronov (1979). Stewart, Lin, and Bodenheimer (1984).
13 Alfvén (1983). Ferrin (1978). Ip (1978). Houpis and Mendis (1983).

in letters and referees' reports (cf. 3.4.5) or try to extract by oral interviews the reasons why some general theories were ignored or rejected.

In Alfvén's case, the glaring discrepancy between his high reputation in space plasma physics and the slight attention given to his cosmogonic theories since the 1970s seems to call for some explanation.[14] The most extensive critique of Alfvén's planetogony that I have been able to find in the public record is a review by William Kaula (1977a), less than two pages long, of the monograph by Alfvén and Arrhenius (1975).[15] My informal discussions with other scientists suggest that Kaula's critique is not significantly different from their private opinions.

Kaula questioned the validity of Alfvén's extrapolation from selected data on laboratory and space plasmas to phenomena differing in scale by orders of magnitude. He went on to state that Alfvén's model for the origin of the Solar System fails to deal quantitatvely with 10 problems, ranging from the very high density that the interstellar medium must have had for the Sun to acquire planetary material as assumed in Alfvén's model, to a mechanism for the Earth to capture a moon coming from a different part of the Solar System. But he stipulated that the model should not be judged by the number of problems it solved or failed to solve, nor did he mention any other model that gave a more satisfactory treatment of these 10 problems. Instead, he suggested that the theorist has an obligation to recognize defects pointed out by others and revise his model in a way that is responsive to the concerns of the rest of the community. "All scenarios of solar system origin are imperfect, but most scenario writers are readier to admit their imperfections and to try to remedy them" (Kaula 1977a: 182). Moreover, Alfvén lacks influence because of his "scornful but vague criticism of others, ignoring others' work on similar problems," and failure to revise his work by responding to "subsequent findings and speculations." Presumably it would be useless for Alfvén to reply that this is just the way his own work had been treated by the scientific community, and that other models also fail to deal quantitatively with many aspects of planetogony.

It seems clear that Alfvén's personal style of interaction with other scientists is partly responsible for the community's resistance to his work. The same is

14 According to Witting (1966: 83), "Alfvén's theory fits a large number of dynamical boundary conditions, even Bode's law, and leads naturally to the Jovian–terrestrial classification. It requires a very hot nebula during at least the start of planetesimal formation, and it is difficult to reconcile these high temperatures with the observed absence of differentiation of substances volatile at these temperatures. . . . Furthermore, the theory requires a large number of ad hoc assumptions, which has led later theorists to reject most of Alfvén's theory, keeping only the hydromagnetic aspects which led to a reasonable solution of the angular momentum problem."
 I do not discuss here Alfvén's views on the evolution of the universe, which are even further removed from the mainstream of cosmological opinion; see, e.g., Alfvén (1966), Lerner (1991).

15 Gustaf Olof Svante Arrhenius (b. 1922) worked as a geologist and oceanographer in Sweden from 1946 to 1948 and received his D.Sc. from the University of Stockholm in 1953. He has been at the Scripps Institute of Oceanography at La Jolla, California, since 1952 and was appointed professor of oceanography in 1959.

probably true for the Estonian-Irish astronomer E. J. Öpik, who is privately credited by scientists with having made important contributions to the theories of the evolution of the Solar System but whose work is not adequately cited in the published literature.[16]

Weidenschilling pointed out that scientists rarely say explicitly in print that they "accept" a theory, whereas they may state that they reject it (or at least one aspect of it) in order to justify a different course of investigation. But they do "vote with their feet" by addressing questions relevant to a particular model. Thus, one "accepts" a theory if (a) it poses interesting questions for further work; (b) the questions are relevant to one's own expertise; (c) "some combination of data, analytical techniques, and/or computational ability must allow progress toward answering those questions"; and (d) funding is available. From this point of view it is risky to accept any of Cameron's theories because they have probably been revised by the time they appear in print; only those working directly with him have access to the latest version. Alfvén's theory by contrast is too rigid. Moreover, it fails on points (b) and (c) because most workers in cosmogony lack the expertise in plasma physics to develop it (letter to S.G.B., 15 December 1988).

Safronov's theory, according to Weidenschilling,

provides a "golden mean" with a content sufficiently stable for meaningful work, but with many areas for progress. The dynamical questions are accessible to the rapidly growing power of computers. This in itself has kept the Safronov model dynamic; without computers it would have reached a dead end at the limits of analytic modeling in the early 1970's . . . I regard the existence of computers to be the greatest single factor in "acceptance" of Safronov's general model in the sense of inspiring further work. The same can be said of the Kyoto model . . . (letter to S.G.B., 15 December 1988),

which is less popular perhaps because there is less contact between Japanese and Western scientists than between Soviet and Western scientists.

If we look at what some experts considered to be the "established" theory (cf. Cameron's "central paradigm," 2.2.4), it is hard to find much agreement on *content*. E. H. Levy, introducing an authoritative compendium of review articles (Black and Matthews 1985), asserted that "today, to a first approximation, there exist no competing theories for the origin of the solar system," and thus, unlike previous conference proceedings that began by reviewing

16 Schwarzschild (D 1984). DeGroot et al. (1986). Ernst Julius Öpik was born in Estonia in 1895, studied at the University of Moscow from 1912 to 1916, then became an assistant at the Moscow Observatory. He escaped the Bolshevik Revolution (Chapman 1982: 24) and was appointed director of the Astronomy Department at the University of Turkestan in 1919. In 1921 he returned to Estonia as astronomer and lecturer at the University of Tartu. After extended visits at Harvard in the 1930s and Hamburg in the 1940s, he settled at the Armagh Observatory in Ireland in 1948. Starting in 1956 he was a visiting professor at the University of Maryland for several months each year. He retired at the age of 88 and died in 1985. Further details of his life and work may be found in the articles by De Groot et al. (1986).

different theories, only one theory need be presented (Levy 1985: 3). As far as I can determine, none of the theories presented in the compendium even claimed to give satisfactory quantitative explanations of more than one or two of the facts listed in Table 2. The authors of different chapters of the book disagreed on such fundamental points as whether the mass of the nebula is about that of the present Sun or only 1/50th as great (Hayashi, Nakazawa, and Nakagawa 1985: 1107), whether angular momentum was transferred from the Sun to the planets primarily by magnetic forces or by turbulent viscosity (Cameron 1985a, Safronov and Ruzmaikina 1985), and whether giant planets are formed by gravitational instability of a gaseous nebula or nucleation (Pollack 1985). The apparent consensus on other points seems rather precarious in view of the rapid swings of opinion we saw during the preceding three decades on hypotheses such as the supernova trigger and high-temperature condensation of the terrestrial planets.

Closer examination of Levy's (1985) statement shows that by "origin of the solar system" he did not mean the actual formation of planets as we now observe them, but only the beginning of the process of forming a star that will "inevitably" be accompanied by some kind of planetary system. This is a much more modest definition than was used in the past. In particular, it gives little weight to "naturalistic" attempts to explain the present features of the system as the outcome of somewhat earlier stages (2.2.3). It gives primary importance to astrophysics and puts considerable pressure on other disciplines, such as meteoritics, to make their models consistent with astrophysics. Meteoriticists may say that their data provide "constraints" or "evidence" for astrophysical models of the solar nebula (Boynton 1985; Wood 1985), but what the astrophysicists really want from meteoritics is isotopic abundances that will indicate where the atoms came from *before* they belonged to the solar nebula (R. N. Clayton, Mayeda, and Molini-Velsko 1985; Kerridge and Chang 1985; Wasserburg 1985).

It is thus the physics of the prenebular epoch (molecular clouds, galactic density waves, supernovae, red giants, etc.) that wags the body of planetogony. The word "tail" does not seem quite appropriate here, but the metaphor may help to explain why there have been such radical changes in the theories used to explain a largely unchanging set of planetary parameters. Planetogonic theories have been evaluated not by their success in accounting for the properties of planets (with a few exceptions), but rather by their consistency with accepted theories of star formation and models of the early Sun. The decision as to whether the Solar System could have formed monistically or required an external stimulus was made not by observing the Solar System itself, but by measuring the distribution of ^{26}Al in the Galaxy and determining the rates of certain nuclear reactions in a laboratory at Yale (2.5.3).

From the historical perspective it is not surprising that planetogonic theories are judged by criteria that have no direct connection with planets. The great popularity of the Nebular Hypothesis in the 19th century was due in no small part to the connection it made between real nebulae that one could see in the heavens, stars, and the origin of the Solar System. It held out the promise (as does today's cloud-collapse model) that if our telescopes could be made just a little more powerful, we could actually see new planetary systems being born.

It seems clear that physical arguments and data (including isotopic anomalies) will continue to provide the most important tests for planetogonic theories. Even in selenogony, where immense quantities of chemical and geological data were collected at considerable expense, calculations of angular momentum turned out to be crucial (4.4). The only thing that is likely to change this situation is the discovery of other planetary systems. "Astronomical" evidence might then become more important in evaluating theories – provided that those theories had been developed sufficiently to yield specific deductions about the observable orbits and sizes of planets. Alfvén himself once tried to play that game (1943b); it's time for another round with more players.

PART 4
Whence the Moon?

4.1

Introduction

Until about 20 years ago, nearly all theories of the origin of the Moon belonged to one of three categories. "Capture" theories assumed that the Moon was formed "somewhere else" and later captured into orbit around the Earth. "Co-accretion" theories postulated that the Moon was gradually formed from material circulating around the Earth during the same time that the Earth itself was forming from similar material. "Fission" theories assumed that the Moon was ejected from an already-formed Earth. It was one of the most remarkable – though for years unrecognized – achievements of the U.S. *Apollo* program to force scientists to abandon all three theories, thus allowing the development of a fourth hypothesis. The "giant-impact" theory, which was widely if only provisionally accepted after 1980, imagines an actual collision between the young Earth and a Mars-size planet, with the Moon being formed from a mixture of material from the impacting planet and Earth's mantle.

4.1.1 Out of the Earth

Modern theories of the origin of the Moon go back to 1878. In that year George Howard Darwin, son of the evolutionist Charles Darwin, proposed that the Moon had once been part of the Earth and was ejected from it by an instability triggered by the action of the Sun's tidal force.[1] The additional hypothesis that the scar left by the Moon's departure became the Pacific Ocean basin was proposed by Osmond Fisher (1882).

Darwin's theory was not based on any direct evidence that the Moon was once part of the Earth, but rather on an interpretation of the observed "secular acceleration" of the Moon. In the 18th century, astronomers thought that the Moon was gradually moving faster in its orbit around the Earth. That would imply (by Kepler's Third Law) that it is approaching the Earth. But in the 19th century, quantitative analysis of the gravitational actions of the other

---

1 Darwin (1878, 1879). G. H. Darwin (1845–1912) studied at Cambridge University; he was Second Wrangler in the Mathematical Tripos and Smith's Prizeman in 1868. He studied law and was admitted to the bar but never practiced. Lord Kelvin inspired him to become a scientist. In 1873 he became a fellow of Trinity College at Cambridge and in 1883 was elected Plumian Professor of Astronomy and Experimental Philosophy (succeeding James Challis). He was knighted in 1905. On G. H. Darwin as the leader of a "research school in geophysics" see Kushner (1993).

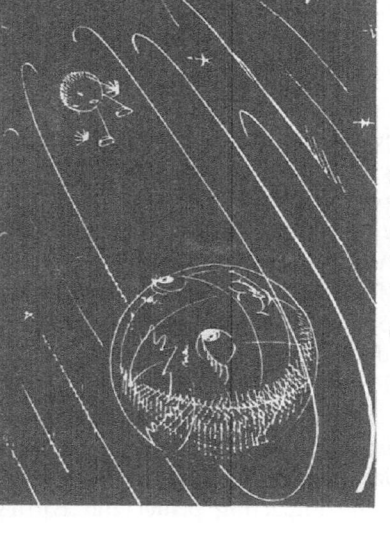

Fission origin of the Moon

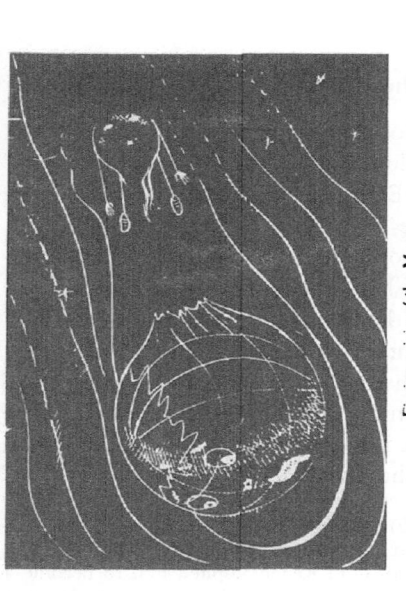

Capture origin of the Moon

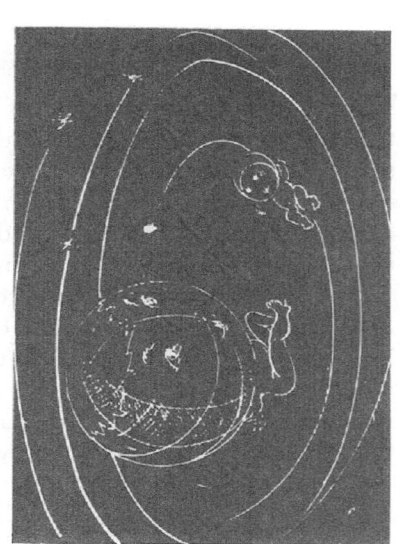

Common origin of the Earth and Moon

Figure 20. Hypotheses of the origin of the Moon – illustrated by Evgenia Ruskol, from the film, *Chronicles of Lunar Nights*, Moscow, Centrnauchfilm, 1983 (E. Ruskol).

planets on the Earth and the Moon indicated that the Moon is actually moving more slowly than in the past and that the apparent acceleration is due to a slowing down of the Earth's rotation. The physical cause was identified as dissipation by lunar tides in the Earth's oceans. Darwin pointed out that since the angular momentum of the Earth–Moon system is conserved, the angular momentum lost by the Earth must be transferred to the Moon. As a result the Moon's orbit is gradually receding from the Earth; conversely it must have been closer in the past. Making specific assumptions about the mechanical properties of the Earth, he traced the lunar orbit back to a state in which the Moon moved around the Earth as if rigidly fixed to it, in a period of 5 hours 36 minutes, with its center about 6000 miles from the Earth's surface. Before that state there was no unique solution.

The major objection to the hypothesis that the Moon was ejected from the Earth was dynamical: A body with the combined mass and angular momentum of the Earth and the Moon, rotating in about 5 hours, would not be unstable against spontaneous fission. Darwin was aware of this objection but proposed to circumvent it by invoking a resonance of the Sun's tidal action with the free oscillations of the proto-Earth.

Another objection was that the Moon could not go into orbit as a single body because it would initially be inside the Roche limit and would therefore be broken up into many smaller bodies by the Earth's tidal force. Darwin argued that this flock of small bodies would still produce tidal dissipation that would expand its orbit out beyond the Roche limit, so it could eventually recombine into a single satellite.

Roche himself was the major proponent of the hypothesis that the Moon was condensed from a ring spun off by the rotating gaseous proto-Earth, just as the Earth was condensed from a ring spun off by the solar nebula in the nebular hypothesis (Roche, 1873). This became known as the co-accretion or "sister" hypothesis.

A third hypothesis, advocated by T. J. J. See (1909d) and others, created the Moon in some other part of the Solar System and later brought it to be captured by the Earth; it was thus Earth's "wife."

Darwin's hypothesis, which described the Moon as Earth's "daughter," remained the most popular until 1930, when Harold Jeffreys criticized it, arguing that viscosity in the Earth's mantle would damp out the motions required to build up the postulated resonant vibration (Jeffreys 1930).

4.1.2 Capture

In the 1950s the capture hypothesis was revived by Harold Urey in the United States and Horst Gerstenkorn in Germany. Urey's theory was developed as part of a general chemical theory of the origin of the Earth, the other planets, and meteorites, described in section 3.3.1. Gerstenkorn (1955, 1969) worked out a quantitative dynamical theory of the capture process, following the approach of G. H. Darwin but with different assumptions about the initial state and the mechanical properties of Earth and Moon.

Urey's theory (1960a, 1960b, 1960/1962, 1962, 1967) was largely qualitative; he was less interested in the dynamics of the capture process than in the nature of the Moon itself as a key to the early history of the Solar System. He argued that the Moon is a frozen relic, a surviving example of bodies that used to populate the Solar System. For example, the abundance of iron in the Moon is much less than that in the Earth (as inferred from the densities of the two bodies) but comparable to that in the Sun (according to solar measurements before 1969). It has always been cold since its capture and thus preserves on its face a record of events that left no trace on the surface of the geologically active Earth. This was a powerful argument for manned exploration of the Moon: Analysis of the lunar surface not only should be able to tell us the conditions prevailing at the time and place of the Moon's formation, but might reveal facts about the Earth's history that cannot be learned by studying the Earth itself. If, on the other hand, Darwin was right and the Moon is just a piece of the Earth, it would not be worth the trouble to go there.

Gerstenkorn's theory raised different kinds of questions. What is the range of initial conditions for which capture is dynamically possible? Could the Moon have been captured from a retrograde orbit? How could the lunar orbit have acquired its present eccentricity and inclination? How much energy had to be dissipated during the capture process, and would this energy have been enough to melt the Earth or at least produce some effects that could be detected today? Extensive calculations by G. J. F MacDonald, Peter Goldreich, S. F. Singer, and others in the 1960s indicated that while capture of the Moon is not dynamically impossible, it would be extremely difficult to satisfy all the conditions necessary to produce the present lunar orbit.

It might appear that the only way to test the capture theory is by mathematics: to see if the known astronomical facts about the Moon can be deduced from a plausible initial state. The larger the set of possible initial states that can be shown to lead to the given final state, the more likely that the hypothesis is correct.

From Urey's point of view this kind of test was irrelevant. Even if the probability that any given Moon-sized body could have been captured by the Earth is very small, there were so many such bodies in the early Solar System that there was a reasonable chance of capturing one of them. In any case this cannot be used to compare capture with other hypotheses since different kinds of adjustable parameters are involved in those hypotheses (viscosity of proto-Earth, conditions in the primeval nebula, etc.). Instead, the real test must be chemical: If the Moon is unlike the Earth it must have been formed elsewhere; if it is like the Earth it was at least formed in the same part of the nebula (co-accretion) if not actually inside the Earth (fission).

4.1.3 Are the Earth and the Moon chemically similar?

To say that the Moon is "like the Earth" does not mean it has the same chemical composition throughout. Cosmochemists were of course aware that the average density of the Moon is significantly lower than that of the Earth. This was generally explained by assuming that the Earth has a substantial high-density iron core, whereas the Moon has little or no core. The fission theory would predict that the Moon is similar to Earth's mantle, if fission occurred after the Earth's core had already been formed. The co-accretion theory needs some kind of fractionation mechanism to get rid of the iron from the material that will form the Moon. Such a mechanism was suggested by Orowan (1969): Iron particles stick together because of plastic deformation when they collide, whereas silicates are brittle and break up in collisions. Thus, the Earth will collect the iron in its region, leaving a shell of silicate particles around it to aggregate into the Moon. (The possibility that iron particles stick together by magnetic forces was considered by Harris and Tozer [1967] but rejected by Banerjee [1967].)

Another way to preserve lunar–terrestrial similarity is to adopt the hypothesis of Lodochnikov (1939) and Ramsey (1948, 1949) that the Earth's core is *not* iron but a silicate compound that has undergone a phase transition to a high-density fluid metallic state. If this hypothesis is adopted, one may suppose that the greater average density of the Earth is due merely to its greater total bulk and consequent higher initial pressure; the pressure inside the Moon is insufficient to produce the phase transition. The co-accretion theory would then be able to get along without any mechanism for separating iron from silicates (Ruskol 1966a). But the Lodochnikov–Ramsey hypothesis was disproved by experiments and theoretical calculations in the 1960s, so this alternative was eliminated. This was not, of course, a serious problem before 1969 since there was no direct evidence for lunar–terrestrial similarity. Nevertheless there were hardly any advocates of co-accretion except in the USSR, where Evgenia Ruskol took the lead in developing this theory.[2]

A few scientists revived the fission hypothesis in the 1960s, in some cases because it was congruent with certain theories about the early development of the Earth. Thus, Ringwood (1960, 1966a) and Cameron (1963c) used the ejection of the Moon to get rid of Earth's primeval atmosphere. Donald Wise argued that the traditional objections to fission had been weakened by recent developments, while Soviet photographs suggesting that the far side of the Moon differs from the near side provided new evidence in favor of the hy-

---

2 Ruskol (1960/1962, 1963a, 1963b, 1975). Ruskol graduated from Moscow University in 1949 and did postgraduate work from 1950 to 1953 in Otto Schmidt's department at the Institute of Earth Physics. Her early research was on galactic dust clouds, dark globulae, the collapse of protostars, and the protoplanetary nebula. She began her research on selenogony in 1958. Ruskol is married to Victor Safronov.

pothesis.[3] John A. O'Keefe argued that if (as he believed) tektites come from the Moon, then the Moon came from the Earth; he emphasized the idea that the fission process involved high temperatures and resulting loss of a substantial amount of volatile substances from the Moon.[4]

Both capture and fission hypotheses, being based on the evolution of the lunar orbit through tidal dissipation, ran into difficulties because the time scale for this evolution was estimated to be only one or two thousand million years. Since the age of the Earth had been determined to be about 4500 m.y. and the oldest rocks were about 3500 m.y., the question arose: Why is there no sign of such a catastrophic event as the ejection or capture of the Moon in the geological record? If one accepted Hartmann's (1965) estimate of 3600 m.y. for the age of lunar maria, this time scale would rule out the fission theory and present the capture theory with the problem of "storing" the Moon outside the Earth's zone for more than a thousand million years after its formation. But the time scale estimate was quite dubious; it depended on the assumption that the dissipative force coefficient had remained constant through the entire period of evolution. Since dissipation was thought to take place primarily in shallow seas, it would depend on the arrangement of land masses on the Earth's surface; with the acceptance of plate tectonics in the 1960s, that became a highly variable quantity. In the 1980s the time scale problem was no more an objection to capture and fission than it was to any other modern theory.

So it was quite difficult to find any conclusive test of selenogonies, before the return of the first lunar samples. On the other hand few theorists actually published specific predictions about what would be found in those samples. I have found only four: Urey's discussion based on his capture theory, and Ringwood's, O'Keefe's, and Wise's based on fission hypotheses. Urey expected to find evidence of water on or near the surface. O'Keefe argued that the Moon would be poorer than Earth in water and other volatile substances, and would also be deficient in siderophile elements such as nickel. (Roughly speaking these predictions reflected the consequences of a cold or hot origin, respectively.) Wise predicted that the near side of the Moon should have the same composition as the Earth's mantle, while the far side should be less dense and the same as the proto-Earth's crust. Ringwood predicted a thermal history in which a temperature maximum starts near the surface and gradually

3 Wise (1963, 1964/1966). Donald U. Wise (b. 1931) received his Ph.D. in geology at Princeton University in 1957. He was on the faculty of Franklin and Marshall College from 1957 to 1968 and was professor of geology at the University of Massachusetts, Amherst, from 1969 to 1980. He was briefly (1968–9) chief scientist and deputy director of the NASA Lunar Exploration Office (see O'Leary 1970: 211).

4 O'Keefe (1963, 1964/1966, 1969a, 1969b). John Aloysius O'Keefe (b. 1916) received his Ph.D. in astronomy at the University of Chicago in 1941. He was a mathematician with the Army Mapping Service from 1945 to 1958, then joined NASA's Goddard Space Flight Center in Greenbelt, Maryland. In addition to selenogony he has done research on tektites and planetary rings.

moves toward the center, possibly exceeding the melting point for a brief period about 10^9 years after formation. This implies loss of volatiles from the crust but not from the deep interior, and in fact the density should decrease with depth (Ringwood 1966a: 90).

4.1.4 *Apollo*'s impact: All theories are refuted by the data

In July 1969 the *Apollo 11* mission brought back the first lunar samples from Mare Tranquillitatis. Preliminary analysis of these samples indicated a high concentration of refractory elements (Ti, Zr, etc.); low concentration of volatiles (Pb, Bi, Tl); strong depletion of siderophile elements, especially Ni and Co; and an absence of hydrated minerals, showing a scarcity of surface water.

The Anders group at Chicago quickly concluded that these results, together with a strong depletion of Au and Ag, provided good evidence against the fission theory (Ganapathy et al. 1970; Anders et al. 1971). Ringwood and Essene (1970a, 1970b) argued that the scarcity of volatile metals, siderophiles, and water did rule out the original fission hypothesis but not the high-temperature version of Ringwood (1960, 1966a) and O'Keefe (1969a, 1969b).

Urey's capture theory seemed to be refuted by the *Apollo* data, in particular the scarcity of water and the evidence for an early high-temperature stage. Also, new measurements of the solar iron abundance (Garz et al. 1969) showed that it was greater than previously believed, thereby removing one of Urey's arguments that the Moon is more like primordial Solar System material than the Earth. After extensive discussions with O'Keefe, Urey decided to abandon his capture theory and eventually leaned toward fission, though he was not very enthusiastic about that or any other theory. Other versions of the capture theory that had relied on a time scale of one to two thousand million years for evolution of the lunar orbit seemed to be refuted by evidence that the Moon is more than 4500 m.y. old and has not undergone significant heating or other catastrophic events more recently than 3500 m.y. ago.

Although it was frequently stated during the 1970s that the fission theory had been refuted, cosmochemists were finding increasing similarities between lunar and terrestrial composition. The early conclusions about excess refractory abundance and depletion of siderophiles were later judged to have been somewhat exaggerated. Moreover, oxygen isotope abundances were found to be the same in lunar and terrestrial material.

The major opposition to lunar–terrestrial similarity came from the Anders group, which favored co-accretion after fractionation in the solar nebula; they argued that the Moon was formed in a circumterrestrial orbit from material that had condensed at higher temperatures than the Earth (Ganapathy and Anders 1974). Another version of the co-accretion theory, developed by

Ruskol (1971/1972a, 1971/1972b, 1972) attributed compositional differences to processing of incoming planetesimals by collisions with the circumterrestrial swarm over a long period of time (10^8 years); volatiles would be removed from the outer edge by the solar wind, silicates would be broken up and remain in the swarm, while iron passed through and was accreted by the growing Earth. Elaborations of this model were proposed by Harris and Kaula (1975).

In the late 1970s and early 1980s, the consensus of the lunar science community was that none of the three pre-*Apollo* theories offered a convincing explanation of the origin of the Moon.

Fission

In addition to the original angular momentum difficulties of this theory, new calculations on viscous rotating fluids indicated that they cannot be spun fast enough to cause fission, instead they simply lose matter from equatorial regions; rotational instability can be produced by planetesimal accretion only if one planetesimal is about one-tenth of the mass of the proto-Earth, in which case the fission model goes over to the impact model. Fission models were deemed incapable of explaining why the Moon is substantially richer in both iron and refractory elements than the Earth's mantle.

Capture

This hypothesis lost its original advantage of being able to explain Earth–Moon compositional differences when it was shown that capture is dynamically impossible unless the Moon was formed at about the same heliocentric distance as the Earth, and even then is rather unlikely. Disintegrative capture was also unlikely. On the other hand, even if the Earth could have captured a Moon formed far away from the Earth (in order to account for chemical differences) one would then have difficulty accounting for the *similarity* of oxygen isotope composition.

Co-accretion

This hypothesis had difficulty in explaining the compositional differences between the Earth and the Moon, even with a postulated "composition filter" to separate iron from silicates; moreover, it couldn't account for the angular momentum of the Earth–Moon system.

Selenogony seemed to have reached an impasse. Other areas of planetary science were also slowing down. Cutbacks in funding for space science, especially in the United States, made it difficult to acquire new data except from the *Voyager* missions to the giant planets, which began to occupy the attention of planetary scientists in the early 1980s.

4.1.5 Giant impact is proposed

What happened next – the emergence of the giant-impact hypothesis – bears a superficial resemblance to a Kuhnian revolution. Selenogony before 1969 had been dominated not by a single theory but by a paradigm: the evolutionary cosmogony exemplified by the 19th-century nebular hypothesis, supplemented by relevant results of physics, chemistry, astronomy, and geology. Within this paradigm, cosmogonic processes had to be deterministic and Uniformitarian, even if their net result was the formation of a qualitatively new system. Thus, fission, a catastrophic event, could occur only when certain physical conditions were present, and its result was predetermined. Two-body interactions, as in the capture theory, or the earlier tidal theory of the origin of the Solar System, should be treated as deterministically as possible; actual collisions or extremely improbable initial states should be avoided. Mainstream cosmogonists were unwilling to postulate random catastrophic events, for reasons that may be called philosophical.

Then, when the accepted paradigm was afflicted with insuperable difficulties, so that the very existence of the Moon became an "anomaly" in the Kuhnian sense, the constraints of the old paradigm were discarded and the first steps were taken toward a new one. The new hypothesis, which was not yet a fully developed theory, suddenly attracted the enthusiasm of many scientists in what even its proponents describe as a "bandwagon" effect (Stevenson 1987: 271). Since the hypothesis explicitly invoked a random catastrophe it was difficult to show that it is objectively superior to theories that exclude such catastrophes on philosophical grounds; if the criteria for testing hypotheses change, the paradigms are at least partially incommensurable (Kuhn 1970). This is not to say that the new criteria are less strict; on the contrary, because of the availability of better computers, proponents of any hypothesis are now expected to demonstrate quantitatively that their mechanism will actually work with physical assumptions, where previously one could get away with qualitative arguments.

I used the phrase "superficial resemblance" to warn the reader that the Kuhnian revolution is only an abstract historiographic model. One cannot expect to find a real historical event that is accurately described by the model, any more than one can expect to find a perfectly rigid sphere in nature. Moreover, most historians and philosophers of science insist that Kuhnian revolutions have (or should have, respectively) nothing to do with how science works. Nevertheless, many earth scientists affirm that the establishment of plate tectonics in the 1960s was a Kuhnian revolution, and the issue will inevitably arise whenever any radical change in accepted theories occurs. It is therefore worthwhile to point out some Kuhnian and non-Kuhnian aspects of the rise of the giant-impact hypothesis.

The most obvious non-Kuhnian feature is that all discussions and calculations on the giant-impact hypothesis employ the same established principles of physics that were used to develop the previous theories, and the major dynamical problem that the new hypothesis was designed to solve is precisely the one that was considered of paramount importance in traditional cosmogony. As Howard Baker pointed out more than 30 years ago, "that the Moon was forcibly separated from the Earth by some extraneous force is indicated by its excess angular momentum about the Earth," and this force must have been exerted by a close gravitational encounter, if not an actual collision, with some large heavenly body (1954: 12, 16).

But Howard Baker's hypothesis, published as a pamphlet by the Detroit Academy of Sciences, was completely ignored by the scientific community; as far as I can determine it was unknown to mainstream selenogonists in the 1970s. Aside from the fact that *most* scientific papers, even those published in respectable journals, are never cited by anyone except their authors (Menard 1971: 96–103), one may attribute the neglect of this work to the general dislike of scientists in the 1950s for catastrophic theories of Solar System history, as shown by their reaction to Immanuel Velikovsky's books. (There were many objective reasons for rejecting Velikovsky, but the emotional tone of the criticism indicates that the argument was partly on a metascientific level; see Bauer 1985.) The dominant paradigm defined such theories as unscientific.

By 1973 the situation had changed enough for another Baker to win serious consideration, though not actual publication, of a giant-impact hypothesis.[5] James Baker proposed that Mars and Earth suffered a grazing collision, ejected material from Mars that eventually formed the Moon (J. G. Baker 1974).[6]

James Baker's hypothesis was praised by the eminent Harvard astronomer D. H. Menzel and mentioned in at least two papers by other scientists, but was rejected for publication in July 1974 and never became part of the recognized literature. It was just on the borderline between "crackpot" and "respectable," and contained several features that even today would be considered unacceptable. Nevertheless, I suspect that if Baker had been a well-known scientist, or if he had marshalled the existing evidence to support his ideas more effectively, he might now be regarded as the inventor of the giant-impact hypothesis. (It is not my function as a historian to give him that title, and experts who have recently looked at his theory decline to do so.)

Before the epoch of planetary exploration, a giant impact on the Earth might have seemed unlikely. But Mercury's cratered surface, revealed by *Mariner 10* in 1974, suggested that the terrestrial planets were bombarded by somewhat smaller bodies for hundreds of millions of years after their formation (3.1.2). This made it much more plausible than before that the Earth could

5 Acceptance of the meteoritic in place of the volcanic hypothesis for the origin of lunar and terrestrial craters helped to foster the recognition of impact as a widespread cosmic process (Hoyt 1987). By the 1980s scientists were willing to give serious consideration to the suggestion that large impacts had a major effect on geological and biological history.

6 James Gilbert Baker (b. 1914) received his Ph.D. in astronomy and astrophysics at Harvard University in 1942. A research associate at Harvard College Observatory since 1949, he is also affiliated with the Lick Observatory and is a consultant on optical physics and aerial photography to the U.S. Air Force. See also 4.4, note 20.

have been struck by an object large enough to knock a substantial amount of material from its mantle.

In August 1974 William K. Hartmann, building on his earlier studies of lunar craters, which suggested bombardment by larger bodies in earlier epochs, proposed that the Moon was formed from material ejected into a circumterrestrial disk by a large body (greater than 1000 km radius) that struck the Earth.[7] He worked with D. R. Davis to develop a theory, incorporating ideas of Ringwood (1970), and published the following year (Hartmann and Davis 1975).

Soon afterward, A. G. W. Cameron and W. R. Ward (1976), motivated by a desire to account for the angular momentum of the Earth–Moon system (Cameron 1985b), suggested a similar hypothesis with an even larger impacting body, comparable in size to Mars. In a later paper (Ward and Cameron 1978) they applied the accretion disk theory of Lynden-Bell and Pringle (1974) to describe the later stages of the lunar formation process.

The Hartmann–Davis and Cameron–Ward hypotheses, though somewhat different in detail, together became known as the foundation of the "giant-impact theory" for the origin of the Moon.

4.1.6 Giant impact is accepted

William Kaula was one of the earliest supporters of the impact theory.[8] In 1977 he mentioned it as a promising explanation for the early differentiation needed to account for the Moon's bulk composition, although the probability of such an impact still seemed quite low (Kaula 1977b). Subsequent calculations of thermal evolution based on Safronov's model seemed to tip the balance in favor of impact (Kaula 1979).

Additional support for the theory was provided by George Wetherill's calculations on the accretion of planetesimals (3.1.2). Wetherill found that a terrestrial planet is likely to be hit by a Mars-sized object; giant impact is not such a rare event after all (Wetherill 1986).

By the mid-1980s giant impact had become the most popular theory of the origin of the Moon and was starting to reach the public through popular magazines and books (Kerr 1984; Abell, Morrison, and Wolff 1987; Cooper 1987; Frazier 1987; Stevenson 1987). Additional theoretical and empirical evi-

---

7 William K. Hartmann (b. 1939) received his Ph.D. in astronomy at the University of Arizona in 1966; he has been a member of the Planetary Science Institute in Tucson since 1970. In addition to his research on lunar and planetary science, he has published several popular articles and books on astronomy, illustrated by his own paintings. See Chapman (1982: 21–2) on his dispute with geochronologists about the validity of the crater-count method of estimating lunar age.

8 William Mason Kaula was born in 1926 in Sydney, Australia. He was educated at the U.S. Military Academy and Ohio State University (M.S. 1953). He served as a geodesist at the Army Mapping Service and at Goddard Space Flight Center; he has been professor of geophysics at the University of California, Los Angeles, since 1963. He was chief of the National Geodetic Survey from 1984 to 1987.

dence against the three earlier theories (fission, capture, co-accretion) had accumulated. Further analysis of the lunar samples convinced many scientists that the similarities between the Earth and the Moon outweighed the differences. Detailed computer simulations of the impact process (e.g., Cameron 1985b) indicated that a significant part of the Moon-forming material would have come from the impactor rather than from the Earth, so it was possible to attribute residual lunar–terrestrial differences to an admixture of nonterrestrial material (Taylor 1987c). Opponents of giant impact could not ignore the hypothesis; instead they had to use it as a basis for arguments and calculations showing that it entailed other consequences (e.g., for Earth's thermal history) that might be tested by observation (Kato, Ringwood, and Irifune 1988; Garwin 1989; Hartmann 1989; Kerr 1989; Newsome and Taylor 1989). Impact itself was now the target to shoot at.

4.2

Early history of selenogony

Speculations about the origin of the Moon must have begun almost as soon as human consciousness, but the kind of evidence needed to develop a quantitative scientific theory has been available for less than 300 years; the evidence required for a rigorous test of competing theories was obtained only with the manned lunar landings beginning in 1969. The first major conference devoted exclusively to the subject was not held until 1984. In this chapter I review the history of selenogony – the study of the origin of the moon – up to the early 20th century.

4.2.1 Earth's Moon and planetary satellites

Since Galileo's discovery of four small bodies orbiting the planet Jupiter, it has been known that the Earth's Moon is only one of several satellites in the Solar System. One might therefore expect that selenogony would be only a special case of the theory of formation of satellites, and indeed one of the most popular theories ("binary accretion" or "sister" hypothesis) treats the Moon this way. But many scientists thought that the Moon deserves her[1] own special hypothesis: First, because she is unusual in being a single and relatively large companion of her primary, unlike the Jovian and Saturnian systems that could be described as miniature planetary systems (it was learned only recently that Pluto's satellite may be comparable to our Moon in this respect); second, because we have much more detailed information about her and thus presumably an opportunity to construct a more reliable quantitative theory.

One of the first attempts to explain the Moon's formation in the framework of the new heliocentric astronomy of the 17th century is found in Descartes's *Le Monde*. This work was apparently written around 1630 but withheld from publication because of the condemnation of the heliocentric system in the notorious trial of Galileo in 1633; it was published posthumously in 1664. Descartes imagines a universe filled with pieces of matter of various sizes, shapes, and motions, evolving into a system of numerous vortices rotating around stars. Large pieces of matter – planets – can move in orbits at definite distances from the central star, depending on the "force" of their motion (mass or quantity of matter, multiplied by speed); they are kept

---

1 In keeping with historical tradition, as well as for grammatical convenience, I refer to the Moon as female in this chapter; for further discussion of this point see Mitroff (1974a; 1974b).

in dynamic equilibrium at those distances by collisions with smaller particles. Each planet then develops its own vortex of these small particles, revolving around it in the same direction that the planet moves around the star. Another planet may move in the same orbit as long as its force is the same, but if it is smaller it will have to move faster and will soon overtake the larger one. It will then be trapped by the vortex of that larger planet and be forced to move around it. Descartes proposed this scheme to account for the Earth–Moon system but left it to his readers to extend it to the other planetary satellites, confessing that "I have not undertaken to explain everything" (Descartes 1664/1824–6: **4,** 246–88).

During the 19th century the hypothesis that the Moon formed from a Laplacian ring around the Earth (Roche 1873; Ennis 1867) was challenged by the proposal that she had been ejected from the Earth. In Britain that idea was expressed poetically by Erasmus Darwin (1790/1825: 28), grandfather of Charles Darwin. An American, Richard Owen, professor of geology and chemistry at the University of Nashville, Tennessee, suggested that the Moon came from the region of the Mediterranean (Owen 1857: 66). The American mathematician Benjamin Peirce hinted at a terrestrial origin in his statement that the Moon when formed revolved in contact with the Earth (Peirce 1871, 1873).

4.2.2 Secular acceleration

Special theories of the Moon's origin developed from attempts to explain observed variations in the lunar orbit. The periodic time of the Moon's revolution around the Earth is gradually increasing (apart from short-term cyclic changes), a fact that implies (according to Kepler's Third Law) that she is slowly retreating from the Earth. If the average Earth–Moon distance has been continually increasing in the past, either at a constant rate or as a result of a force varying with distance in a known manner, we can extrapolate backward in time to reconstruct the earlier history of the Moon's orbit. This leads us to an epoch when the Moon would have been inside the Roche limit – the distance at which the Earth's tidal force would break up a body held together by gravitational forces. At that point (if not before) the extrapolation becomes invalid, and we must introduce a specific hypothesis about the Moon's earlier motion – and indeed about whether she even existed in her present form.

What I have called a "fact" in the preceding paragraph was not recognized as such before the middle of the 19th century. On the contrary, the first quantitative studies of the Moon's motion indicated that the periodic time is *decreasing;* hence, the term "secular acceleration" has traditionally been applied to this effect.[2] In 1749 the British astronomer Richard Dunthorne reviewed

---

2 Astronomers use the word "secular" to refer to a relatively long-term monotonic variation, as distinct from short-term "cyclic" variations. The significance of secular inequalities in 18th-century science and in the work of Laplace is discussed in *Nebulous Earth,* 1.2.2. On the history of the problem of the Moon's secular acceleration see Kushner (1989).

ancient and modern records of eclipses in order to test Edmond Halley's suggestion (1695) that the Moon has gradually been moving faster in her orbit. He concluded that there is indeed a secular acceleration, about 10 inches per century, and this estimate was confirmed by other astronomers (Dunthorne 1749; Grant 1852: 60–4; Forbes 1972: 11, 20–1, 76–7). Modern values are substantially higher (30 to 50 inches per century).

Laplace attempted to explain the secular acceleration as a combination of the action of the Sun on the Moon and the secular variation of the eccentricity of the Earth's orbit. Perturbations of other planets are slowly decreasing this eccentricity, and since the Sun's action on the Moon's orbit is greatest when the Earth is at perihelion, his effect on her motion will also be slowly diminishing. Taking into account directly only the radial component of the Sun's action, Laplace found that it decreases the angular velocity of the Moon at perihelion, so the decrease in eccentricity of the Earth's orbit will be accompanied by an acceleration of the Moon.[3]

If this explanation is sufficient to account for the entire effect, it implies that the Moon's acceleration is not truly "secular" (changing always in the same direction) but will eventually be reversed, since the change in the eccentricity of the Earth's orbit is itself a cyclic effect. (Indeed, this cycle may be partly responsible for periodic "ice ages.")[4] This cyclic character is consistent with the expectation that gravitational forces should not produce any irreversible effects in a conservative dynamical system.

Since Laplace's theoretical calculations gave a secular acceleration in good agreement with the observed value, it was generally assumed that he had solved the problem. But the British astronomer John Couch Adams (1853, 1859), looking into lunar theory half a century later, noticed an error in Laplace's calculation: The tangential component of the Sun's action does produce a significant effect.[5] When higher-order terms were computed it appeared that the Laplace mechanism could account for only about half of the observed acceleration. This result was announced in the same year as the publication of Charles Darwin's *Origin of Species*.

The explanation for this newly discovered discrepancy was already at hand, though it took a few years before it was recognized by astronomers. The German philosopher Immanuel Kant had pointed out as early as 1754 that tidal dissipation should retard the Earth's rotation.[6] The German physician J.

---

3 For an explanation of Laplace's contributions to this problem see De la Rue (1866) and Gillispie, Fox, and Grattan-Guiness (1978: 332).
4 Milankovich (1920). Imbrie and Imbrie (1979). Imbrie (1982).
5 John Couch Adams (1819–1892) studied mathematics at Cambridge University, graduating First Wrangler and Smith's Prizeman in 1843. He was a fellow of St. John's College, Cambridge, and later Pembroke College, and was appointed professor of astronomy and geometry at Cambridge University in 1859. He is best known for his prediction of a new planet, based on detailed analysis of perturbations of the orbit of Uranus; unfortunately he could not persuade James Challis and G. B. Airy to look for the planet, and the major share of the credit for the discovery of Neptune went instead to U. J. J. Leverrier (who also predicted it from theoretical calculations) and J. G. Galle (who observed it in 1846).
6 Kant (1754a, 1754b), reprinted in Kant (1910: 183–91, 193–213). Kant (1755/1981). See also Wackerbath (1867).

Robert Mayer[7] and the American scientist William Ferrel[8] both attempted quantitative determinations of this effect. But their rough estimates came out much too large, so they assumed that the tidal effect was cancelled by the gradual cooling and contraction of the Earth (which should *increase* its rotation speed, if angular momentum is conserved). Mayer's and Ferrel's papers did not attract much attention at the time, but Hermann von Helmholtz emphasized the significance of tidal dissipation in his famous Koenigsberg lecture of 1854 (Helmholtz 1854/1962). All of these scientists recognized that similar tidal forces acting on the Moon could have increased her rotation period to synchrony with her revolution period, thus explaining why she always keeps the same face to the Earth.

In 1865, the French astronomer Charles-Eugene Delaunay, who had confirmed Adams's calculation of the secular acceleration, proposed that part of the apparent secular acceleration of the Moon is due to a deceleration of the Earth's rotation-produced tidal dissipation.[9] (The rest would still be attributed to Laplace's mechanism, as described earlier.) Ferrel, strictly speaking, deserves priority for this proposal since he made it at a meeting of the American Academy of Arts and Sciences in Boston in December 1864, a few weeks before Delaunay read his paper to the Académie des Sciences in Paris,[10] but it was Delaunay's reputation that persuaded astronomers to adopt it. G. B. Airy, the British Astronomer Royal, was initially skeptical of Delaunay's claim but eventually accepted it after doing his own calculations.[11]

4.2.3 Darwin's fission theory

Although Mayer had noted in 1848 that tidal action would increase the Moon's distance from the Earth, no one seems to have pursued the cos-

---

7 Mayer (1848, 1851/1893). Lindsay (1973: 176–95). Julius Robert Mayer (1814–1878) received his M.D. in 1838 at the University of Tubingen. He is best known as a co-discoverer of the Law of Conservation of Energy, based on his observations as a physician on a Dutch merchant ship on a voyage to the East Indies in 1840–1. After his return from that trip he settled as a physician in his native town, Heilbronn.

8 Ferrel (1853; 1895: 294–97). Burstyn (1971). William Ferrel (1817–1891) served for several years as a school teacher in several states in the United States. He started to publish scientific papers in 1853 and in 1858 joined the staff of the *American Nautical Almanac*. In 1867 he was appointed to the U.S. Coast Survey and was also employed by the U.S. Army Signal Service (predecessor of the Weather Bureau). He was known for his work on the effect of the Earth's rotation on atmospheric and oceanic circulation, and has been called "after Laplace, . . . the chief founder of . . . geophysical fluid dynamics" (*DSB*). He wrote an autobiographical sketch, published after his death (Ferrel 1895).

9 Delaunay (1865, 1866a, 1866b, 1866c, 1866d, 1866e, 1866f). Charles-Eugene Delaunay (1816–1872) graduated from the Ecole Polytechnique (Paris) in 1836. After serving as a mining engineer he became a lecturer on mechanics, mathematics, and astronomy at the University of Paris and elsewhere. In 1846 he published an analytical method for solving problems in celestial mechanics and applied it to the Moon's motion. In 1870 he was appointed director of the Paris Observatory.

10 Ferrel (1866, 1895).

11 Airy (1866). De la Rue (1866). Berry (1898/1961: 308–9, 368–70). Dorling (1979).

Figure 21. G. H. Darwin (*Scientific Papers*, Cambridge University Press).

mogonic implications of this fact for more than two decades. I have not yet determined what led George Howard Darwin to the idea that the history of the lunar orbit should be traced backward to a time when the Moon and Earth were in contact. His papers published in 1877–8 suggest that contemporary discussion of the cause of ice ages inspired him to test the hypothesis, proposed by geologist John Evans, that major disturbances in the Earth's surface could change its axis of rotation. This led him to inquire whether the planets could have acquired their present obliquities at a time when they were large, gaseous, rapidly spinning bodies with pronounced equatorial bulges; satellites might have been spun off at this stage. Having become interested in the process of satellite formation, he realized that mathematical difficulties would prevent him from drawing definite conclusions about the behavior of the system just before separation into two bodies; instead, he decided to follow the two-body system backward in time to the point just after separation. Thus, the son of the biological evolutionist became the first student of the evolution of the Moon's orbit, and the first to work out the quantitative details of the later stages of a specific process by which Mother Earth might have given birth to her satellite.

Darwin announced his theory in 1878 and published the details in a long memoir the next year.[12] He treated the Earth as a homogeneous rotating viscous spheroid and assumed that the Moon moves in a circular orbit in the plane of the ecliptic. Assuming the viscosity of the Earth to be large enough to give the observed (apparent) secular acceleration, he could work back to a state in which the Moon moved around the Earth as if rigidly fixed to it, in a period of 5 hours 36 minutes. This would have been at least 54 million years ago, and the Moon's center would have been no more than 6000 miles from the Earth's surface; both period and distance would be smaller if the Earth were not homogeneous.

"These results point strongly to the conclusion," Darwin declared, "that if the Moon and Earth were ever molten viscous masses, then they once formed parts of a common mass" (Darwin 1879: 536). But how could this mass have broken up? A system rotating in 5 hours, with the combined mass of the present Earth and Moon, would *not* be rotationally unstable. However, Darwin's senior colleague Sir William Thomson, later known as Lord Kelvin, had shown that a fluid spheroid of the same density as the Earth would have a period of free oscillation of about 1 hour 34 minutes.[13] A less dense body would have a longer period. Darwin could then invoke the Sun's tidal action to trigger fission: He proposed that the solar semidiurnal tide, reaching a maximum every 2½ hours at a given place on the Earth's surface, might be in resonance with the free oscillations, thus producing enormous distortion sufficient to disrupt the body.[14]

Darwin undertook further calculations to apply his theory to other possible models, including finite eccentricity and inclination of the Moon's orbit. He was cautious about stating his conclusions on the earliest stages of the Earth–

12 Darwin (1878, 1879). Most of his papers are reprinted in Darwin (1907–16). A nonmathematical survey, originally published in *Atlantic Monthly* (1898), may be found in chapters 15 and 16 of his *The Tides* (1905/1962).

13 William Thomson (1824–1907) became Baron Kelvin of Largs in 1892. His research in heat theory, solar physics, and geochronology is discussed in *Nebulous Earth*, 1.5 and *Transmuted Past*, 1.3. Although he did not reject biological evolution himself, Kelvin's low estimates of the age of the Earth indirectly undermined Charles Darwin's theory, since Darwin himself had stated that evolution by natural selection would take hundreds of millions of years.

14 Love (1889) supported Darwin's theory, confirming that the spheroid has a period of oscillation of the required length.

Moon system, recognizing that when one approaches the point of contact, an infinitesimal disturbance may cause an irreversible finite change. He thought it likely that most of the tidal dissipation at present is due to ocean tides rather than body tides in the solid Earth, but argued that the early Earth was hotter and more plastic, so body tides would have been more important (Darwin 1880).

Darwin's theory was quickly popularized by Robert S. Ball, the Royal Astronomer of Ireland, in a lecture at Birmingham, published in *Nature* in 1881. Ball stressed the possible geological effects of the enormous tides that should have been present in the Earth's early history because of the Moon's proximity. Alluding to Kelvin's arguments that the age of the Earth is much less than geologists had assumed, he pointed out that to compensate for limiting the time available for geological processes, the mathematicians have given geologists a "new and stupendous tidal grinding-engine." On the other hand, the social reformers who are attempting to reduce the working day may find their efforts nullified in the long run by the Moon's action in slowing the Earth's rotation – "Where will the nine-hour's movement be when the day has increased to 1400 hours?" (Ball 1881: 79, 103; 1882a; 1882b; 1889).

Ball's speculations stimulated a round of letters to the editor of *Nature,* the most substantial being a proposal by the geologist Osmond Fisher (1882) that the scar left by the Moon's separation did not completely heal.[15] Fisher suggested that the ocean basins are the holes left in the Earth's crust, after some flow of the remaining solid toward the original cavity. In this way the birth of the Moon would have resulted in both the Pacific Ocean basin and the separation of the American continent from Europe and Africa.

4.2.4 Alternatives to fission

The major alternative to Darwin's fission theory in the 19th century was the explanation based on Laplace's nebular hypothesis: The Moon condensed from a ring spun off from the rotating gaseous proto-Earth, just as the Earth itself condensed from a ring spun off from the rotating solar nebula. Although this might have been a satisfactory explanation for satellites of the giant planets, there seemed to be some difficulty in accounting for the relatively large mass and orbit of the Moon. The best defense of the Laplacian theory of the origin of the Moon, now known as the "sister" (or "binary planet" or "co-accretion") hypothesis in contrast to Darwin's "daughter" theory, was offered in 1873 by Edouard Roche.[16] Roche corrected Laplace's calculation of

15 Osmond Fisher (1817–1914) read mathematics at Cambridge University, graduating in 1841. He did most of his research on planetary science while holding church positions. See *Nebulous Earth,* 2.2.7, for his role in the debate on the solidarity of the Earth, and biographical data in note 6.

16 Eduard Albert Roche (1820–1883) was educated at the University of Montpellier and served on its faculty for his entire career. His contributions to the theory of Saturn's rings

the extent of the Earth's rotating atmosphere. Roche's results showed that the sister theory was tenable though not compelling.

A variation of the sister theory was the hypothesis, proposed by the American geologist Grove Karl Gilbert, that the moon formed from a ring of small solid particles; the final stage of the process would produce the craters on the Moon's face.[17]

Gilbert's hypothesis, though introduced without specific reference to Darwin's theory, could have been combined with it. Indeed, the first major criticism of Darwin's theory, published by James Nolan (1885, 1886, 1887, 1895) in Australia, was that the material spun off from the Earth would not remain intact in a close orbit, but would immediately be torn apart by tidal forces and form a ring of particles. This is because it would initially be inside the "Roche limit," although Nolan did not refer to Roche's theory. Darwin (1886) was forced to admit that the Moon must have been broken into a flock of meteorites as soon as she escaped from the Earth, but he insisted (contrary to Nolan) that this flock could still exert tidal forces on the Earth. Even a symmetrical ring of fragments would raise tides, and the resulting dissipation would expand its orbit beyond the Roche limit, whereupon the fragments could recombine into a single satellite.

Nolan's forgotten pamphlet (1885) also suggested an additional hypothesis that might be used to assist the fission theory: "The earth, which was supposed to have acquired the rapid rotation which caused the moon to separate, from the process of contraction, could only acquire that condition by the contraction of a denser nucleus" (Nolan 1885: 5).

At the beginning of the 20th century Darwin's theory was widely accepted. Lingering doubts about the excessive time required for tidal evolution, as compared with Lord Kelvin's later estimates of only 20 million years for the age of the Earth, were dispelled by the discovery of radioactivity and the resulting multi-billion-year estimates.[18] Henri Poincaré's mathematical studies of the equilibrium figures of rotating fluids seemed to offer a new and more respectable basis for the theory, and Darwin enthusiastically cooperated with Poincaré in working out the details (*Nebulous Earth*, 1.8.3).

Fisher's hypothesis that the Moon came from the Pacific Ocean basin, revived two decades later by the American astronomer W. H. Pickering (1903:

---

and his development of the nebular hypothesis are discussed in *Nebulous Earth*, 1.6.1, 1.7.3, 1.7.8. His 1848 formula for the tidal stability limit of a satellite plays a crucial role in many modern theories.

17   Gilbert (1893). See Hoyt (1982). Harold Urey, who noted that the paper was written between the time of his own conception and birth, admonished other selenologists for ignoring it (D1959).

  Grove Karl Gilbert (1843–1918) was educated at the University of Rochester. After receiving his A.B. in 1862, he worked for a company that sold scientific materials to schools, then participated in geological surveys of Ohio and the Rocky Mountains. He was employed by the U.S. Geological Survey from its beginning in 1879 and was its chief geologist from 1889 to 1892. See the comprehensive biography by Stephen Pyne (1980).

18   Darwin (1907–16, **II:** lv). Eddington (1906). Kelvin himself accepted Darwin's theory (1908).

7; 1907), eventually became a standard addition to Darwin's theory. For some geologists it provided a satisfactory catastrophic explanation of the same geographical features that Alfred Wegener's continental drift hypothesis claimed to interpret in a gradualist fashion.[19]

By 1936 the Darwin–Fisher theory had been translated into popular mythology, as illustrated in the following excerpt from a script prepared by the U.S. Office of Education for broadcast as a children's radio program:

[Start with "weird mysterioso" fanfare]
Friendly Guide. Have you heard that the Moon once occupied the space now filled by the Pacific Ocean?

Once upon a time – a billion or so years ago – when the earth was still young – a remarkable romance developed between the Earth and the Sun – according to some of our ablest scientists. . . . In those days the earth was a spirited maiden who danced about the princely Sun – was charmed by him – yielded to his attraction, and became his bride. . . . The Sun's attraction raised great tides upon the earth's surface . . . the huge crest of a bulge broke away with such momentum that it could not return to the body of mother Earth. And this is the way the Moon was born!
Girl. How exciting!

In 1909 the fission theory was attacked by two American astronomers. Forest Ray Moulton, noting that the Russian mathematician A. M. Lyapunov had disproved Poincaré's conjectures about the stability of his rotating fluids, argued that the Earth–Moon system could not have been produced by Darwin's mechanism.[20] Thomas Jefferson Jackson See proposed that the Moon, like other satellites, was captured through the action of a resisting medium. (Capture had been proposed by other writers but this suggestion was generally ignored.)[21]

According to See, the Moon was originally formed in the outer part of the Solar System, near the present orbit of Neptune. Following Leonhard Euler, he argued that all the planetary orbits have been gradually shrunk and their eccentricities reduced by the resisting medium. The discovery of retrograde satellites of Saturn and Jupiter suggested that at least some satellites must have been captured; hence, we must follow Newton's second rule of reasoning and assign the same causes to the same effects whenever possible – that is, we must assume that *all* satellites were captured. See claimed that the Moon is approaching the Earth and stated that there is no direct evidence that the Earth ever rotated more rapidly than at present; thus, he rejected the major conclusions of Darwin's theory of tidal evolution of the lunar orbit.[22]

19 Patterson (1909), Wegener (1912). Bowie (1929, 1930), Gutenberg (1930). For a negative view see Barrell (1907).
20 Moulton (1909a, 1909b). Lyapunov (1905, 1908). See also Schwarzschild (1898: 231).
21 See (1909d; 1909f; 1910a: ch. 11; 1910b). Mackey (1825). F. B. Taylor (1898: 29).
22 See (1909d: 380; 1909i; 1915). He also insisted that lunar craters are due to meteorite bombardment rather than volcanism: "The Moon's surface can be nothing but fragments of

These attacks did not lead to the rejection of Darwin's theory. See was rapidly losing his earlier scientific reputation because of his eccentric behavior, and no one seems to have taken his capture theory very seriously.[23] A theory based on similar ideas was proposed by the Austrian engineer Hanns Hörbiger as part of a grandiose "glacial cosmogony" (Hörbiger 1913). The curious history of this theory and its popularity in Nazi Germany has been thoroughly reviewed in a monograph by Brigitte Nagel (1991); see also Bowen (1992).

Harold Jeffreys, in England, came to the defense of the fission theory in 1917, pointing out that the different moments of inertia of the Moon indicate that she must have solidified at a time when she was much closer to the Earth (Jeffreys 1917b). Jeffreys suggested that Moulton's objections could be avoided by taking account of the heterogeneity of rotating fluid; resonance could then produce fission even for fairly slow rotation. The experiments of A. A. Michelson and others on the Earth's body tides indicated that it is highly elastic rather than viscous, thereby throwing doubt on the tidal dissipation mechanism used by Darwin.[24] But Jeffreys (1920a, 1920b, 1925a), following G. I. Taylor (1919), estimated that friction in shallow seas could dissipate enough energy to account for the secular acceleration of the Moon, thus reinstating the Darwinian principle for tracing its orbital evolution.

Having become one of the principal advocates of the fission theory, Jeffreys was able to deprive it of most of its support when he rejected it in 1930. His objection — that viscosity in the Earth's mantle would damp the motions required to build up a resonant vibration and thereby prevent fission — was considered conclusive by later researchers, though I find it unconvincing.[25]

During the next 25 years there seems to have been neither any major progress in developing theories of lunar origin nor any clear agreement on adopting one of the previous theories. The most important writings on the subject were those of the German astronomer F. Nölke, who advocated a modified Laplacian hypothesis in which the Moon condensed from the outer parts of the Earth's atmosphere (Nolke 1922; 1924; 1930: 294–5; 1932; 1934). The revival of selenogony came only in the 1950s with the capture theory of Horst Gerstenkorn and the application of the physicochemical approach to planetary science by Harold Urey.

rock filled with finer dust; and it is evident that it has never been molten as a whole and has never shown true volcanic activity" (1910d: 19). Cf. Gold (1955), Urey (1966b).

23 Obituary of See in *New York Times,* 5 July 1962, p. 23. J. Ashbrook (1962). Lankford (1980). Biographical note in 1.2, note 48.

24 Michelson (1914). Moulton (D1914a, D1914b, D1915). Michelson and Gale (1919). Nölke (1924). For further discussion of this topic see *Nebulous Earth,* 2.2.7, 2.2.9, 2.3.4.

25 Jeffreys estimated the frictional force for the mantle flowing over the (presumably liquid) core from a formula that appears to pertain to liquids flowing over solids; he does not explain why it would be valid in this case. Dicke (1957) reported Wheeler's (unpublished) argument against this objection. McKinnon and Mueller (1984) reexamined the effect of the solar resonance and concluded that Jeffrey's conclusion was correct even though based on oversimplified assumptions.

4.3

Harold Urey and the origin of the Moon

"Oh, I'd love to go to the moon," said Harold Urey in July 1969. "I wish I could go rock-hunting with the astronauts this month. . . . I think I'd go to the moon . . . even if I knew I could never get back" (Urey 1969b). The business executives who read this statement in *Forbes* magazine on the eve of the *Apollo 11* lunar mission may have laughed at the dreams of an eccentric scientist, but they could not ignore the impact of the biggest technological project of the decade – a project on which Harold Urey had a major influence.

4.3.1 Scientific rationale for lunar exploration

Of course, the primary reason for going to the Moon was political – we needed a space spectacular to demonstrate our superiority over the Soviets, and a manned lunar landing was the most feasible way to do this from an engineering standpoint. But why did we undertake such an extensive program of lunar exploration rather than shifting resources quickly to planetary missions as some scientists would have preferred? It was Urey who provided a scientific rationale for *Apollo* and persuaded the leaders of the U.S. space program that lunar exploration would yield more valuable information than other missions under consideration.[1]

Even without the brilliant success of *Apollo,* Urey would still deserve much of the credit for reviving interest in Solar System research during the past four decades. His 1952 book *The Planets* helped to persuade physical scientists to enter a field that had been largely abandoned or downgraded by astronomers in favor of the more spectacular realm of stars and galaxies.[2] Urey showed that chemistry and physics could greatly enrich planetary science, previously dom-

---

1 Jastrow (1959: 6–8, 90–3; 1960; 1981a; 1981b). Newell (D1959; 1973; 1981: 113, 237). Urey (D1961a). Emme (1968). R. C. Hall (1977), see index, under Urey. Brooks, Grimwood, and Swenson (1979: 125–6). The scientific justification for a manned lunar landing is largely ignored in most accounts, which focus instead on the political and technological aspects of the *Apollo* program. See, e.g., Van Dyke (1964), Mandelbaum (1969), Logsdon (1970), Bruno (1979).
2 Urey (1952). De Vaucouleurs (1962: 142–51). This is a special case of the general downgrading of planetary science in the early 20th century discussed in *Transmuted Past*, 1.4.

inated by astronomy and geology. Many of those now active in the field acknowledge the stimulus Urey gave to their own careers even if they disagree with his hypotheses and methods: As one of them said, those hypotheses and methods "defined the paradigm from which we all started."[3]

Astronomers and geologists who had undertaken careful scrutiny of the Moon's surface were curious about the origin of the craters and other features, but did not expect to learn much about the origin of the Moon itself from such study. Urey convinced himself that the Moon is a primordial object, almost unchanged since her formation during the infancy of the Solar System; unlike the Earth, she should have preserved a record of conditions billions of years ago, and thus a close examination should give us valuable clues to the formation of our own and other planets. In particular, Urey concluded that the Moon was never part of the Earth but was formed independently and later captured.

Urey was well aware of the temptation to reinterpret any new piece of data in a way favorable to one's own hypothesis.[4] It must have been hard for him, as an octogenarian, to abandon his hypothesis that the Moon is a primordial body captured by the Earth – but he did. His struggle is the topic of the present chapter; the parallel and subsequent work of the rest of the planetary science community on the problem of selenogony (lunar origin) will be surveyed in the following chapter.

4.3.2 Revival of the capture theory

In the 1950s Gerard Kuiper proposed a gas-dust protoplanet theory of the origin of the Solar System, based on Weizsacker's revival of the nebular hypothesis, and assumed (without much elaboration) that the Moon had formed inside the same protoplanet as the Earth (Kuiper 1959). According to Alter, this was the most popular view as of 1960.

Aside from Urey's brief remark in his 1952 book, quoted later, the first modern proposal of a capture theory is that of Horst Gerstenkorn (1955).[5] Gerstenkorn published a calculation of the history of the lunar orbit, going back to an epoch 2500 million years ago when it coincided with the Roche limit at a distance of $2.9 R_\odot$. The Earth then rotated in a period of 4.79 hours,

3 O'Keefe (D1978). In this letter O'Keefe called Urey "the greatest planetologist of our times. He is a great man; that means that he defines the questions which are important. It does not mean that by some miracle he always gets the right answer, more often than not, he was wrong. But it does mean that he showed us all what the real questions were." See also Ringwood (1979: v).
4 Urey (1969a; D19XX: 15). The tendency of "almost any theorist to explain a new observation in a way that supports his particular theory" is sometimes called "Urey's law" according to French (1977: 60).
5 Ruskol (1960) mentioned a proposal of a capture theory by Savchenko (1953), also by Radzievsky (1952) and Razbitnaya (1954). See also references to theories of Jeffreys and Lyttleton in the paper by Wise (1963). All of these papers seem to have been forgotten after 1960.

Figure 22. Harold Urey and the lunar globe (Von Del Chamberlain).

while the Moon revolved in 6.86 hours. He suggested that the orbit was retrograde before that epoch.

In another paper (1957) Gerstenkorn considered the hypothesis of Dirac and Jordan, that the gravitational constant varies inversely with the age of the universe. Tidal evolution would have been more rapid in the past if gravity was stronger, but the qualitative behavior of the orbit was the same. The "initial" state (Moon at the Roche limit) now occurred only 600 to 800 million years ago, in what Gerstenkorn called "geological time."

E. J. Öpik (1955b) commented favorably on Gerstenkorn's work soon after its first publication. He noted that the Moon could have "jumped into the picture" at any stage near or before "zero hour" when the orbit was at the Roche limit; thus, fragments already existing inside that limit could have gathered into one body as they drifted beyond it. Such fragments might have been ejected by the sudden phase transition recently postulated by Bullen to produce the Earth's core, though Öpik considered this postulate "rather un-promising."[6]

Gerstenkorn's theory became the basis for a revival of the capture theory in the 1960s, though at first it was ignored by everyone except Öpik. Later work did not support all the details of Gerstenkorn's calculations, but most of his qualitative conclusions were found to be valid (Gerstenkorn 1967a, 1967b, 1968; Goldreich 1968). Hannes Alfvén, the Swedish physicist-cosmogonist who achieved recognition for his own work only after considerable difficulty and opposition, pointed out that at the time when the two superpowers were spending billions of dollars and rubles on lunar exploration and had sup-posedly put much effort into compiling and exchanging scientific informa-tion, a high school teacher could make a major contribution with no re-sources other than his own brains, yet this contribution could go unnoticed for nearly a decade (Alfvén 1965).

Dinsmore Alter (1960) was one of the first scientists to present a systematic comparison of the three major theories in the light of modern knowledge. He favored capture but did not cite Gerstenkorn's papers, noting instead that it was an idea that Urey "casually mentions" in The Planets (1952: 97). Alter reviewed the physical arguments against fission and co-accretion theories – for example, according to both theories the Moon's orbit should be in the Earth's equatorial plane and her shape should be that produced by hydrostatic equi-librium. He claimed that the capture theory avoided these objections, but he did not try to develop it.

In his book The Planets (1952) Urey mentioned both the sister and wife theories for the origin of the Moon:

The qualitative difference in density of the Moon and Earth can be explained by the assumption of two stages of growth, the first during a period in which low-density silicates collected into a primordial Moon and Earth, and the second subse-quent to this in which metallic iron–nickel phase was an important ingredient in the material collected, together with the assumption that the rate of growth of the Earth in this second phase was much more rapid than that of the Moon. . . .

We should explore the possibility that the Moon was formed from its own proto-planet and not from a secondary nucleus within the Earth's protoplanet, and that the Moon or its protoplanet was captured by the Earth. So far as any evidence presented here is concerned this may well have been the case. In this event there is no difficulty in accounting for the rate of growth of the Earth relative to the Moon. . . . The comparatively large angular momentum of the system arose then from the details of the capture collision. Such a capture would be aided by the presence of gas. (Urey 1952: 97)

In spite of this early remark Urey did not propose a definite capture theory in any other publications until 1959, and he seems to have been led to it

6 Bullen (1951). See also Ramsey (1948, 1949). Bullen (1967: 261). Öpik (1962b) rejected capture in favor of co-accretion in 1962.

indirectly through his studies of meteorites rather than directly by consideration of the Moon herself. But he did admonish colleagues who continued to mention Darwin's theory as a possible mode of lunar origin, insisting that this theory was untenable.[7]

In the mid-1950s Urey had severe scientific and personal disagreements with Kuiper; in 1958 he left Chicago to accept an appointment at the new San Diego (La Jolla) campus of the University of California. In 1957 he abandoned Kuiper's hypothesis that protoplanets (large masses of gas and dust of solar composition) had been involved in the formation of the terrestrial planets. Instead he postulated that two sets of objects of asteroidal and lunar size, called "primary" and "secondary" objects, were accumulated and destroyed during the history of the Solar System.[8] The primary objects were suddenly heated to the melting point of silicates and iron, then cooled for a few million years and broken into fragments of less than centimeter size. "The secondary objects accumulated from these about 4.3×10^9 years ago, and they were at least of asteroidal size. These objects were broken up, and the fragments are the meteorites" (Urey 1956b: 625).

The reason for constructing this scheme was to explain the presence of diamonds in meteorites, which Urey thought required an earlier high-pressure environment that could have been provided inside larger bodies. Subsequently it was found that the diamonds could have been formed by impact, but the hypothesis of "lunar-sized bodies" kept its hold on Urey's mind.[9]

At the first Symposium on the Exploration of Space in April 1959, Urey suggested that "the Moon may be one of these primary objects, as I realized after devising what seemed to me a reasonable model for the *grandparents* of meteorites" (1959: 1727). As the *New York Times* headlined one of his speeches two years later, "Urey holds moon predated earth" and is one of the few relics of an early state of the Solar System (Sullivan 1961). Urey could therefore prescribe a set of chemical and physical observations to be made from the Moon's surface to give information not only about meteorites but also the formation of the planets. He concluded his 1959 paper with the remark: "It is hoped that such observations will be forthcoming during the immediate years ahead" (1959: 1736).

Urey discussed the nature of the Moon's capture in more detail in 1960, attributing the necessary energy dissipation either to tidal effects or to collisions with small bodies remaining in orbit around the Earth. "The very short period of time for the formation of the maria indicated by the surface features of the moon is quite in accord with the hypothesis that the moon was captured by the earth late in the process of the formation of the earth by the

---

7 Letter to G. Gamow (Urey D1953). Letter to D. H. Menzel (Urey D1955b). In a letter to D. Alter, Urey (D1954) supported the binary-accretion theory.
8 Urey (1955b, 1956a, 1956c, 1958, 1959). Letter to Slichter (Urey D1955a).
9 In 1963 Urey wrote to S. Chandrasekhar that his (Urey's) ideas about diamonds in meteorites, apparently refuted by Lipschutz and Anders, seemed to be confirmed by more recent results of George Kennedy (Urey D1963).

capture of smaller objects" (1960a: 502; see also 1960b, 1962). He stated that this "obvious explanation" of the remarkably short duration of bombardment of the lunar surface occurred to him during the past year.

Urey paid little attention to the detailed calculations on the possible mechanisms for capture of the Moon and subsequent evolution of her orbit. At the 1960 International Astronomical Union Symposium at Pulkovo Observatory, he admitted that capture of an object like the Moon from a heliocentric orbit by the Earth is quite improbable; thus, the justification of his theory had to depend on the more general argument that the early Solar System was populated by a large number of such objects, one of which happened to have been captured and survived (Urey, Elsasser, and Rochester 1959; Urey 1962). But since the capture may well have involved collisions with other bodies near the Earth, it would be difficult to develop a precise dynamical model for it (Urey 1967).

Based on the low solar iron abundance then popular with solar astronomers, Urey surmised that the iron content of the Moon is only about half that of the Earth or Mars, but is comparable to that of nonvolatile solar material. From this viewpoint, the problem was not to explain why the Moon has so little iron, but why the Earth has so much. Later upward revisions of the solar iron abundance were to undermine this aspect of Urey's theory.[10]

Urey also concluded that the Moon must have been cold ever since she was captured, for if she had been melted or undergone the same kind of processing as the surface of the Earth, she would have nothing interesting to say about conditions in the early days of the Solar System. One consequence of this line of reasoning was Urey's suggestion that the lunar maria were not formed by lava flows (a sign of undesirable heating) but by water. The water may have come from the primordial stuff of which the Moon was formed (e.g., carbonaceous chondrites) or even splashed from Earth during the capture process. "If indeed the surface of the Moon carries a residue of the ancient oceans of the Earth at about the time that life was evolving, the Apollo program should bring back fascinating samples which will teach us much in regard to the early history of the Solar System, and in particular with regard to the origin of life." Others who saw the Moon as Earth's wife described the wedding as an event involving much generation of heat; according to Urey, the bride was so frigid and infertile that she could only preserve for eternity, unused, her consort's seed.[11]

Some NASA leaders tried to gain public support for the *Apollo* program by using an extrapolation of Urey's cold-moon theory: If the Solar System was formed by an encounter of two stars, as suggested earlier in the 20th century by T. C. Chamberlin, F. R. Moulton, J. H. Jeans, and H. Jeffreys, then the

10 Urey (1960/1962). On the iron abundance problem see 3.3.3.
11 Urey (1956c, 1966c, 1969c). Letter to J. A. O'Keefe, 27 July 1967. Extensive press coverage of his 30 December 1963 speech at AAAS meeting in Cleveland. For the remark about residues of ancient oceans see Urey (1966a: 166) and a statement to a congressional committee (Urey D1965: 10).

Moon and planets must have been hot, so all would have separated their iron and other heavy elements into central cores. But if they had condensed from cold gas and dust, the iron might not flow to the center except in bodies as large (or with as much radioactive minerals to provide heating) as the Earth. Thus, the Moon might be found to have bits of iron scattered throughout her interior. In that case we could accept the nebular hypothesis, which implies that planets are normally formed by the same process that builds stars. Hence, there is likely to be other intelligent life in the galaxy, whereas the encounter theory implies that planetary formation is a very improbable process and life is rare.[12] Although Urey participated actively in the discussions of the National Academy of Science's Space Science Board, which advised NASA in the early 1960s, and was probably responsible for increasing the magnitude of the resulting lunar science effort, his influence on how the space program actually operated was minimal. He argued successfully that the last *Ranger* mission (*Ranger IX*) should be sent to Alphonsus, partly because of Russian reports of gaseous eruptions there and partly because it was thought to be a very old area, but he had no influence on the choice of landing sites for later missions. He insisted on the need for manned missions to bring back carefully selected samples for detailed analysis, rejecting the view of other investigators that the use of television images and unmanned sample-return missions would be a more effective use of available funds.[13]

4.3.3 Revival of the fission theory

According to the Russian astronomer B. Yu. Levin, at the Pulkovo symposium "all participants unanimously agreed that revival of the hypothesis of the separation of the moon from the earth is impossible" (Levin 1966). Levin seemed to consider it almost a personal insult that certain scientists, such as A. E. Ringwood in Australia and A. G. W. Cameron and Donald Wise in the

12 Jastrow and Newell (1963), reprinted by U.S. Government Printing Office for mass distribution. The article does not mention the fact that the encounter theory had been abandoned by most scientists at least 20 years earlier. For examples of statements to Congress on the relevance of *Apollo* to studies of the origin of the Solar System, see Jastrow (1961), Newell (1961), Mueller (1964: 68–9). An example of a popular account of the "moon core test" of theories of the origin of the Solar System may be found in Alexander (1964: 190–1). A somewhat vaguer promise that the Moon would tell us "something of the Earth's early evolution" was made by Lowman (1966).

13 N. W. Hinners (1972); personal communications to S.G.B. from D. E. Wilhelms, E. M. Shoemaker, and W. H. Pickering, and anonymous referee's report on version of this paper for *Science*. Urey's suggestions for landing sites may be found in a letter to H. E. Newell (Urey D1961a) and other correspondence at the NASA archives; his complaints about NASA's site selection are in letters to C. H. Townes, J. Findlay, and G. E. Mueller (Urey D1969a, D1969b, D1969c). Compare Urey's interview with E. M. Emme and R. C. Hall (D1976). For Urey's activities on the Space Science Board see Space Science Board (D1959: 3–4, 8–10; D1960: 8); Urey (1961a, 1963b). In the last item (p. 51) Urey recalls how he first gained his enthusiasm for landing a man on the Moon.

Alan Binder says that it was Kuiper rather than Urey who influenced the landing site for *Ranger IX* (interview with S.G.B., 14 November 1984).

United States, had nevertheless dared to revive the idea. (His compatriot Lyapunov was one of those who had proved that the Darwinian fission theory was incorrect.) Even Urey, who decisively "repudiated" this theory in favor of capture a few years earlier, had (he thought) recently shown sympathy for it. But here Levin was a little premature; Urey was not yet ready to abandon the capture theory.

Ringwood's hypothesis (1960), which he originally described as a return to the "ancient fission hypothesis," eventually developed into a sophisticated version of the Laplace–Nölke theory that the Moon has precipitated from the outer parts of the proto-Earth's atmosphere (Ringwood 1966a; Ringwood and Essene 1970a, 1970b). Ringwood rejected Urey's approach – explaining density differences between terrestrial planets and their satellites in terms of varying degrees of mechanical removal of silicates from iron – in favor of explanations involving different redox states. He considered this view to be confirmed by later findings (Garz et al. 1969) that the iron abundance in the Sun is comparable to that in the planets, contrary to the assumption that the iron/silicate ratio varies with distance from the Sun. "It was this basic opinion on the limited role of iron/silicate fractionation," he recalled (letter to S.G.B., 5 December 1984), "which obliged me to interpret the moon as having an entirely different origin to the planets. This led to my revival of the fission theory in 1960 and in 1966 and later papers, to more complex models whereby the material in the moon was derived from the earth's mantle." Any differences in composition between the Moon and the Earth's mantle should be accounted for by the process of separation. This philosophy guided his later contributions (1978, 1979, 1984).

Wise's paper (1963) included an explicit defense of the fission theory in the light of recent changes in the earth sciences, and a critique of capture hypotheses. A major feature was the postulate (briefly mentioned by Ringwood and before that by Nolan, as quoted earlier) that formation of the Earth's core would increase the rotational speed beyond the critical value and trigger fission. Wise admitted that his mechanism was still quantitatively insufficient to account for the necessary angular momentum, but thought that modern ideas like magnetic braking or a decreasing gravitational constant might resolve this difficulty.[14]

---

14 Wise sent me the following account of his encounter with Urey: "I met Urey in 1961 as a result of an early version of the fission paper. It had already been rejected once by J[ournal of] G[eophysical] R[esearch] but Harry Hess suggested I show it to Urey while I was on sabbatical at Scripps. After making an appointment I was ushered into his office. As soon as he heard I was a geologist he began a tirade against half baked geologists – he had just finished a verbal battle with one who wanted to form the world's oceans as giant impacts. After cooling down, he asked me what I wanted. He took the paper to review & said to return in a week. I came back with fear & trembling. Much to my relief he was all smiles, shook my hand and said that next to his own theory for the origin of the moon, he liked this best. With his support I sought opinions and reviews by other big names until finally enough of a case could be made to force publication of the fission paper in JGR in 1963. He remained a kindly father figure after that, even though he never really supported the method as the most probable" (Letter to S.G.B., 1981). See Urey's correspondence with S. Chandrasekhar about Wise's paper (Urey D1961b, D1962a).

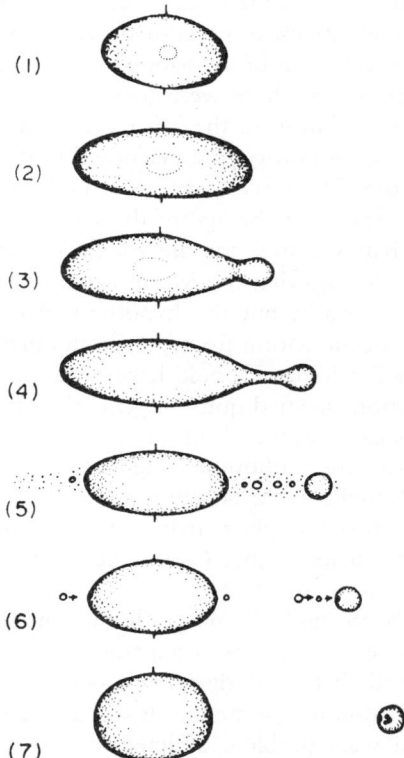

Figure 23. Sequence of forms in lunar origin by fission from the Earth during formation of the Earth's core (D. Wise, "Origin of the Moon by Fission," in *The Earth–Moon System*, ed. Marsden and Cameron, Plenum Press, 1964/1966, p. 215).

A. G. W. Cameron (1963c) supported the hypothesis that fission resulted from sudden formation of the Earth's core; he noted that, according to Elsasser (1963), this would be a catastrophic process, so that rotational instability would suddenly set in at the equator rather than the slow plastic deformation suggested by Wise.

John A. O'Keefe (1963) suggested that the fission theory could be revived with the help of the magnetic-braking mechanism; he was led to the idea that the Moon came from the Earth by his theory that tektites come from the Moon. In those days Urey did not consider O'Keefe a "reliable scientist" (Urey D1962b).

The revival of the fission hypothesis by Ringwood, Wise, Cameron, and O'Keefe did not generate much enthusiasm in the scientific community.[15] In meeting the old objections to Darwin's theory, its modern advocates seemed

15 MacDonald (1964a, 1964b). Marsden and Cameron (1964/1966). Baldwin (1965: 37).

to have introduced ad hoc hypotheses that destroyed the simplicity that had made the original model attractive, without providing a satisfactory explanation of other facts such as the tilt of the Moon's orbit. In addition to problems about the initial fission event, there were new difficulties in reconciling the time scale for the tidal evolution of the lunar orbit with modern knowledge about the Earth's geological history; the period of 1000 to 2000 m.y. implied by current values of the tidal force[16] was now significantly shorter than the value of 4500 m.y. accepted for the age of the Earth but significantly longer than the age of less than 500 m.y. for the Pacific Ocean basin according to plate tectonics.[17] The old idea that the Moon came out of the Pacific Ocean was no longer taken seriously, but the hypothesis that a catastrophic event such as the formation of the Moon from the Earth's mantle, accompanied by the segregation of the Earth's core, could have occurred as long as 2000 m.y. *after* the Earth's formation, seemed quite implausible.[18] It was difficult enough to imagine how a less catastrophic event like the capture of the Moon could have taken place at that time without leaving obvious traces in the geological record (Field 1963; Pannella 1972; Scrutton and Hipkin 1973). Of course, if one looked hard enough one might convince oneself that such traces do exist (Cooper, Richards, and Stacey 1967; Cloud 1968; Turcotte, Nordmann, and Cisne 1974).

The time scale problem might be put aside by arguing that the magnitude of tidal friction is very sensitive to the arrangement of continents and oceans – it is sometimes claimed that most dissipation occurs in shallow seas – and therefore there is no reason to assume that it was as great in the past as now.[19] In any case the time scale problem afflicted any theory that assumed the Moon (or the parts from which she formed) had once been in orbit around the Earth near the Roche limit and later moved outward because of tidal friction; thus, chemical or physical properties that could be related to the mode of formation would be more likely to discriminate between alternative theories.

Shortly before the first lunar landing John O'Keefe, one of the small group of supporters of the fission origin of the Moon, proposed to modify Darwin's theory by assuming that both the Earth and Moon would have been so hot, immediately after fission, that they would have been largely vaporized, and thus would have suffered substantial mass loss. The Moon, composed of mantle material, would have received more heat per gram and would have contained a larger proportion of volatile substances, thus she would have lost a larger fraction of her mass. O'Keefe predicted that the Moon would be poorer in water and other volatile substances than the Earth. Moreover, a crucial test

16 MacDonald (1964a). Gerstenkorn (1967a, 1967b).
17 See any comprehensive work on plate tectonics, e.g., Tarling (1978: ch. 10).
18 Cameron (1964/1966) noted that the results of Ostic, Russell, and Reynolds (1963) indicate that the core formed 4½ billion years ago.
19 Singer (1967, 1970). Burns (1977: 132). Other models that avoid the problem have been proposed by Hansen (1982) and Finch (1982).

would be provided by the abundances of siderophile elements such as gold, platinum, and nickel: These are depleted in the Earth's crust (compared with Solar System abundance) presumably because they followed iron into the core. If the Moon was formed by fission after core formation, her crust should also be deficient in siderophiles; since she probably does not have an iron core, we would not be able to attribute this deficiency to a purely lunar process. On the other hand if siderophiles are found to be abundant in the Moon's crust the capture theory would be favored.[20]

4.3.4 Siderophile abundance: A crucial test

As soon as the early results from the July 1969 *Apollo 11* mission were announced, O'Keefe pointed out to Urey that the depletion of siderophile elements could be considered an argument in favor of the fission theory. The fact that these elements are scarce in the Earth's crust as compared with their Solar System abundances is usually explained by assuming that they went down into the core of the Earth. We cannot assume that the Moon's nickel went down into her core since the Moon seems to be nearly homogeneous, and hence has little or no core. Therefore, the Moon must have come from the Earth's mantle after the Earth's core had been formed. O'Keefe reminded Urey that he had predicted the deficiency in volatiles from his fission theory (letter, 30 September 1969; see also O'Keefe 1970).

Urey replied that he was still skeptical about fission but recognized that the alternative theories were in trouble.[21] It was becoming clear that several of Urey's theses about the Moon – the presence of water and a completely cold history, for example – were contradicted by the evidence from *Apollo 11* (Lyons 1969; Cooper 1970: 26). His assumption that the Moon's low concentration of iron matched the solar abundance (thus making the Moon a primordial object) was refuted by redeterminations of the solar iron abundance (3.4.3), On the other hand the determination that the Moon's surface is at least as old as the Earth's (though subject to some melting episodes) confirmed another aspect of Urey's theory.

Early in 1971 Urey suggested to O'Keefe that depletion of siderophiles

---

20 O'Keefe (1968, 1969a, 1969b, 1969c). Wolfe (1969).
21 Urey's letter to O'Keefe (11 November 1969) said that he was finishing a chapter for Kopal's book, co-authored by Gordon MacDonald: "Neither Gordon or I are able to see any way by which the Moon could have gotten out of the Earth and be consistent with all the evidence. As I have consistently said I do not think the evidence is conclusive for any of the models for the origin of the Moon. . . . I find insurmountable problems for the present escape hypothesis, but also no conclusive evidence for the capture or the accumulation theory."

    The Urey–MacDonald chapter was published in the second edition of *Physics and Astronomy of the Moon*, edited by Z. Kopal (Urey and MacDonald 1971), but articles based on preliminary versions were published elsewhere as early as 1969.

    In a speech on 10 November 1971, Urey said: "I do not know the origin of the moon. I'm not sure of my own or any other's models. I'd lay odds against any of the models being correct" (Treash 1972: 21).

might be explained by postulating that they had been removed by iron that now formed a layer 200 km below the surface. O'Keefe argued that such a layer would be unstable because it would be denser than the material below it. Urey agreed that there were difficulties with his idea, and wished that "I had stuck to my model for the meteorites and claimed that the Moon was one of these objects." The problem was, as O'Keefe said, that "it is necessary to have removed the siderophile elements from the surface of the Moon by means of liquid iron." O'Keefe was "very encouraged that Urey saw the problem this way."[22]

In November 1971, after Urey had given a talk at the University of Maryland, O'Keefe reminded him that they both agreed that "we have this much at least in common; we think there must be a more volatile-rich region in the interior of the Moon" (letter from O'Keefe to Urey, 12 November 1971). Urey replied that "unless the interior of the moon contains the volatiles, my model cannot be correct and, therefore, I will go back to another model and it will be Darwin's model." But he would be disappointed if that model turned out to be right, "for the moon then will be an incidental object and not of fundamental importance. We can decide that it escaped from the earth and then 'to hell with it'" (letter from Urey to O'Keefe, 15 December 1971). He expressed similar views in a paper published in 1972 (Urey 1972: 322).

During 1972 and 1973, Urey and O'Keefe continued to discuss the problem of the Moon's deficiency in nickel and other siderophile elements. O'Keefe argued that the nickel must have been extracted by molten iron that is no longer to be found in the Moon; hence, the extraction must have occurred while the Moon was still part of the Earth's mantle. Urey proposed that they collaborate on a quantitative study of the extraction process and O'Keefe agreed. Urey became convinced that the siderophile problem must be solved, but he continued to look for places within the Moon where the iron could go after it had extracted nickel from the surface.[23]

The contrast between public and private views during the period 1970–4 is remarkable. Those who attempted to determine the "consensus" of lunar scientists concluded that the fission theory was being rejected in this period, while capture, binary accretion, and various modifications of these hypotheses gained ground (Marvin 1973; R. Lewis 1974b: 52; Mitroff 1974a: 152–6). Yet during this same period Ringwood was developing a modified fission theory while Urey was moving toward the fission theory and reworking with

22 O'Keefe–Urey correspondence (May–June 1971). Urey favored capture over fission and suggested that iron had carried siderophile elements down to the bottom of a melted surface layer, in a joint paper with Marti, Hawkins, and Liu (Urey et al. 1971).
23 Urey–O'Keefe correspondence (February 1973). In an article published in November 1973, Urey stated that the fission theory was hard to accept because of the Moon's depletion in volatiles and siderophiles, and favored a modified version of his earlier capture hypothesis (Urey 1973: 5). O'Keefe elaborated his arguments for fission in three papers published in 1972 (O'Keefe 1972a, 1972b, 1972c). Urey called attention to new research on the spectral lines of iron, which suggested that the "revised" higher abundance of iron in the Sun might have to be revised downwards again, thus removing one objection to the idea that the Moon is a primitive object (Urey 1974: 475).

O'Keefe the paper that would present (in their opinion) conclusive arguments for that theory. O'Keefe (1973) wrote that he was astonished to be told that the *Apollo* results excluded the fission theory because lunar materials are poor in volatiles, since he had predicted just that from his fission theory in 1969.

While the dynamical objections to Darwin's original hypothesis were still an obstacle, geochemical evidence (at least as interpreted by many scientists) was providing support for substantial similarity between the Moon and the Earth's mantle. In these years, also, Robert Clayton and his group at the University of Chicago were refining their measurements of oxygen isotope abundances, leading to the conclusion that the Earth and Moon are related by blood, not merely by marriage (Onuma, Clayton, and Mayeda 1970, 1972; Grossman, Clayton, and Mayeda 1974; Clayton and Mayeda 1975; Hammond 1976).

In 1974 Heinrich Wänke at the Max-Planck-Institut für Chemie in Mainz, Germany, published an authoritative review of the chemistry of the Moon in which he emphasized the difference between the FeO/MnO ratios in the lunar samples and in terrestrial rocks. He asserted that this difference "rules out the possibility that the moon was once part of the earth" (1974: 142).

O'Keefe wrote to Wänke that the high FeO/MnO ratio for the Moon was a consequence of her general depletion in volatiles, which had been predicted by Wise and himself before *Apollo 11*, since they had concluded that the Moon must have lost most of her initial mass immediately after fission (letter, file date 16 October 1974). Wänke replied that "if you can find the physical conditions for the evaporation of about 90 percent of the original mass without disruption of the moon and with a strict fractionation of the compounds according to their boiling points a fission origin of the moon could be brought into agreement with the geochemical observations" (letter from Wänke to O'Keefe, 23 October 1974). O'Keefe responded that this "agrees well with the mass losses which I calculated from dynamical arguments" (letter from O'Keefe to Wänke, file date 1 November 1974).

A few days later O'Keefe wrote to Urey, "We really seem to be making progress on the problem of the moon's origin," since Wanke's 90 percent depletion factor

agrees unexpectedly well with the depletion that one finds by dynamical methods. A big loss of mass is needed because the earth breaks up via a pear-shaped configuration (piroid, as Poincaré called it) and the small end of the pear has around $\frac{1}{5}$ of the total volume. Hence the moon probably began life with a mass $\frac{1}{10}$ or more of the earth's mass. It now has $\frac{1}{81.3}$. So there has been a mass change of one order of magnitude. Maybe we are on the right track after all. (letter from O'Keefe to Urey, 4 November 1974)

Wänke eventually leaned toward fission, not so much because of O'Keefe's arguments, but because his own experiments began to suggest a strong genetic relationship between the Earth and Moon, in agreement with Ring-

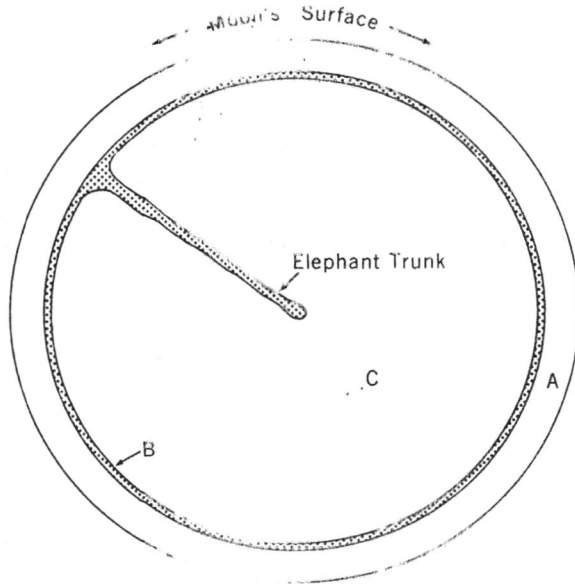

Figure 24. "Elephant trunk:' shaped flow of iron–nickel liquid toward the Moon's center, as shown in original version of the O'Keefe–Urey essay "The deficiency of siderophile elements in the Moon," *Phil. Trans.*, **A285**, 1977, (John A. O'Keefe)

wood's ideas. His early results had indicated an excess of refractory elements (consistent with a high fractional loss of volatiles), but this turned out to be incorrect. He also concluded that Ni is not strongly depleted in the Moon but that the high concentrations found in some samples should be attributed to a primary component of the Moon rather than to meteorites (Wänke et al. 1975). Further study of terrestrial minerals convinced him that the Earth's mantle is strikingly similar to the bulk composition of the Moon, especially in the elements V, Cr, and Mn (Wänke et al. 1975, 1977; Wänke and Dreibus 1977/1979; Wänke 1981a, 1981b; interview with S.G.B., October 1984).

In the version of their paper "The Deficiency of Siderophile Elements in the Moon," submitted to *Geochimica et Cosmochimica Acta* in 1974, O'Keefe and Urey did not try to make a strong case for the fission theory. Instead, they emphasized their conclusion that the depletion was due to liquid–liquid extraction – the "blast furnace" process described in 1921 by V. M. Goldschmidt – rather than to gaseous fractionation in the solar nebula as recently suggested by E. Anders and others.[24] They estimated that the minimum amount of

24 1974 draft of Urey–O'Keefe paper, copy provided by J. A. O'Keefe. Goldschmidt (1958). Larimer (1967). Anders (1971). Anders told me that this statement, as published in my *Science* article (Brush 1982c), misrepresents his view: "You have me and others claim that the

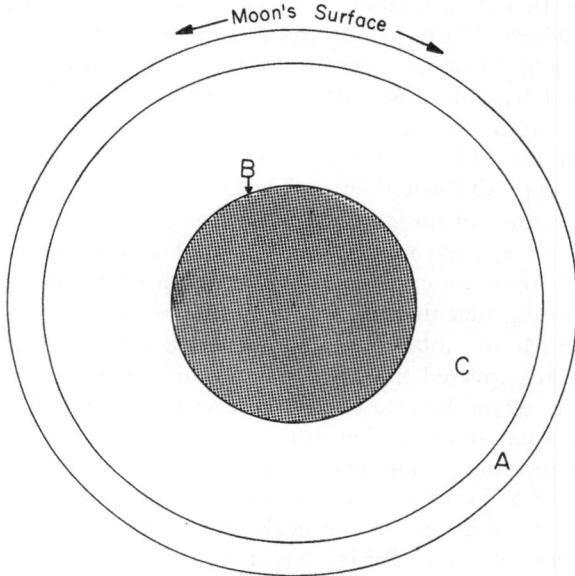

Figure 25. Iron–nickel liquid collected to form lunar core (hypothetical), as shown in original version of O'Keefe–Urey paper "The deficiency of siderophile elements in the Moon," *Phil Trans.*, **A285**, 1977, (John A. O'Keefe).

metallic liquid required to produce the observed depletion of nickel in lunar crystal rocks is about 2 percent; this liquid would eventually form the Moon's core. The only definite argument they offered against this possible alternative to fission was that "it does not predict the high heatflow values which are suggested by the scanty measurements available." Those values could be accounted for by assuming a substantial amount of uranium and thorium – refractory elements that would presumably be enriched in a high-temperature fission process.

The paper encountered considerable criticism from the referees for *Geochimica et Cosmochimica Acta*.[25] During the long-drawn-out process of responding to criticisms and rewriting the paper, other scientists published new ideas that led Urey to reconsider his decision to support fission (letter to O'Keefe, 8

depletion of siderophiles occurred in the solar nebula, not by segregation of molten iron in a planetary body. In all my papers on this subject, I have drawn a clear distinction between metal–silicate fractionation in the solar nebula, which Urey and I hold responsible for *small* variations in the Fe/Si of chondrites (up to $\sim 2\times$) and planets (up to 4 to $5\times$), and metal segregation in solid bodies, which can lead to depletions of 10^5 or even 10^6. See pp. 1133–1134 and 1138 of my Apollo 11 paper" (letter from Anders to S.G.B., 17 September 1987).

25 Urey papers at University of California, San Diego; copies provided by W. M. Kaula; interviews with J. A. O'Keefe.

November 1974). But in October 1975, when Urey was awarded the V. M. Goldschmidt Medal of the Geochemical Society, he stated in his published acceptance speech, "I rather favor the view that the Moon escaped from the Earth, though I have no good idea as to how the separation of the Moon from the Earth occurred" (Urey 1976: 570).

By this time the O'Keefe–Urey paper had been rejected by *Geochimica et Cosmochimica Acta*.[26] O'Keefe then took advantage of the opportunity to publish a shorter version of the paper in the *Philosophical Transactions of the Royal Society of London,* as part of the proceedings of a conference (O'Keefe and Urey 1977). By then the case for fission had been strengthened by new seismic data indicating that the mass of metal in the lunar core is less than 0.5 percent of the Moon's total mass.[27] O'Keefe and Urey estimated that if the nickel had been extracted from the crust by liquid iron that stayed in the Moon, this core would have to contain at least 1 percent of the Moon's mass; otherwise the nickel has to be left in the Earth.

The seismic recording equipment that might have determined the size of the Moon's core had to be turned off because of lack of funding – thus keeping scientists from attaining one of the goals of the *Apollo* program (Nakamura, Latham, and Dorman 1982). Participants at the 1984 Hawaii conference on the origin of the Moon agreed that seismic experiments to determine the size of the lunar core should be given high priority in any future lunar mission.[28]

Some evidence for a lunar core has been obtained from magnetic measurements. It appears that the Moon used to have a magnetic field, probably generated by an internal dynamo (Runcorn 1978, 1979; Runcorn, Collinson, and Stephenson 1985). A magnetic moment can be induced in it by the geomagnetic tail (Russell, Coleman, and Goldstein 1981; Russell 1984). Christopher Russell suggested that "tidal forces associated with orbital evolution drove a dynamo until the moon moved sufficiently far from the earth" (letter to S.G.B., 31 October 1984). The early magnetic history of the Moon may therefore provide clues to her origin.

When the O'Keefe–Urey paper finally appeared in 1977, it had been rewritten to stress from the beginning the case for the fission hypothesis:

1. By comparison with cosmic composition, the Earth's crust is deficient in siderophiles.

---

26 The paper was rejected on 2 December, according to a letter from Denis M. Shaw, the editor, to S.G.B. (1981), but correspondence about it continued into the following year (e.g., letter from Shaw to O'Keefe, 13 February 1975). Copies of some of the referees' reports and related correspondence are in the Urey papers at UCSD, but I have not yet received permission to quote them.

27 Nakamura et al. (1974). Urey–O'Keefe correspondence, April–July 1974; letter from O'Keefe to J. T. Wasson (17 September 1974). For further discussion of evidence for a small lunar core see Wiskerchen and Sonnett (1977), Runcorn (1984), Russell (1984), Yoder (1984).

28 Newsom (1984a, 1984b) discussed the relation between a lunar core and the depletion of molybdenum and rhenium in lunar samples; in the discussion of possible crucial experiments at the 1984 Hawaii meeting, he stated that a core as large as 5 percent would be strong evidence that the Moon didn't come from the Earth.

2. This is usually explained by the migration of these elements into metals at a time when the metal and silicate portions were mixed, then the metal sank to the core.
3. The Moon's crust is also deficient in siderophiles.
4. But the Moon now has no core; hence, bodily separation of metal from silicate mass is suggested; hence, the Moon originated by fission.
5. Fission cannot be rejected on the grounds that the Moon is deficient in volatiles compared with the Earth, since dynamical calculations predict a loss of 90 percent of the Moon's mass after fission.

These arguments would also support Ringwood's hypothesis that the Moon precipitated from an extended terrestrial atmosphere, passing through an intermediate "sediment ring" as proposed earlier by Gilbert and Öpik; conversely, Ringwood's continued advocacy of his own model during the 1970s did no harm to the fission theory since many of his arguments merely supported the general proposition that the Moon had come from the Earth's mantle.[29]

4.3.5 Science and subjectivity

Robert Jastrow (1981) recalled that, as a physicist, he was attracted by Urey's deductive approach to planetary science. From the single fact that the Moon has an irregular shape, significantly different from what would be expected for hydrostatic equilibrium, Urey concluded that the Moon had been frozen and dead for billions of years, and it was this conviction that the Moon preserves a record of the earliest days of the Solar System that helped him convince NASA to give high priority to lunar exploration. The same deductive approach forced him to change his views, once he had agreed with O'Keefe that rigorous conclusions could be drawn from the deficiency of siderophiles:

Because of its similarity to iron in many respects, the marked fractionation between iron and nickel in the Earth and Moon is a critical phenomenon for understanding the geochemistry of the Moon. (O'Keefe and Urey 1977: 572)

As will be seen in the next chapter, several planetary scientists did not accept this reasoning and pointed to other "critical phenomena" implying a different origin for the Moon. The existence and size of the Moon's core was still a subject of controversy. Some critics may even have suspected that Urey's abandonment of his capture theory was the act of an exhausted old man under pressure by a vigorous advocate of another theory (O'Keefe). That

---

29 O'Keefe and Urey (1977: 569). Ringwood (1979). Nickel depletion was also used as an argument for fission by G. M. Brown (1976/1978), as was tungsten depletion, by Rammensee and Wanke (1977). For other versions of the fission hypothesis see Binder (1974, 1975, 1977), Shoemaker (1977). A news report by Waldrop (1982) indicated that five years later the implication of siderophile abundances for selenogony was still an active research topic, but the contribution of O'Keefe and Urey had been forgotten.

suspicion should be quickly dispelled by a glance at Urey's letters during the period 1972–6; it is clear that he was critical of almost all of O'Keefe's ideas and rejected at least as many as he accepted.

A letter from O'Keefe to Urey (17 September 1974), written at a time when Urey was wondering whether to withdraw their paper in the face of adverse criticism, is relevant to this point and also to recent discussions of the alleged failure of scientists to adhere to certain ideal standards of behavior:

For heavens sake don't drop out of the paper. All of the chemical thinking is yours, and is full of solutions to problems that I failed to solve. The main idea is yours. If you drop out, I'll never be able, alone, to break down those characters. I would never have undertaken this paper alone.

Do you remember Ian I. Mitroff, the sociologist from the University of Pittsburgh, who came around during the Apollo program with a tape recorder and asked every-one lots of questions about the origin of the moon, whether the moon was hot or cold, and so on? Well, he's coming out this fall with a book on the subject of our interactions with one another. I haven't seen the book, but I have seen four papers covering the subject; they probably give a good idea of what is in the book.

The drift of Mitroff's findings is that the lunar scientists, especially the theorists, don't give up their ideas in the face of facts and don't think logically. He says we allow personal considerations to dominate our thinking. It's a real attack on science as such.

In this situation, this paper by you and me will do a lot of good because of its clear anti-Mitroffian message that scientists can get together on the meaning of the facts, no matter how different their ideas may have been.

Mitroff did not claim that science is "completely subjective, irrational, rela-tivistic" (1974a: 268), but rather that subjective factors are more important than scientists are willing to admit. Since Urey himself had often remarked on the tendency of scientists to interpret data as confirming their own theories (4.3.1, note 4) he might not have objected to this conclusion, although a sociological monograph bearing the title *The Subjective Side of Science* is some-what less palatable than occasional remarks intended to needle one's col-leagues. Nevertheless, Urey himself did abandon his theory when he decided that the facts no longer supported it. Don E. Wilhelms, a geologist who was involved in the *Apollo* program wrote (letter to S.G.B., 1981):

In general he [Urey] was surprised that his theories of the completely primitive and cold Moon were wrong, and he gave them up grudgingly. But give them up he did, and it was very refreshing to observe his willingness to change his mind so drastically in later years. Other pioneers of his era did not prove so flexible.

My own informal discussions with lunar scientists indicate that Urey's shortcomings may have been in just the opposite direction from that por-trayed by Mitroff: He was too ready to change his ideas and was likely sooner or later to take both sides on any issue. This tendency does show up in later correspondence between O'Keefe and Urey. The qualitative conclusion that the Moon came from the Earth was not sufficient; Urey demanded that O'Keefe produce a detailed theory for the separation process, and in May 1977 he wrote, "I'm doubtful about your origin of the Moon, though there are certain features of the moon that would be fitted by your suggestion that the moon comes from the earth." Eight months later O'Keefe was able to produce a theory that did please Urey (O'Keefe and Sullivan 1978). But shortly afterwards, when they discussed the new giant-impact hypothesis, Urey saw some merits in this idea too.

I conclude that Urey was never satisfied that the problem of the Moon's origin had been definitely solved. He left us with this remark (1976: 570):

But what can one expect? One must always leave something for the young people to solve. It would be most disappointing, I am sure, if we older people solved all the problems of science, which, of course, none of us will ever do.

4.4

History of modern selenogony

4.4.1 The co-accretion theory

During the 1960s and 1970s, one of the most active selenogonists was the Russian scientist Evgenia Ruskol. Although her theory has been classified under the heading "binary accretion" (Wood 1977), she rejected Kuiper's protoplanet theory (Ruskol 1958/1964) and concluded as early as 1960 that "the Earth–Moon system did not originate as a double planet from some double embryo" (Ruskol 1960/1962: 154). Her hypothesis, based on the cosmogonic ideas of O. Yu. Schmidt and V. Safronov, was that a swarm of particles formed around the growing Earth, with a radius of 100 to 200 times the radius of the Earth. Other particles were captured from heliocentric orbits (in a "supply zone" ranging from 0.8 to 1.3 AU from the Sun) into the circumterrestrial swarm by inelastic collisions. It is this specific mechanism – inelastic collisions of particles most of which are less than 100 km in radius – that distinguished Ruskol's hypothesis from others that postulated a circumterrestrial swarm destined to evolve into the Moon.

The swarm grew most rapidly, according to Ruskol, when the Earth's mass was 0.3 to 0.5 its present mass. Since, according to Safronov's (1958/1964) calculations, the rates of growth of the largest body in a system and the second largest body start to diverge when their sizes reach a critical value, the Moon could have started growing very early without having grown as rapidly as the Earth and hence could still be much less massive while being almost as old. The difference between the ages of the Moon and Earth should be less than 100 million years (Ruskol 1960/1962: 155).

Ruskol (1960/1962) suggested that the Moon was initially formed at a distance of 5 to 10 Earth radii; its subsequent recession was due to tidal evolution (Ruskol 1963a, 1963b, 1966a, 1966b). In order to extend the time scale to more than 4 thousand million years, it was necessary to assume that the effective lag angle of the Earth–Moon tidal interaction (tidal effective Q) was smaller in the past than at present. This assumption was justified, according to Ruskol (1975: 41), by evidence that the Earth's interior "was gradually heated from a lower temperature, while releasing the water of the oceans to the surface."

An obvious objection to all binary accretion theories is the apparent difference in composition between the Earth and Moon: Since the Earth is denser it presumably has a substantial iron core, whereas the Moon has little or no core. Ruskol (1966a: 225) suggested that this difficulty could be overcome "if

Figure 26. Evgenia Ruskol (Evgenia Ruskol).

we admit that the Earth's core is composed of metallized silicates," so that Earth and Moon have essentially the same chemical composition; the greater density of the Earth is due to a phase transition occurring at high pressures – the "Lodochnikov–Ramsey hypothesis" (*Nebulous Earth*, 2.4.4). That idea had briefly attracted Urey's attention, but his physics colleagues at Chicago (Enrico Fermi and Edward Teller) persuaded him that a phase transition that could squeeze silicates to the density of iron at core pressures was quite un-likely.[1] Their skepticism was confirmed by equation-of-state calculations done by Walter Elsasser and shock-wave compression experiments done in the USSR and at Los Alamos.

Ruskol informed me that her "agreement" with the Lodochnikov–Ramsey hypothesis in 1962–6 "may better be referred to Prof. B. Yu Levin who was [an] active proponent of it [at] that time . . . in explicit form I 'abandoned' this hypothesis only later, in 1971. But I had never written that I hold this hypothesis" (letter to S.G.B., 18 January 1985).

Ruskol's theory was not widely known in English-speaking countries in the 1960s, although it was vigorously supported by Levin at a Caltech–Jet Propulsion Laboratory conference in 1965 (Levin 1965/1966), and several of her papers appeared in cover-to-cover translation journals. Following the publication of an English translation of Safronov's 1969 book in 1972, Western scientists took up Safronov's theory and its application to selenogony by Ruskol. Soviet contributions to understanding the origin of the Moon, as well as the origin of the Solar System, thus became fairly well known during the 1970s and 1980s (Levin and Brush 1995).

[1] Urey (D1949, D1950a, D1950b, D1950c, D1952a, D1952b, D1957). Bullen (D1952a, D1952b, D1952c). DeMarcus (D1956).

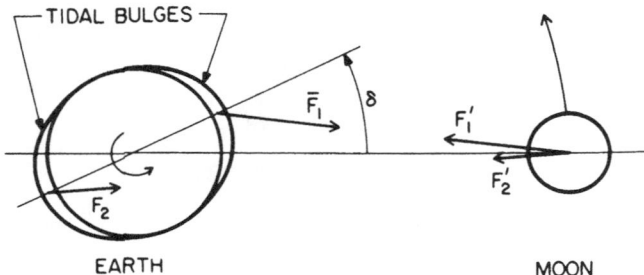

Figure 27. Schematic of tidal bulges raised on the Earth by the differential gravitational acceleration of the Moon. Because the Earth rotates faster than the Moon revolves and dissipation leads to a time lag in the response of the Earth to the tide raising potential (i.e., high tide occurs after the Moon has passed overhead), the tidal bulge leads the Moon by a phase angle δ. F_1 and F_2 are the lunar attractions for the near and far bulges, respectively, and F_1' and F_2' are the corresponding equal and opposite reactions on the Moon. Since $|F_1| > |F_2|$ a net torque retards the Earth's spin and increases the orbital angular momentum of the Moon, thereby increasing the Earth–Moon separation (A. P. Boss and S. J. Peale, "Dynamic constraints on the origin of the Moon," in *Origin of the Moon*, ed. W. K. Hartmann et al., 1986, p. 64; copyright 1986 by Lunar and Planetary Institute).

4.4.2 Evolution of the lunar orbit

In January 1964 a conference was held at the Goddard Institute for Space Studies in New York, to discuss the Earth–Moon system. Its proceedings provide a good overview of the status of selenogony at that time (Marsden and Cameron 1964/1966). Extensive calculations of the past evolution of the lunar orbit had been done by L. B. Slichter (1963), G. J. F. MacDonald (1964a, 1964b, 1965), and W. Kaula (1964). This and later work did not support all the results of Horst Gerstenkorn's calculations (4.3.2), in particular his claim that the distance of closest approach is very nearly the same as the Roche limit (Sorokin 1965; Ruskol 1966a), and Gerstenkorn admitted that his calculation of the evolution of orbital eccentricity would have to be changed to take account of dissipation (Gerstenkorn 1967b); but some of his qualitative conclusions were found to be valid (Gerstenkorn 1967b, 1968; Goldreich 1968). MacDonald (1964a: 535) estimated that the rate of energy dissipation required in Gerstenkorn's model was so high, because of the flip-over from a retrograde to a prograde orbit, that the entire Earth would have been melted by the capture event. According to S. F. Singer (1986: 472), "This criticism effectively discredited the capture theory" until he produced a new model, based on capture from a prograde orbit (Singer 1968). Goldreich concluded that the capture theory also suffered from serious difficulties in

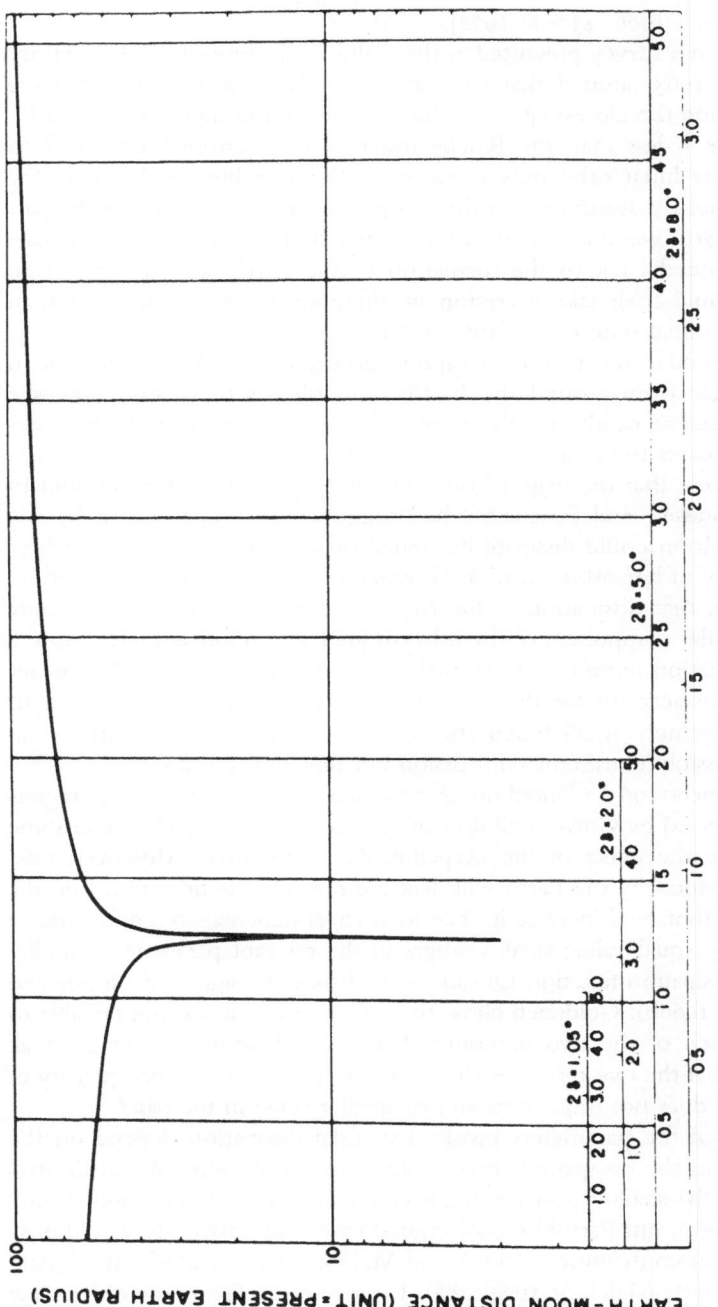

Figure 28. The variation of the Earth–Moon distance with time, according to the capture theory. The various time scales correspond to different values for the phase lag δ. In the left-hand branch of the curve, the Moon is in a direct orbit approaching the Earth. In the right-hand branch of the curve, the Moon is in a retrograde orbit receding from the Earth (G. J. F. MacDonald, in *The Earth–Moon System*, ed. Marsden and Cameron, Plenum Press, 1966, p. 182).

reconciling the numerical values of angular momentum, eccentricity, and inclination, and he favored co-accretion even though that theory also had difficulties (Goldreich 1966: 437–8; 1972).

B. J. Levin, in a survey presented at the Caltech–Jet Propulsion Laboratory conference in 1965, argued that calculations of the tidal evolution for the branch preceding the closest approach have no real meaning since the Earth–Moon distance is less than the Roche distance; the capture theory fails to explain why the lunar orbit now coincides with the ecliptic rather than the equatorial plane. By assuming that the dissipation rate was smaller in the past (because the Earth was colder and had less water on its surface), the time scale could be extended back to the formation of the Earth, and the evolution calculations could then take accretion in circumterrestrial orbit rather than capture as the initial state (Levin 1965/1966).

Levin also pointed out that, after capture occurred, the Moon would be in a very eccentric orbit around the Earth; one still has to explain how she reached her present nearly circular orbit. (This is even more difficult to explain if one accepts the consequences of the hypothesis of Gerstenkorn, supported by Alfvén, that the original orbit was retrograde just after capture.) In 1959 Urey, Elsasser, and Rochester had suggested that tides raised by the Earth in the Moon could dissipate her radial orbital motion and thus reduce the eccentricity of her orbit. Gordon Groves (1960), in a calculation based on the assumption that dissipation in the Moon can be neglected and that only the perpendicular component of the tidal torque has a significant effect, found that tidal dissipation *increases* the eccentricity of the Moon's orbit. This meant that in order to account for the present eccentricity one must assume it to have been extremely small when the Moon was close to the Earth – an assumption possibly consistent with fission but not with capture.

Peter Goldreich (1963) pointed out that radial forces and some of the tangential terms neglected by Groves will decrease the eccentricity e and may in some cases overcome the effect of the perpendicular component. Moreover, tides raised on the Moon by the Earth will decrease e, while the tides raised on the Earth by the Moon will increase it. The total effect depends on a difference of terms of nearly equal value; small changes in the relevant parameters (rigidity and specific dissipation function Q) can easily change the sign of de/dt (see also Ruskol 1966a, 1966b). Goldreich chose to neglect e since it was not possible to determine which of the two opposing effects would dominate. Yoder et al. (1984) found that the two terms nearly cancel so that the present eccentricity of the lunar orbit does not imply a greater or smaller value in the past.[2]

The values of the parameters involved in tidal dissipation depend on the physical cause of the dissipation. Previously it had been assumed that dissipation in the Earth's shallow seas was the dominant cause (Jeffreys 1920b; Munk 1964/1966: 66–7), but Pariiski (1960) argued that solid Earth tides could provide a significant contribution (Munk and McDonald 1960: 208f; MacDonald 1964a; Kaula 1971; Melchior 1974: 288). Later research favors the older view

---

2 I thank Alan Harris for information on this point.

that oceanic dissipation is the most important, perhaps because of a near-resonance in the present epoch.[3]

To further complicate matters, some physicists thought that the gravitational "constant" is changing, which would produce an apparent *increase* in the Earth's rotation rate (Dicke 1964/1966).

In addition to the change in eccentricity of the Moon's orbit one has to account for the change in its inclination to the Earth's axis of rotation, or to the plane of the ecliptic if one assumes that her origin was not related to the Earth's rotation. Goldreich (1966), in what he calls "the answer to an exam question posed to me by Thomas Gold in June 1963," concluded that if the Moon came out of the Earth or was formed in an equatorial orbit she would now have to be much closer to the ecliptic than she actually is (5°). This result was cited by Urey (1967) and Öpik (1972) as a major argument against all fission theories. O'Keefe (1972c) and Rubincam (1975) argued that the theoretical prediction of evolution toward very small eccentricity would not apply if the Earth were a viscous liquid, as it might have been during the period when tidal stresses were greatest; hence, this objection to the fission theory may not be decisive. (But their own calculations still made idealized assumptions about the viscoelastic properties of the Earth.)

Almost all versions of the capture and fission theories postulated that the Moon is composed of material that had at one time been broken into smaller particles orbiting around the Earth at or inside the Roche limit (Öpik 1967). Given a circumterrestrial ring or disk of small particles, it is not too difficult to see how they might accumulate into a moon while moving outward because of tidal dissipation; but it is rather hard to determine by dynamical arguments what was the previous state of this system – whether (1) it was originally a single body moving outside the Earth's gravitational influence, (2) the material came from inside the Earth itself, (3) the particles were captured separately from heliocentric orbits, (4) they condensed from the same gas–dust protoplanet as the Earth, or (5) some combination of all of these.

4.4.3 First results from the lunar samples

The Mare Tranquillitatis rocks returned by the *Apollo 11* mission in July 1969 provided a flood of new data; I will mention here only the results that seemed to be directly relevant to selenogony.

S. R. Taylor, an Australian geochemist, was put in charge of the initial chemical analyses of the first lunar samples.[4] He described his experiences in Houston in an unpublished essay (Taylor D1980):

3 Lambeck (1975, 1977). Hansen (1982). Sündermann (1982). Kerr (1983).
4 Stuart Ross Taylor (b. 1925) was educated in New Zealand, then obtained his Ph.D. in geology at Indiana University in 1954. He was a lecturer in mineralogy at Oxford University in 1956 when Urey was visiting professor, and was inspired by Urey's lectures to devote his efforts to cosmochemistry and the study of the Moon in particular. He settled in Australia in 1961 and has been on the faculty of the Research School of Earth Sciences at the Australian National University in Canberra since then.

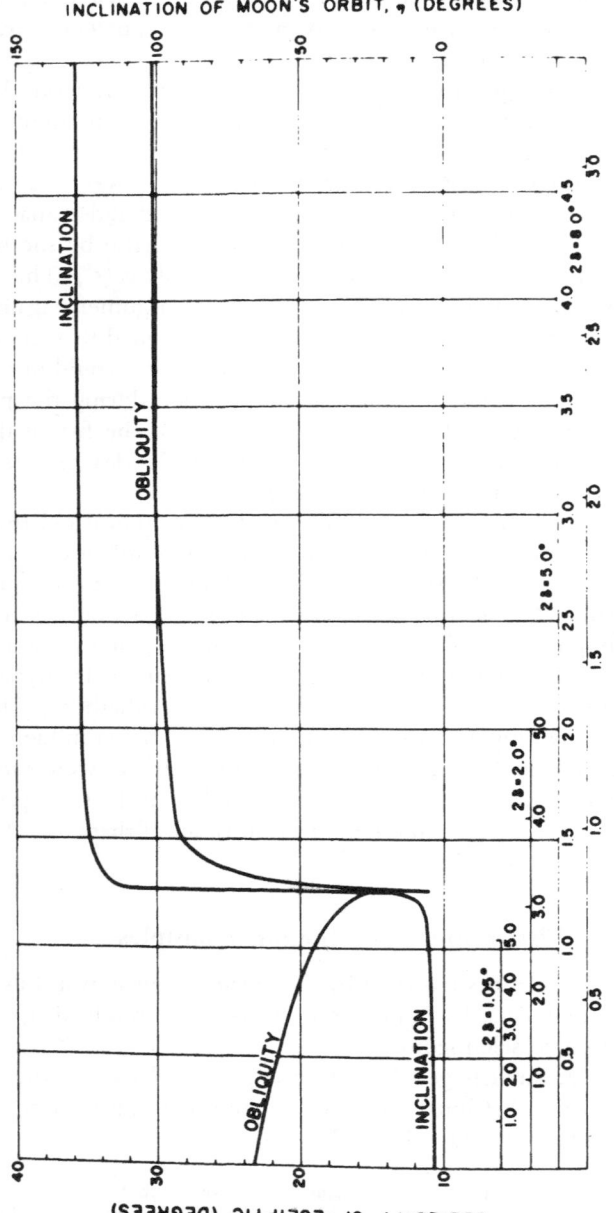

Figure 29. The variation of the obliquity of the ecliptic (left-hand scale) and the inclination of the lunar orbit (right-hand scale) according to the capture theory (G. J. F. MacDonald, "Origin of the Moon: Dynamical considerations," in *The Earth–Moon system*, ed. B. G. Marsden and A. G. W. Cameron, Plenum Press, 1966, p. 184).

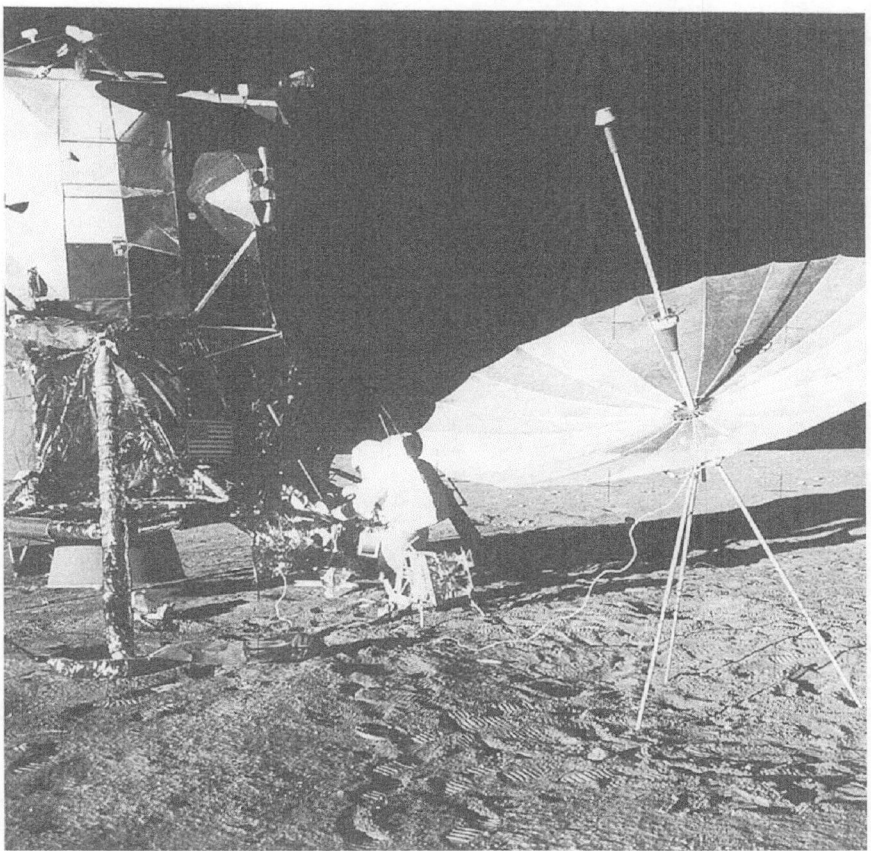

Figure 30. *Apollo 12* extravehicular activity (EVA). Astronaut Charles Conrad, Jr., commander of the *Apollo* 12 lunar landing mission, stands at the Module Equipment Stowage Assembly on the Lunar Module during the first *Apollo 12* EVA on the lunar surface. The erectable S-band antenna is already deployed at right. The carrier for the Apollo Lunar Hand Tools is near Conrad (NASA, Lyndon B. Johnson Space Center, Houston).

The Lunar Sample Preliminary Examination Team (LSPET or PET in the fashionable acronyms of the period) was an interesting group. It comprised fifteen scientists of assorted specialities from a variety of universities and organisations such as the U.S. Geological Survey, aided by a group of eleven "in-house" NASA scientists. Each member had his own unique skill. There were geologists who had devoted their professional careers to investigating volcanic rocks, minerals and glasses produced at meteorite impact craters on earth; chemists who had specialised in the mass spectrometric determination of the noble gases; biochemists who might detect signs of life

Figure 31. *Apollo 17* EVA. Scientist-astronaut Harrison H. Schmitt, *Apollo 17* lunar module pilot, with his adjustable sampling scoop, heads for a selected rock on the lunar surface to retrieve the sample for study (NASA, Lyndon B. Johnson Space Center, Houston).

in the lunar soil; physicists who had specialised in detecting low levels of radio-activity, a bacteriologist, and a micropalaeontologist. No lunar religion ever had a more devoted band of adherents.

I started to organise the spectrographic laboratory. The instrument had been installed so that the samples could be handled through the biological barrier of stainless steel and glass, in an atmosphere of dry argon and nitrogen, using portholes fitted with rubber gloves. Weighing and manipulating milligram amounts of powder was very difficult because of the high static charge set up in the dry atmosphere.

The first step was to prepare atomic emission spectra for the 70 or so chemical elements detectable on the instrument. About 100,000 spectral lines of elements in the ultraviolet and visible portions of the spectrum, are potentially detectable. Many lines are coincident, and the possibilities of erroneous identification of an element, or of missing one that was present, haunted me. The analytical technique depends heavily on precise calibration with known samples. I scrambled together a set, partly from Canberra and partly from phone calls to colleagues. By the same procedures, a variety of likely analogues of lunar samples was assembled. These ranged from ocean floor volcanic rocks to several varieties of meteorites. It was possible in the days before the landing to make a case for the existence of nearly any chemical element on the lunar surface. . . .

The lunar landing was on July 20. . . . On July 26, the sample box was opened. The dust coated rocks resembled charcoal briquettes used for barbecues. The Greeks had imagined the heavenly bodies to be composed of shining crystal, the quintessence or fifth element of the Alexandrine cosmologists. The reality looked less exotic. . . .

By 12.28 p.m. [July 28], the [first] sample was prepared and loaded into the cup shaped electrode, which formed the anode of a direct current arc. The excitement of the moment and the tendency of the lunar dust to adhere to everything made this routine operation difficult. The 10 amp direct current arc was struck. It was designed to volatilise refractory silicates and produce atomic emission spectra, characteristic of the elements present. The sample flared red, indicative of calcium. A little later (the burn took 2½ minutes) a white fringe to the flame suggested that titanium might be a major component.

Early in August 1969, Taylor wrote a preliminary report for the *Lunar Science Institute Newsletter,* summarizing the data and their implications (Taylor 1969):

The geochemistry in these samples is of unusual interest. They are distinct both from terrestrial rocks and meteorites in many respects. In particular, they possess a high concentration relative to "cosmic" abundances of refractory elements, such as Ti, Zr, Y, etc. The enrichment factors over cosmic abundances are 50–100 times indicating that these samples have undergone severe fractionation, probably at high temperatures. "Volatile" elements, e.g., Na, K, Rb, Pb, Tl, etc., and most of the "toxic" elements are low in abundance supporting the concept of high temperature processes with temperatures of the order of 1500°C or greater. Nickel and cobalt are very low in abundance and the Fe/Ni and Fe/Co ratios are high compared to chondritic or iron meteorites.

Although the data are preliminary, they already place some constraints on theories of lunar origin. Large scale chemical fractionation has occurred to raise the concentration levels of several elements by 1–2 orders of magnitude over primitive nebula compositions (based on our current ideas of these derived from solar and meteoritic abundances). Such processes can take place more readily during condensation of a double-planet system, with interchange of material between primitive moon and earth, than in a small independent moon-sized body during accretion. Significant differences (in Ti, Mg, Zr) from our ideas of mantle composition do not favor deriving the moon from an already formed earth. Thus, of the various alternatives for lunar origin, the initial chemical data may be interpreted as favoring the double-planet hypothesis.

This summary should not be taken to mean that concentrations of all the elements mentioned had actually been measured; rather, "the volatile ele-

ments (Pb, Bi, Tl, and the like) in general are below the limits of detection of
the spectrographic techniques employed; also the elements of the Pt group,
Ag and Au, were not detected" (Lunar Sample Preliminary Examination
Team 1969: 1221). The low abundances of Bi, Tl, Ir, Ag, and Au were established by the Anders group at Chicago, using radiochemical neutron activation analysis. They inferred that the Moon probably formed as an original
satellite of the Earth. They also argued that the depletion of Co, Ir, Ni, Pd,
and especially Au contradicts the fission hypothesis: "If the moon had lost its
siderophiles while still part of the earth, it should show a terrestrial siderophile
pattern. The fact that it has a characteristic pattern of its own suggests that
removal of siderophiles from the moon was a separate event, that proceeded
under different physico-chemical conditions" (Ganapathy et al. 1970: 1133).

Taylor recalled that "the depletion in nickel was one of the great surprises
in looking into the chemistry of the lunar samples" (letter to S.G.B., 17 December 1984):

The first sample which we analysed (a soil, 10084) had about 250 ppm Ni (from
meteoritic contamination). However, when we analysed the rocks, the nickel spectral
lines were missing. This was extraordinary for material with about 15 percent Fe. The
emission spectrographic technique used had a detection limit of about 2 ppm. We
were barely able to see the strongest Ni lines in terrestrial granitic standards at about
that level. The Apollo 11 high-Ti basalts in general had no detectable Ni. I recall
searching through the spectrum, checking the alignment of master and sample spectra,
and eventually concluding that rather than an expected value of several hundred ppm,
nickel was absent. In contrast, chromium was present at several thousand ppm, an
order of magnitude more enriched than in terrestrial samples. This and similar experiences alerted me to the hazard of using our terrestrial geochemical experience as a
basis for dealing with a different planet. This ultimately led to the concept of extensive early lunar differentiation, in apparent contrast to the Earth. (Taylor and Jakeŝ
(1974)

The formal report by the Lunar Sample Preliminary Examination Team
(1969) included the same results:

1. the high concentration of refractory elements (Ti, Zr, etc.),
2. the low concentration of volatiles (Pb, Bi, Tl),
3. strong depletion of siderophile elements, especially nickel and cobalt,
4. absence of secondary hydrated minerals, indicating that there had been no
   surface water at Tranquility base at any time since the rocks were exposed,
5. the great age of some rocks, perhaps greater than any known on Earth.

But the report carefully refrained from speculating about the implications of
these data for theories of lunar origin.[5]

5 The team consisted of D. H. Anderson (Manned Spacecraft Center, Houston, denoted hereafter by MSC), E. E. Anderson (Brown and Root–Northrop), P. R. Bell (MSC), Klaus

Similarly, a summary of the *Apollo 11* Lunar Science Conference (held at Houston, 5–8 January 1970) cautiously stated:

The results reported do not resolve the problem of the origin of the moon. However, the number of constraints that must be met by any theory have been greatly increased. For example, if the moon formed from the earth, it can now be stated with some confidence that this separation took place prior to 4.3×10^9 years ago. Furthermore, such a hypothesis must now take account of certain definite differences in chemical composition. (Arnold et al. 1970: 451)

Initial determinations of the ratio of ^{18}O and ^{16}O in lunar rocks indicated that they are generally within the range of terrestrial rocks or slightly higher. If one assumes on the basis of meteorite data that the $^{18}O/^{16}O$ ratio decreases as one goes outward from the Sun, then these results "indicate an initial condensation either at the same position as terrestrial material, or somewhat nearer the sun," according to Naoki Onuma, Robert N. Clayton, and Toshiko K. Mayeda (1970: 537). (See also Epstein and Taylor 1970; Friedman et al. 1970.)

A subsequent analysis of *Apollo 11* and *Apollo 12* samples by the group led by S. R. Taylor reinforced the conclusion that refractory elements (Ti, Zr, Y) are enriched in comparison with terrestrial rocks, while Ni and Co are depleted. However, they also noticed that Ni is more abundant in lunar breccias and fines; "The high concentration of nickel in the fines is attributed to a meteoric contribution."[6]

Those lunar scientists who found high concentrations of nickel and other siderophile elements in some lunar rocks and soils generally assumed that they were dealing with "extralunar" material and that the intrinsic abundance of siderophiles in the Moon is less than that in the Earth (Baedecker, Chou, and Wasson 1972: 1351; Cuttita et al. 1973: 1090; Ganapathy et al. 1973). The

Bieman (MIT), D. D. Bogard (MSC), Robin Brett (MSC), A. L. Burlingame (Berkeley), W. D. Carrier (MSC), E. C. T. Chao (U.S. Geological Survey = USGS), N. C. Costes (Marshall Space Flight Center), D. H. Dahlem (USGS), G. B. Dalrymple (USGS), R. Doell (USGS), J. S. Eldridge (Oak Ridge), M. S. Favaro (U.S. Department of Agriculture = USDA), D. A. Flory (MSC), C. Frondel (Harvard), R. Fryxell (Washington State), J. Funkhouser (SUNY, Stony Brook), P. W. Gast (Columbia), W. R. Greenwood (MSC), C. S. Gromme (USGS), G. H. Heiken (MSC), W. N. Hess (MSC), P. H. Johnson (Brown and Root–Northrop), Richard Johnson (Ames), E. A. King Jr. (MSC), N. Mancuso (MIT), J. D. Menzies (USDA), J. K. Mitchell (Berkeley), D. A. Morrison (MSC), R. Murphy (MIT), G. D. O'Kelley (Oak Ridge), G. G. Schaber (USGS), O. A. Schaeffer (SUNY, Stony Brook), D. Schleicher (USGS), H. H. Schmitt (MSC), E. Schonfield (MSC), J. W. Schopf (UCLA), R. F. Scott (Caltech), E. M. Shoemaker (Caltech), B. R. Simoneit (Berkeley), D. H. Smith (Berkeley), R. L. Smith (USGS), R. L. Sutton (USGS), S. R. Taylor (Australian National University), F. C. Walls (Berkeley), J. Warner (MSC), R. E. Wilcox (USGS), V. R. Wilmarth (MSC), J. Zähringer (Max-Planck-Institute, Heidelberg).

  For further details see special issue of *Sciences*, **30** (January 1970), and the proceedings of the *Apollo 11* Lunar Science Conference.

6 Taylor et al. (1970: 1633). The other authors were P. H. Johnson, R. Martin, D. Bennett, J. Allen, and W. Nance, all affiliated with Brown and Root–Northrop, at the Lunar Receiving Laboratory in Houston. The report on the *Apollo 12* samples by the Lunar Sample Preliminary Examination Team (1970) stated that nickel is "strikingly depleted" in a way that is to be expected as a result of fractional crystallization involving olivine and pyroxene separation.

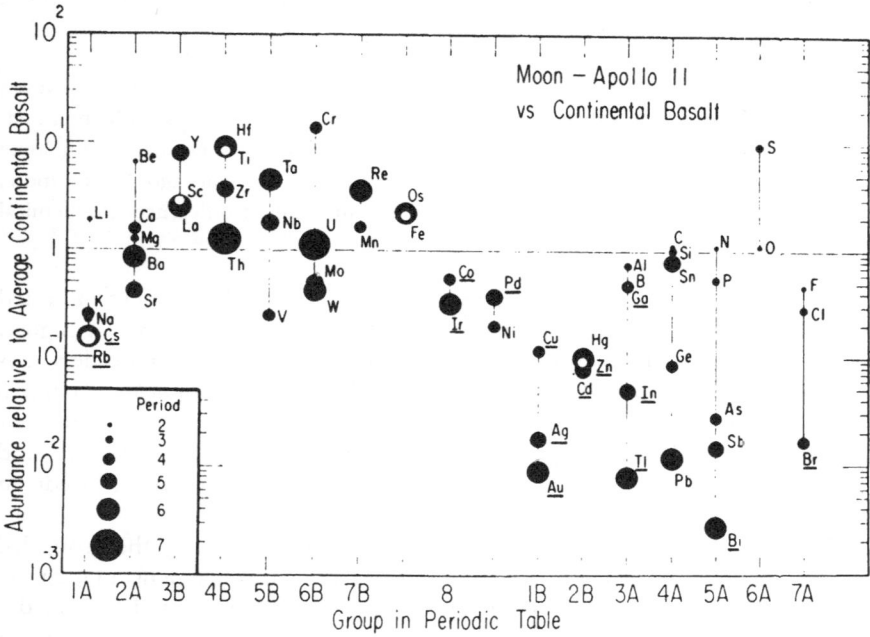

Figure 32. Abundances in *Apollo 11* rocks, normalized to average continental basalt (R. Ganapathy et al., "Trace elements in Apollo 11 lunar
rocks," in *Proceedings of the Apollo 11 Science Conference*, ed. A. A. Levison,
vol. 2 [1970], p. 1133).

reasons for this assumption were stated (with respect to highland breccias) by
the Chicago group as follows: First, the siderophiles are greatly overabundant
in breccias compared with "genuine" igneous rocks; second, the lowest abundances of siderophiles in highland rocks are consistent with measured metal–
silicate distribution coefficients, thus these values represent true indigeneous
abundances (Morgan et al. 1974: 1704). Other investigators accepted this assumption and used it as a basis for determining the amount of extralunar
(meteoritic) material in their samples.[7] In general it was believed that siderophiles would be strongly depleted on the surface of any differentiated
planet.

As noted by Anders et al. (1973: 4), a similar assumption had earlier been
used in attributing meteoritic origins to material found in terrestrial oceanic
sediments and polar ice (Barker and Anders 1968; Hanappe et al. 1968).
Barker and Anders (1968) concluded that a more reliable, and much lower,
limit on the rate of influx of cosmic matter can be obtained from Ir and Os
than from Ni. Anders (letter to S.G.B., 17 September 1987) said that earlier

7 Ganapathy et al. (1972). Morgan et al. (1972: 1377). Baedecker et al. (1974: 1629). Wasson et
   al. (1975). Morgan et al. (1976: 2190).

results of Picciotto's group (Hanappe et al. 1968) were often two orders of magnitude too high, and that the Barker and Anders results had been confirmed by recent measurements.

Anders's group used a radiochemical neutron activation procedure to analyze trace elements in the lunar samples and presented some "highly tentative conclusions":

It is surprising that siderophile elements (Co, Ir, Au) are more depleted in A, B lunar rocks than in their terrestrial analogs. Apparently the removal of siderophiles from the crust was more efficient on the moon, although its metal-phase content is at most $\frac{1}{m}$ that of the earth. Perhaps conditions were more reducing, or the scarcity of water and sulfur depressed the solubility of siderophiles in the silicate phase.

The remaining elements (Ag, Cd, Zn, In, Tl, and Bi) are volatile and have large ionic radii; they are therefore expected to concentrate rather strongly in the crust [Larimer (1967), Larimer & Anders (1967), Anders (1968), Ringwood (1966)]. The most straightforward inference to be drawn from their low abundance is that the moon accreted at higher temperatures than did the earth (that is, above 600° to 700°K), and acquired a lesser proportion of low-temperature volatile-rich material in the terminal stages of accretion: perhaps only 10^{-4} to 10^{-5} lunar masses. The scarcity of water and carbon certainly agrees with this view. (Keays et al. 1970: 492)

The Anders group soon concluded that the detailed pattern of depletion in siderophiles was evidence *against* the fission theory, as indicated in the passage quoted earlier from Ganapathy et al. (1970: 1133): "Re and Os are less strongly depleted than in terrestrial rocks, but Co, Ir, Ni, Pd and especially Au are more strongly depleted."

Further data strengthened their conclusion; it was the strong depletion of Au and Ag rather than the mild depletion of Ir, Re, and Ni that seemed to be the best evidence against fission (Anders et al. 1971: 1026). In a letter to me (17 September 1987), Anders stated:

The depletion of Ir was greater than we realized at that time. In a later paper (Wolf et al. 1979), we disowned 7 of our 1971 Apollo 12 Ir analyses, having realized by comparison with later analysis that the samples were contaminated. . . . In 1971, we weren't yet measuring Re and hence used the data of Herr, without realizing that they were high by about $30\times$. Our own data later showed that the Earth–Moon difference was $370\times$ for Re and $300\times$ for Ge (Wolf and Anders, 1980).

The final summary of the Anders group results was presented in a paper by Wolf and Anders (1980):

Volatiles (Ag, Bi, Cr, Cd, In, Sb, Sn, Tl, Zn) are depleted in lunar basalts by a nearly constant factor of 0.026 ± 0.013, relative to terrestrial basalts. Given the differences in volatility among these elements, this constancy is not consistent with models that derive the Moon's volatiles from partial recondensation of the Earth's mantle [Ringwood and Kesson 1977; Binder 1978b] or from partial degassing of a captured body [Kaula 1974/1977; Smith 1977a]. It is consistent with models that derive planetary volatiles from a thin veneer (or a residuum) of C-chondrite material [Anders 1968 and later papers; Turekian and Clark 1969; Wasson 1971]; apparently the Moon received only 2.6 percent of the Earth's endowment of such material per unit mass. . . . Ni is relatively abundant in lunar basalts ($4 \times 10^{-3} \times$ C1-chondrites), whereas Ir, Re, Ge, Au are depleted to $10^{-4}–10^{-5} \times$ C1.

A. E. Ringwood and E. Essene argued that the lunar rocks showed important similarities to terrestrial rocks; in particular, "The abundances of most of the relatively nonvolatile, oxyphile major and trace elements in Apollo 11 basalts fall within the ranges observed for these elements in terrestrial basalts." The similarities indicate "some kind of genetic relationship" between the Earth and Moon (Ringwood and Essene 1970a: 794). At the same time they noted significant differences, such as the Moon's depletion in volatile metals, siderophiles, and water, which rule out the original fission hypothesis. "Recent developments of this hypothesis by O'Keefe (1969b) and Wise (1969) suggest that fission may be accompanied by formation and then loss of a massive hot terrestrial atmosphere and that the moon may be the residue of this atmosphere" – a model similar to that proposed by Ringwood (1960, 1966a).

Ringwood and Essene also concluded that the binary-planet and capture hypotheses were not consistent with the *Apollo* data. The former, which they attributed to W. M. Latimer (1950) and Orowan (1969),[8] assumed that iron would be preferentially concentrated in the Earth during condensation from the nebula, but implied that the Earth's mantle and the Moon would have the same composition; thus, it could not explain the observed chemical differences between lunar and terrestrial basalts. The latter, which assumes that the capture occurred as recently as 2 thousand million years ago, implies that both the Earth and Moon were strongly heated and deformed at that time; but "the ages of the Apollo 11 basalts and the older ages inferred for the lunar highlands are in direct conflict with the assumed timescale of lunar tidal evolution." Moreover, Urey's argument that the Moon is a "primitive object" was based in part on the belief that there is a discrepancy between the solar and terrestrial iron abundance, but this has been refuted by the results of Garz et al. (1969). On the contrary, the capture hypothesis fails to account for the deficiency of iron in the Moon compared with the Earth and explains very few of the known facts while postulating an event of low intrinsic probability (Ringwood and Essene 1970a: 795). (Defenders of the co-accretion and cap-

8 Latimer's brief remarks (1950: 102) can hardly be called a binary-planet theory of the Moon's origin, especially since he suggests that the Moon was formed when tidal waves in the "loose material" of the condensation caused "rupture" to produce the Moon after iron had been concentrated toward the center of the Earth. Ringwood, like most Western scientists, ignored the more recent papers by Ruskol on the binary-planet theory.

ture theories could argue that their hypotheses did not necessarily entail those consequences, as indicated elsewhere in this chapter.)

4.4.4 The post-*Apollo* stalemate

The multi-billion-dollar Apollo program increased our knowledge of the Moon and early history of the Earth's environment by several orders of magnitude but throughout the 1970s could not provide a definitive answer to the No. 1 scientific question: Where did she come from? As noted in 4.3.4, O'Keefe, Binder, and a few other scientists (including Urey somewhat tentatively) concluded that the fission hypothesis had been confirmed. Ruskol in the USSR and some Americans supported binary accretion.[9] Others advocated disintegrative capture.[10]

Most advocates of capture in the 1960s assumed that the Moon was broken up during the capture process, either by the Earth's tidal action inside the Roche limit or by collision with smaller bodies orbiting the Earth, so one could argue that disintegrative capture was not really a new theory. On the other hand the hypothesis did have considerable influence and helped lead W. K. Hartmann to consider the plausibility of a major collision instead of a near miss (private communication, August 1987).

A capture theory that does *not* assume disintegration but invokes gas drag to trap the Moon whole was proposed by Nakazawa Komuro, and Hayashi (1983). According to Stevenson, Harris, and Lunine (1986: 81–2) this theory is rather less suitable for the Earth's Moon than for other satellites.

But the consensus of the lunar science community (as I perceive it) was that *none* of the three pre-*Apollo* theories was tenable in its original form.[11] Of course, the hypotheses proposed after *Apollo* bore some resemblance to the earlier ones, but they included catastrophic events that seem a priori so unlikely that they would not have been seriously considered as long as the simpler theories were still in the running.

Ian Mitroff (1974a) quantified the failure of the *Apollo* missions to yield any immediate selenogonic consensus, on the basis of interviews with 42 lunar scientists from late 1969 through mid-1972. He asked them to indicate their evaluation of five hypotheses of lunar origin: *fission, capture,* and three versions of simultaneous formation: *double planet, condensation* from a hot silicate atmosphere of the primordial Earth, and *accretion* from planetesimals of Type I carbonaceous chondrites. On a scale from 1 ("agree strongly") to 7 ("disagree

9 Ruskol (1972, 1973, 1974/1977, 1975, 1982). Harris and Kaula (1975). Safronov and Ruskol (1977). Wood (1977). Harris (1978b).
10 Singer (1970, 1977). Wood and Mitler (1974). Mitler (1975). J. V. Smith (1977b, 1982).
11 Schmitt (1975). Barshay and Lewis (1976). French (1977: 56, 247; 1981: 78). Kaula (1977b: 329). Urey (D1977). Drobyshevskii (1978). Dorschner (1979). Murray, Malin, and Greeley (1981). Sullivan (1983). Smoluchowski (1983a). Wood (1983: 79). Blanchard (1985). Couper and Henbest (1985). Friedlander (1985). Lowman (1985).

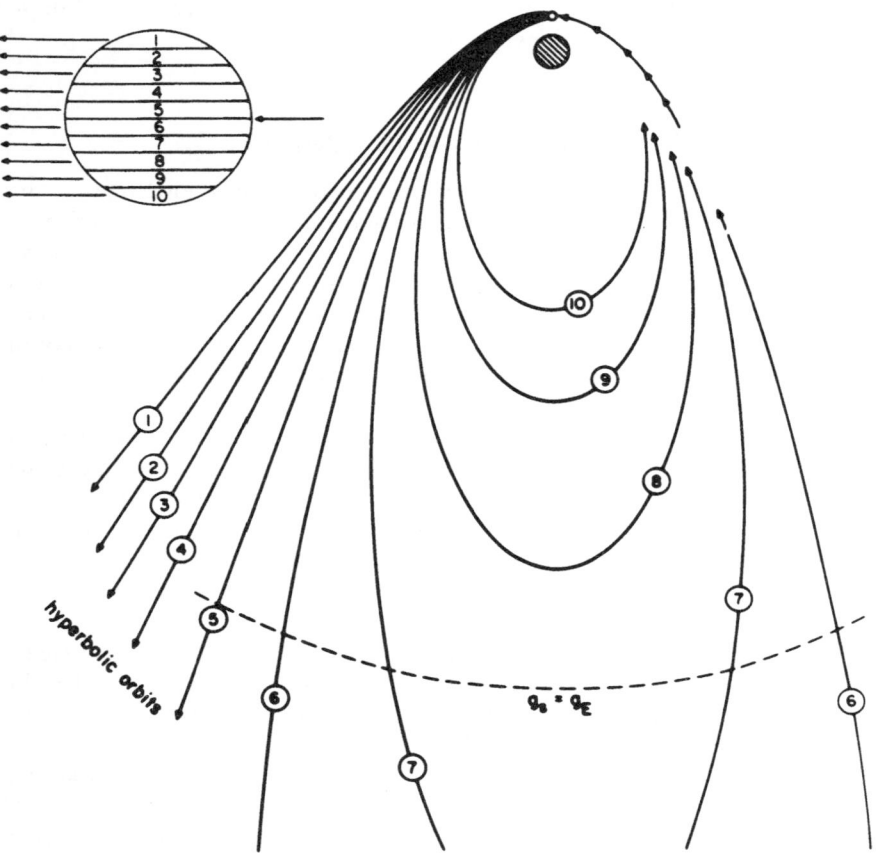

Figure 33. An idealized example of disintegrative capture. A planetesimal at the top of the figure is tidally disrupted during passage at parabolic velocity through Earth's Roche zone. The subsequent trajectories of material from ten parallel slabs in the origin planetesimal (inset, upper left) are followed. Debris from slabs 1 to 5 (farthest from Earth) is returned to heliocentric orbits; debris from slabs 6 to 10 is captured in geocentric orbits. In the case of encounters at greater than parabolic velocity, a smaller number of slabs, those that pass closest to Earth, are retained in geocentric orbits. These would contain a disproportionately large amount of crustal and mantle material if a differentiated planetesimal was disrupted; most of the core material would return to heliocentric orbits. Reprinted from Wood (1986: 46).

strongly"), no theory had an average score of more than about 4 ("neither agree nor disagree") before *Apollo 11*; *accretion* was slightly more favored than the others. "*No theory enjoyed a statistical significant improvement in its average rating; fission and accretion suffered significant losses*" (Mitroff 1974a: 152). The double-planet theory received the best rating, 3.73, but even that was weaker than "agree moderately." Fission dropped to 5.56 ("disagree moderately"), leading Mitroff to conclude that this theory "was the first to fall by the wayside" (1974a; 156).

Mitroff's finding that co-accretion was more popular than fission or capture (though not enthusiastically supported) is confirmed by occasional remarks in popular writings by scientists (Goldsmith and Owen 1980: 249; Abell 1985).

Although *Apollo* had apparently produced little information about the origin of the Moon as measured by Mitroff's technique, the project did yield significant changes in opinion on related issues, especially in reversing the previous consensus on the nature of mascons (Mitroff 1974a: 153–9).

There was one definite conclusion directly relevant to lunar origin: The age of the Moon is about the same as that of the Earth.[12] This fact knocked out some older versions of the fission and capture theories and made futile the search for evidence on the present Earth's surface of the Moon's birth. But it left open the important question: Precisely how old is the Moon *relative to the Earth?* (Swindle et al. 1986).

A major question, halfway between observation and theory, soon emerged from analysis of the lunar samples: Is the composition of the Moon so similar to that of the Earth as to imply a genetic relationship? In spite of Öpik's (1972) view that dynamical criteria should be given priority, chemical arguments were to play an increasingly important role in selenogony.

Ringwood's first answer was that the "differences between lunar and terrestrial basalts [depletion of lunar basalts in volatile metals and siderophile elements] preclude the moon's having once been fissioned directly from the earth's mantle." But they did not exclude the "sediment-ring hypothesis" (attributed to Öpik, 1961) as modified by Ringwood (1966a) to produce the ring by precipitation from the Earth's massive primitive atmosphere; thus, "The moon was ultimately derived from the earth" (Ringwood and Essene, 1970b: 609–10). Of course, the sediment-ring hypothesis (like all the others) has a dynamical defect: The present inclination of the lunar orbit could not have evolved from a ring in the Earth's equatorial plane (Goldreich 1966; Urey and MacDonald 1971: 271–2). Ringwood proposed to avoid the difficulty by adopting Wise's suggestion (1969) that the Earth's rotation axis changed its relation to the ecliptic after the Moon was formed (Ringwood 1970). In response to a criticism by Singer (1971), Ringwood (1971) admitted that the inclination problem could not be satisfactorily handled in this way, but cited other mechanisms that might work instead.

Cameron (1970) supported Ringwood's views with the additional hypothesis that the hot proto-Earth was formed by the collision of two bodies of comparable mass. Kaula, in an extensive review (1971), also favored Ringwood's theory.

12 Albee et al. (1970). Gopalan et al. (1970). Murthy, Schmitt, and Rey (1970). Silver (1970). Tatsumoto (1970). Tatsumoto and Rosholt (1970). Wanless, Loveridge, and Stevens (1970).

For the Anders group at Chicago (Ganapathy et al. 1970), the depletion of siderophile and volatile elements meant that the Moon accreted in a circumterrestrial orbit at a distance of 5 to 20 Earth radii, probably from material condensed at higher temperatures than that which formed the Earth. Depletion of siderophiles, especially Au, was attributed to metal segregation within the Earth and Moon, the lunar–terrestrial differences being ascribed to more reducing conditions on the Moon, lower pressure, and a scarcity of water. Depletion of volatiles was attributed to the lower cross-section for accretion by a Moon moving in orbit around the Earth because of her higher velocity (as compared with the low relative velocity she would have had in order to be captured). The fractionation that produced this metal–silicate separation took place in the solar nebula as described in the model of Larimer and Anders (1967), rather than in the Earth. "The principal conclusion thus is that the geochemistry of the Moon does not require an exotic mode of origin, such as capture or fission. On the contrary, the data favor a simultaneous origin, as a planet–satellite system" (Ganapathy et al. 1970: 1138; see also Ganapathy and Anders 1974). Hartmann (1973: 127) agreed that the differences between lunar and terrestrial materials are too great for the former to be derived from the latter – leaving only capture or binary accretion as acceptable origins.

Don L. Anderson (1972, 1973) also concluded that the Moon formed as a high-temperature condensate enriched in refractory elements, but in an earth-crossing orbit highly inclined to the median plane of the solar nebula. Anderson's interpretation of the chemical composition of the Moon led Cameron (1972, 1973a) to propose that she was born inside the orbit of Mercury, where only the highly refractory substances could condense according to the Cameron–Pine model of the solar nebula. The interaction between the Moon and Mercury ejected the former to the Earth's vicinity and gave the latter its large orbital eccentricity. While Cameron and Anderson disagreed on where the Moon was formed, they agreed that she was captured by the Earth. If capture involved collision with other bodies in geocentric orbit and subsequent assembly of the present Moon from fragments, not unlike the later stages of the binary-accretion model, one might also be able to explain the depletion of volatile elements.

Cameron's Mercurial Moon was shot down by Kaula and Harris (1975), who pointed out severe dynamical difficulties with the idea. In general it seems more plausible that the Moon was formed, if not in orbit around the Earth, at least in a nearby heliocentric orbit; a slow relative approach velocity would make capture much easier, and the chemical composition of the Moon is close enough to that of the Earth (apart from the lack of metallic iron) to make an origin in a very distant region of the nebula seem improbable. Capture theories went out of favor after most lunar scientists decided that the Moon is not (as had earlier been thought) enriched in refractory elements, and therefore did not have to be formed in a different part of the solar nebula (cf. Kaula and Harris 1973; Wood 1986: 22, 31). As S. R. Taylor noted (1975: 328–9), much of the lunar research since 1969 focussed on detailed analysis of geochemical and geophysical phenomena that could demonstrate specific differences and similarities between the Earth and Moon, "and it is esthetically unsatisfying to relegate these processes to a remote and unspecified part of the solar system."

The assertion that the lunar crust is depleted in siderophiles, on which the O'Keefe–Urey and Ganapathy–Anders theories depended, was disputed by Ringwood and others on the basis of results from later lunar missions. Basalts and fines from *Apollo 12* and *15* contained much higher amounts of nickel, comparable to terrestrial basalts,[13] although many scientists attributed this to extralunar sources. They also contained much lower amounts of titanium and other refractory elements. According to Ringwood:

The high-titanium basalts have had a more complex petrogenetic history than the low-titanium types. It is the latter which have most to tell us about the composition of the lunar interior. . . . The initial observations of very low siderophiles in Apollo 11 rocks received considerable attention and were responsible for the widely shared belief that the moon is exceptionally depleted in siderophiles. The significance of the much higher siderophile abundances in low-Ti basalts was largely ignored until the papers of Wänke and colleagues and Ringwood, Delano, and Kesson in the late 1970s. (letter to S.G.B., 5 December 1984)

Moreover, Ringwood argued:

The earth's mantle is actually much richer in siderophiles than would be expected from segregation of an iron core of terrestrial size. [The siderophile patterns] are unique to the earth or to a planetary-size body similar to the earth. The fact that these patterns (modified in some cases by the separation of a very small lunar core) are present in lunar mare basalts and also in lunar highland rocks (when corrected for meteoritic contamination) provides a very strong set of arguments for derivation of the moon from the earth's mantle. (letter to S.G.B., 5 December 1984; see Ringwood and Kesson 1977, 1981; Delano and Ringwood 1978; Ringwood 1978)

Edward Anders and his co-workers at Chicago continued to argue that the Moon is essentially different from the Earth in chemical composition. They pointed out strong similarities in the siderophile–volatile patterns of lunar basalts and eucrite meteorites. They proposed that the Moon is similar to a parent body of eucritic meteorites, perhaps similar to the asteroid Vesta. This

calls into question an entire class of models that invoke ad hoc processes to explain the Moon by a unique chance event, e.g., disintegrative capture [J. V. Smith; Mitler]. Such an event would have to happen twice, with similar chemical but different dynamical consequences: once leading to a satellite and next time to an independent

13 Compston et al. (1971). S. R. Taylor et al. (1971). Wänke et al. (1971). Ringwood and Kesson (1976). Ringwood (1979: 199–202).

planet. It may be better to look for ways of producing a multitude of small volatile-poor and metal-poor bodies, as first attempted by Urey.[14]

Anders accused Delano and Ringwood (1978) of practicing "Procrustean science" by rejecting low-siderophile lunar rocks in order to conclude that the Moon is *not* substantially depleted in siderophiles and therefore could have come from the Earth (Anders 1978). A similar critique of Delano and Ringwood's interpretation of lunar rocks, directed at the problem of telling which ones are "pristine" and which formed as "secondary differentiates," was published by Warren and Wasson (1978).

I will not try to judge whether Anders or Ringwood is correct in their controversy, but simply report what they said. In his letter to me Anders stated, first, that the evidence from "fines" should be disregarded; "There is universal consensus that they are crushed surface rock contaminated with micrometeorites, as indicated by the enrichment not only of Ni but of a dozen other siderophiles and volatiles in C-chondrite proportions." Second, as for Ni in basalts, Anders cites the arguments he has already published (1978: 167; Wolf and Anders, 1980: 2113). Third, concerning other siderophiles in basalts, he states:

Our data on lunar basalts and pristine rocks were summarized in Figs. 1–3 of Wolf et al., 1979. In response to Ringwood's point, I deleted all high-Ti basalts, but there still remained substantial differences between Earth and Moon, as shown in Fig. 1 of Wolf and Anders, 1980. Ni is a fairly abundant element that is easy to measure by rather inaccurate methods, and hence one can find occasional high values that are not confirmed by more accurate methods. Trends are cleaner for gold, which is more noble than Ni and hence a better test case. (Moreover, it is less abundant and therefore is generally measured by radiochemical neutron activation (RNAA), a labor-intensive but accurate method). . . . The controversy broke out in the open at the 1978 Lunar Science Conference, where Ringwood and I clashed in a 1-hr debate. By most people's account, I won. The essentials of that debate are preserved in my 1978 paper, "Procrustean Science" and Ringwood's rejoinder. . . . time has vindicated me on most points.

More recent results of Warren and Wasson on pristine rocks should be taken into account. "The alleged similarity of the Earth and Moon patterns is totally contradicted by Fig. 1 of Wolf's and my 1980 paper" (letter to S.G.B., 5 December 1984).

Ringwood continued to claim that the abundances of siderophiles, rare earth elements, oxygen isotopes, xenon isotopes, and other characteristic properties of the Moon were so similar to those of the Earth's mantle that she must be a child of Earth; any differences could be explained by the high-

14 Vizgirda and Anders (1976). See also Gros et al. (1976), Anders (1977), Hertogen et al. (1977), Morgan, Hertogen, and Anders (1978), Wolf, Woodrow, and Anders (1979), Wolf and Anders (1980).

temperature separation process and by subsequent development of an iron oxide core in the Earth.[15] He criticized binary accretion, the other major contender, as unable to account satisfactorily for the removal of iron from Moon-forming planetesimals.

Ringwood gained a strong supporter when Heinrich Wänke, the leader of the most active European group of lunar scientists, concluded that the chemical similarities between Earth and Moon outweighed the differences. Wänke was one of those who decided that the real meaning of the lunar samples could not be understood without further investigation of the chemistry of the Earth and meteorites (interview with S.G.B., 15 October 1984). Having ruled out "the possibility that the Moon was once part of the Earth" on the basis of FeO/MnO ratios in 1974, his experimental studies of the W/La ratios led him to accept the idea that metal–silicate fractionation of lunar materials took place within the proto-Earth and that these materials were subsequently removed as vapor from the Earth (Dreibus et al. 1977; Rammensee and Wänke 1977). Contrary to inferences from the first lunar samples the Moon is not enriched in refractory elements compared with the Earth's mantle (cf. Warren and Wasson 1979; Wasson 1985: 194).

According to Dreibus and Wänke (1979), the elements Mn, Cr, and V are depleted in both the Earth's mantle and the Moon, but not in the parent bodies of eucrite and shergottite meteorites.[16] They regarded this similar depletion as "the strongest argument for a close genetic relationship of Earth and Moon. The Moon must have formed somehow from the material of the Earth's mantle after core formation" (Wänke and Dreibus 1982: 340).

According to Anders (letter to S.G.B., 17 September 1987) this argument is refuted by the results of Newsom and Drake (1987) on the partitioning of V, Cr, and Mn into S-rich metallic liquids. There is a similar disagreement between Newsom and Drake (1983) and Wänke et al. (1983) on the significance of the difference in P content between the Moon and the Earth's mantle: the former claiming that it indicates independent formation, the latter arguing that since the Co/P ratio is similar for lunar and terrestrial rocks the P depletion can be explained by segregation into a metallic core.[17]

Ringwood's view in the late 1980s (letter to S.G.B., 11 November 1987) was that "there is general agreement among all those now working in the field concerning the siderophile patterns in terrestrial and lunar (low-titanium) basalts. Several comparative diagrams have been published independently by Newsom, Drake, Jones, Delano, Wänke, and myself. All of them show that moderately siderophile elements are indeed depleted in lunar basalts. All of these workers agree that this latter depletion is due to their seg-

15 Ringwood (1977, 1978, 1979, 1984, 1986a, 1986b). Ringwood and Kesson (1977, 1981). Lightner and Marti (1974). R. N. Clayton, Onuma, and Mayeda (1976). Ringwood et al. (1980).
16 The latter may have come from Mars; see Bogard, Nyqvist, and Johnson (1984) and references cited therein.
17 See also the discussion of the significance of alkali volatile depletion by Jones and Drake (1986), Kreutzberger, Drake, and Jones (1986), and Ringwood (1986c).

regation into a small lunar core. There is no serious disagreement among any workers concerning the nickel abundance. With the benefit of later data, we all agree that it is slightly depleted, by a factor of about three, in the source region of mare basalts, due to segregation into the lunar core. The nickel abundances of lunar low-Titanium basalts are nevertheless remarkably high as compared with expectations based upon equilibrium between metallic iron and silicates, as displayed in the meteoritic parent body. That is, they are much more "Earthlike" than "meteoritelike." Wänke and I and others have written at length on the significance of these relationships for lunar origin. Our arguments have yet to be answered. I continue to believe that they demonstrate quite definitively that a large proportion (> 50 percent) of the material now in the Moon was derived from the Earth's mantle after the Earth's core had formed.

Ringwood stated that there is still substantial disagreement on the siderophile chemistry of the lunar highlands:

It has been acknowledged from the beginning that there has been substantial siderophile contamination of lunar highland breccias. Wänke, Dreibus, Delano and I assumed the contamination to be caused by the most common class of meteoritic material – chondrites. After correcting for this contamination using a standard geochemical procedure, we found that the residual siderophile patterns in the highlands were similar to mare basalt and to terrestrial patterns. . . . I remain confident of our position on this topic, but should emphasize that the highland chemistry data are not essential to our thesis concerning similarities between Earth and Moon which are securely based upon the data from basalts.

He objects to Anders's statement (earlier)

implying that our conclusion rests upon the interpretation of information from lunar "fines" and that we had overlooked the problem of meteoritic contamination. Our conclusion rests primarily on the evidence of lunar basalts. When we used highland materials, we took great care to apply corrections for meteoritic contamination. (letter to S.G.B., 11 November 1987)

Finally one should note the following information from another lunar scientist, Paul H. Warren:

Very recent results from lunar meteorites have further weakened Ringwood's case for a "striking" similarity between the siderophile patterns of the Moon and the Earth. Ringwood and Wänke have made much of the surprisingly high Ni/Ir ratios found for breccias from Apollo 16, the one "representative" highlands Apollo site. By mentally stripping away a meteoritic component of supposedly known Ni/Ir ratio, Ringwood and Wänke claim that they can prove that the indigenous or "residual" highlands component must be Ni-rich. Provided one bears in mind the salient weakness that by no means do all meteorites have identical Ni/Ir ratios, this approach is worthwhile. However, we now have several lunar meteorites (four that have thus far been analyzed) that appear to be from at least two additional "representative" highlands sites, and these meteoritic breccias consistently have much lower Ni/Ir ratios than typical Apollo 16 breccias. Thus the "residual Ni" contents of the meteoritic highlands samples are incontrovertibly far lower than the Apollo 16 "residual Ni" contents. (letter to S.G.B., 1987).

To conclude this discussion, my impression is that most (but not all) planetary scientists think the Moon's chemical similarities to the Earth are more striking that her differences (e.g., Drake 1986; Kaula, Drake, and Head 1986: 625). But could the Moon have been born from the Earth without the action of another body? The fission hypothesis was still strongly supported by O'Keefe, Binder, and others.[18] But most other scientists found it implausible.[19] Something else was needed.

4.4.5 The giant-impact hypothesis

Before discussing the introduction of the giant-impact theory by Hartmann and Davis (1975) and Cameron and Ward (1976), I want to mention what might be called a precursor of that theory, proposed by James G. Baker in 1973.[20]

Baker assumed that Jupiter once had a much larger mass, which produced large gravitational perturbations in the orbits of the other planets. In particular, Mars and the Earth came close enough to experience a grazing collision. Calculations carried out by B. E. Baxter at Aerospace based on Baker's hypothesis indicated that fragments of Mars could be captured into orbit around the Earth; Baker proposed that this material eventually formed the Moon.

18 Binder and Lange (1977, 1980). Binder (1978a, 1978b, 1980, 1986). O'Keefe and Sullivan (1978). Andrews (1982). Drake, as reported by Waldrop (1982). Durisen and Scott (1984).
19 Kaula and Harris (1975). Burns (1977: 129). French (1977: 52–3; 1978). Kaula (1977b). J. V. Smith (1977a). Wood (1977). Cadogan (1981). Condie (1982: 21).
20 Baker was the co-designer of a camera used to track Earth satellites in the 1950s and 1960s and also designed many aerial camera lens systems for the Air Force in World War II and after (letter from A. W. Harris to S.G.B., 23 September 1987). While a consultant to the Aerospace Corporation, living in Winchester, Massachusetts, he developed a theory of the origin of irregular satellites, including the Moon. See also 4.1, note 5.
   James Baker should not be confused with Howard Baker, an earlier precursor, who suggested that the Earth had a gravitational encounter with "some large heavenly body, in close proximity, which removed a part from the Pacific side" (1954: 12). This body, which he named "Pentheus," was more massive than the Earth; it was later broken up by Jupiter into comets, some of which then brought water to the Earth.
   Similarly the geologist R. A. Daly proposed, at first rather vaguely (Daly 1933) and then more specifically (1946), an impact origin of the Moon (Baldwin and Wilhelms 1992; I thank H. S. Yoder for information on Daly's theory).
   M. M. Woolfson, one of the few remaining advocates after World War II of a dualistic origin of the Solar System, proposed (1971) that two planets moving between the Earth and Jupiter collided; the Moon was an orphaned satellite of one of them.
   Such catastrophic speculations were consistently ignored or flatly rejected by the scientific community in the 1950s and 1960s, usually for good reasons (Bauer 1985).

Baker presented his ideas to Donald H. Menzel at Harvard; Menzel reported them to an International Astronomical Union (IAU) Symposium on the Stability of the Solar System in Poland, 5–8 September 1973, but no record of this report appeared in the published proceedings of the symposium. Menzel wrote a foreword to Baker's (unpublished) technical report (1974: iv–viii) stating that the theory "appears to combine the best features of all earlier hypotheses." In 1974 Baker submitted a long manuscript for publication in the *Reviews of Geophysics and Space Physics*. The editor, W. M. Kaula, stated that it was "too speculative" for a review journal, the major criticisms being that the author had incorrectly assumed the orbital eccentricities to increase linearly with Jupiter's mass and had failed to explain how Jupiter lost its original mass (letter from W. M. Kaula to J. G. Baker, 3 July 1974, quoted by permission). But Kaula urged Baker to break his work into smaller parts that could be judged separately on their merits. "Your ideas have quite respectable company," Kaula wrote. "For example, just this March John Wood of the Harvard Center for Astrophysics presented a tidal 'rip-off' hypothesis for origin of the moon similar to yours in several respects."

Baker's theory was briefly mentioned by Kaula and Harris (1975: 368) but cited only as a "personal communication" along with similar ideas of W. K. Hartmann and A. G. W. Cameron. Harris recalled that they regarded Baker's theory as "bordering on crackpot" at the time (letter to S.G.B., 7 November 1984). H. E. Mitler, whose "partial capture" theory of the Moon has some similarities to Baker's, used Baker's formula incorporating the tensile strength of the material into the Roche limit for tidal breakup, citing the preprint as "to be published" (Mitler 1975: 265–7). But since Baker never published the details of this theory it was eventually forgotten (letter from Baker to S.G.B., 8 December 1984). Most of his basic premises, such as the assumption that the impactor survived to become the present planet Mars, are still considered unacceptable, and his ideas had no influence on later developments (letter from W. K. Hartmann to S.G.B., August 1987). This case serves mainly to show that the idea of attributing the Moon's origin to a giant impact, which could not have been taken seriously by scientists 20 years earlier, might enjoy a favorable reception if presented with proper evidence and arguments.

W. K. Hartmann presented his giant-impact theory – ejection of moon-forming material into a circumterrestrial disk by a large (>1000-km radius) planetesimal – at an IAU colloquium at Cornell in August 1974. (The basic idea was very briefly mentioned in a news report by Hammond, 1974.) He worked with D. R. Davis in developing the theory.[21] A. G. W. Cameron, who had been studying a similar hypothesis, collaborated with W. R. Ward to work out another version independently of Hartmann and Davis.

Hartmann had long been interested in lunar craters and the time variation of their size distribution. During the 1960s he had also been impressed by

---

21 Donald Ray Davis (b. 1939) received his Ph.D. in 1967 at the University of Arizona. He has been on the staff of the Planetary Science Institute in Tucson since 1972.

Figure 34. William K. Hartmann (S. G. Brush).

V. S. Safronov's papers on planetary formation by accretion of solid planetesimals, including the hypothesis that the tilt of the Earth's rotation axis ("obliquity of the ecliptic" in astronomical terminology) is primarily due to the impact of the last large body that was added during its accretion.[22] Also, measurement of the flux of impacting bodies as a function of time showed that the pre-mare cratering rate was enormous and included basin-forming bodies much bigger than the crater formers in the last few billion years (Hartmann 1966). Those were debris left after planetary formation. Extrapolation back to an earlier epoch suggested that bodies as large as the Moon itself could plausibly have been moving in the vicinity of the Earth.

Hartmann and Davis attempted a numerical reconstruction of the size distribution of bodies that could have grown during planetesimal accretion and were left behind near the end of planet formation, assuming a process starting with accretion of small particles. They found that "the probability of the planet interacting with a large body is much larger than has been considered in some descriptions of planetary growth." For certain assumptions, they found that among Earth-sized planets the second-largest bodies could be of radius 500 to 3000 km, and there could be tens of bodies larger than 100 km radius. Half of the kinetic energy of a planetesimal about 1200 km in radius, arriving at the Earth's surface at a velocity of 13 km/sec, could eject two lunar masses to near-escape speeds. Assuming that the collision occurs after the

22 Interview with S.G.B., 16 October 1984; letter to S.G.B., 16 April 1985; see Safronov (1958/1964, 1960/1964, 1965/1966, 1969/1972: ch. 11).

Earth's core starts to form, one expects that the ejected material would be depleted in iron, as in the fission theory.

The advantages of impact over fission, according to Hartmann and Davis, are (1) an energy source to raise the material off the Earth is provided; (2) "The theory is not purely evolutionary [i.e. the outcome for a given planet is randomized, not purely deterministic], depending on a chance encounter so that it does not require prediction of similar satellites for Mars or other planets" (Hartmann and Davis 1975: 512). After the material is ejected it forms a cloud of hot dust, enriched in refractory elements and rapidly depleted in volatiles. The subsequent evolution follows the "widely admired" theory of Ringwood (1970).[23]

According to Cameron (1985b: 319), he and W. R. Ward

were led to the suggestion of a collisional origin of the Moon through the following consideration. The angular momentum of the Earth–Moon system is less than sufficient to spin the Earth to rotational instability; we were nevertheless interested to determine the mass of the body which, striking a tangential blow to the protoearth, could impart the angular momentum of the Earth–Moon system to the protoearth. The required projectile turned out to be about the mass of Mars. That defined the basic scenario of our lunar formation process.

Their theory was published as a three-page abstract in 1976.

Although Cameron was present at the original presentation of the Hartmann–Davis hypothesis, he had already begun to develop a similar hypothesis independently and was not directly influenced by the Safronov planetesimal-accretion model for planet formation since he preferred a gaseous protoplanet model instead (interview with S.G.B., 15 October 1984). Cameron says that the difference between his model and that of Hartmann and Davis should be more strongly emphasized. "The HD idea involves *many* collisions and therefore is fatally slain by angular momentum" (note to S.G.B., November 1984); it "did not grapple with the problem of how the material that was thrown up got into orbit" (Cameron, quoted by Cooper, 1987: 79).

Hartmann said that the two theories start from a similar basic assumption, and that his hypothesis includes the idea of a single large impactor that may be nearly as big as Mars:

Each of us considered different aspects of the problem: we, the question of whether such big bodies could have grown and been available, and the neat consequence that stochastic big impacts could explain differences from planet to planet in obliquity, satellites, etc.; they, the angular momentum problem and (to some degree) the orbiting of the debris. This is perhaps the best possible scientific situation: different teams working independently on different aspects of an interesting new idea . . .

---

23 For popular expositions with dramatic illustrations see Hartmann (1976; 1978; 1983; 149–51), Rubin (1984), Brownlee (1985), Frazier (1987).

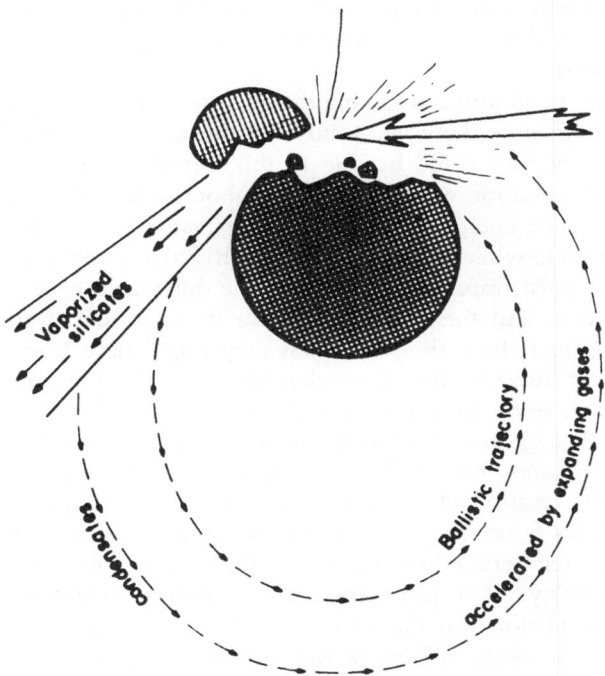

Figure 35. Collisional ejection model: Expansion of ejected silicate vapor accelerates condensates into orbits with perigees that clear the Earth's surface (in contrast to the trajectories of solid debris knocked into simple ballistic orbits, which would reaccrete to the Earth) (J. A. Wood, "Moon over Mauna Loa," in *Origin of the Moon*, ed. W. K. Hartmann et al., 1986, p. 43; copyright by Lunar and Planetary Institute).

No paper on certain aspects of a problem is obliged to treat every aspect. Ours didn't. Cameron & Ward didn't, either. We don't say *theirs* is fatally flawed because it failed to treat growth rates of the planetesimals, which is what we discussed. (letter from Hartmann to S.G.B., August 1987)

Cameron and Ward (1976) postulated that both colliding bodies were differentiated and possibly molten at the time of impact. (This follows from the assumption that they were formed in gaseous protoplanets.) The mantle material of both bodies would be largely vaporized. The early motion of the ejected material would not be a set of ballistic trajectories (which would return it to Earth), but would be governed by gas pressure gradients in the expanding silicate vapor. Later the more refractory silicates will recondense into particles. These will form a thin disk at a distance of 2 to 4 Earth radii; beyond the Roche limit (about 3 Earth radii) gravitational instability will produce clumps that will have substantial tidal interactions with the Earth.

Further suggestions about the later stage of this process, based on the Lynden-Bell and Pringle theory of accretion disks (1974), were offered by Ward and Cameron (1978).

The impact, in addition to solving the angular momentum problem, might explain the differences between Venus and the Earth. The Earth's relatively fast prograde rotation could be due to this fortuitous event, while Venus's small retrograde rotation would be the natural outcome of many small impacts (Wood 1986). The energy for Earth's tectonic activity could also have come from this impact, while Venus's lack of such activity could be due to the absence of a similar impact (Kaula 1981). The differences between the atmospheres of Venus and the Earth might also be explained (Cameron 1983): Since those planets have the same mass they might have been expected to acquire similar atmospheres in condensing from the presolar nebula, but Earth's atmosphere seems to have lost almost all of its primordial noble gases. The impact that ejected the Moon might also have dissipated much of the Earth's original atmosphere. Other consequences have been suggested by Kaula (D1987): "The creation of the Moon was a necessary condition for us to be here today. This event blasted away the excess atmosphere and allowed the Earth's surface temperature to drop below the boiling point of water."

The plausibility of a giant impact was greatly enhanced by George Wetherill's calculations on the accretion of planetesimals, (1976), based on Safronov's model. Wetherill's work was motivated in part by photographs of Mercury's surface taken by the *Mariner 10* spacecraft on 29 March and 21 September 1974, analyzed by Murray et al. (1975). It appeared that not only the Moon but also Mercury had suffered a "late heavy bombardment" after formation (Wetherill 1975a). ("Late" means with respect to the growth process starting from dust and going through small planetesimals. Hartmann [1966] discussed "early intense cratering" of the Moon, meaning more than 3.9 billion years ago. Both phrases refer to the same process occurring at the same time period.) Hence, it was likely that there was a high flux of asteroid- or Moon-sized bodies throughout the inner Solar System, 4 to 4.5 billion years ago, and that the Earth had suffered such a bombardment even though its traces have been removed by geological processes. Wetherill found from his version of Safronov's model that a substantial fraction of the total mass in each region of the nebula would reside in bodies only one order of magnitude smaller than the dominant planetary embryo at a fairly late stage of the process. It is therefore an essential feature of this process that a terrestrial planet will be hit by an object as large as Mars during the final stage of its growth, although Wetherill emphasized (1986) that impact is only one of several processes that can be expected to provide material for the formation of the Moon in his model.

Because of the suggested cosmogonic importance of solid-body collisions, Thomas J. Ahrens and his colleagues at Caltech analyzed the melting and vaporization produced by impacts of silicate compounds (O'Keefe and Ahrens

1977). Early results indicated that collisions at velocities appropriate for planetesimals (11 to 16 km/sec) would not yield enough vaporization to place adequate amounts of material in orbit (Rigden and Ahrens 1981). But in subsequent experiments with anorthosites, Mark B. Boslough and Ahrens (1983) found that about five times the mass of an impacting meteoroid could be vaporized by impact at 15 km/sec. This was an order of magnitude greater than the result of the previous studies and made impact models of lunar origin "much more plausible."

Further support for the impact theory came from A. C. Thompson and D. J. Stevenson (1983), who pointed out that a hot disk of the type postulated in the Cameron–Ward theory could undergo gravitational instability leading to the rapid formation of a Moon-sized body because the speed of sound (which determines the stability criterion) is substantially lower in a two-phase (gas–liquid) system than in either phase by itself.

A consensus in favor of the giant-impact theory emerged at a conference on the origin of the Moon held in Hawaii in October 1984. According to Stevenson (1987) this was "not because of any dramatic new development or infusion of data, but because the hypothesis was given serious and sustained attention for the first time. The resulting bandwagon has picked up speed (and some have hastened to jump aboard)." Hartmann, one of the organizers of the conference, says his idea "had languished" since its publication; "When I went to a planning session for the conference to look over the abstracts for the proposed papers, I found, to my amazement and joy, that eight or ten of the abstracts – independently of each other – were about the impact idea" (quoted by Cooper 1987: 80).

A major reason for the attention given to the giant-impact hypothesis was the increasing evidence against the three pre-*Apollo* theories:

1. In addition to the original angular momentum difficulties of the fission hypothesis, new calculations on viscous rotating fluids indicated that they cannot be spun fast enough to cause fission, but instead simply lose matter from equatorial regions; rotational instability can be produced by planetesimal accretion only if one planetesimal is about one-tenth of the mass of the proto-Earth, in which case the fission model goes over to the giant-impact model (Boss 1985; Boss and Mizuno 1985; Boss and Peale 1986). Fission models were deemed incapable of explaining why the Moon is substantially richer in both iron and refractory elements than the Earth's mantle (Drake 1986; Taylor 1987a).

2. The capture hypothesis lost its original advantage of being able to explain Earth–Moon compositional differences when it was shown that capture is dynamically impossible, unless the Moon was formed at about the same heliocentric distance as the Earth (Wood 1986), and even then is rather unlikely (Boss and Peale 1986). Disintegrative capture is also unlikely (Mizuno and Boss 1985). On the other hand even if the Earth could have

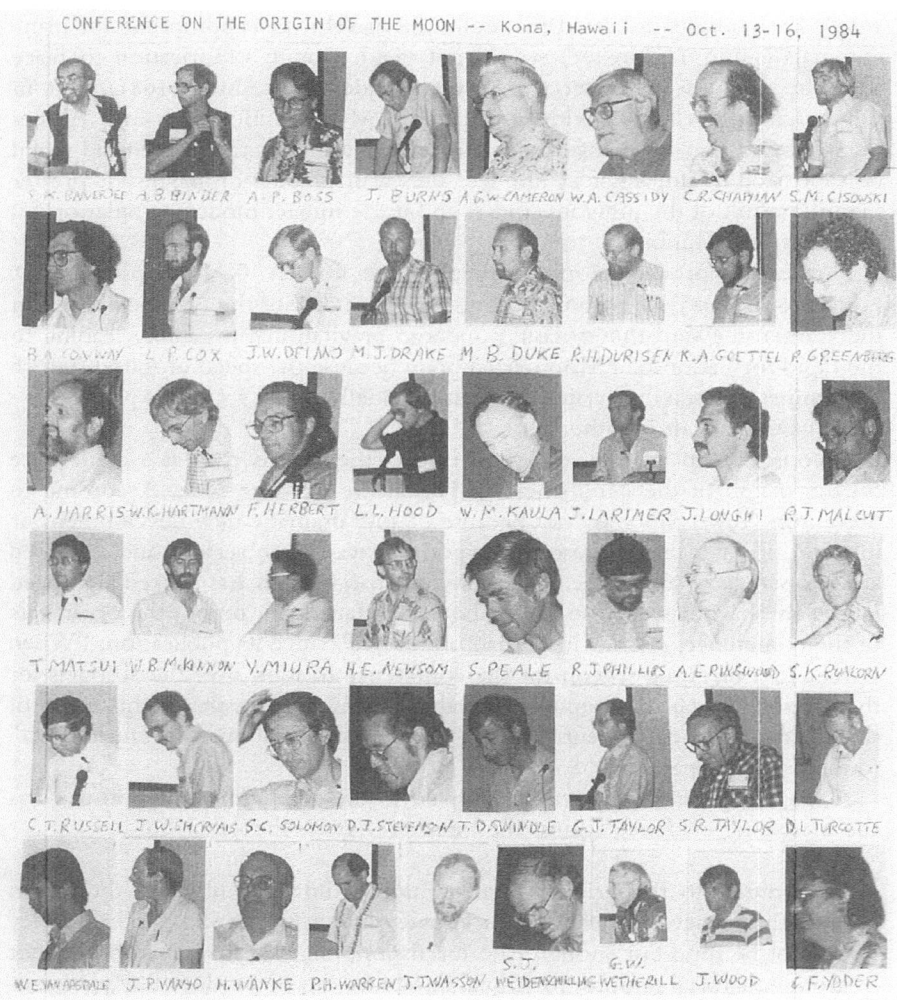

Figure 36. Some of the participants in the Conference on the Origin of the Moon at Kona, Hawaii, 13–16 October 1984 (S. G. Brush).

captured a Moon formed far away from the Earth (in order to account for chemical differences), one would then have difficulty accounting for the *similarity* of oxygen isotopic composition.

3. Co-accretion had difficulty explaining the composition differences between the Earth and Moon, even with a postulated "composition filter" to separate iron from silicates; moreover, it cannot account for the angular momentum of the Earth–Moon system (Boss and Peale 1986; Wood 1986).

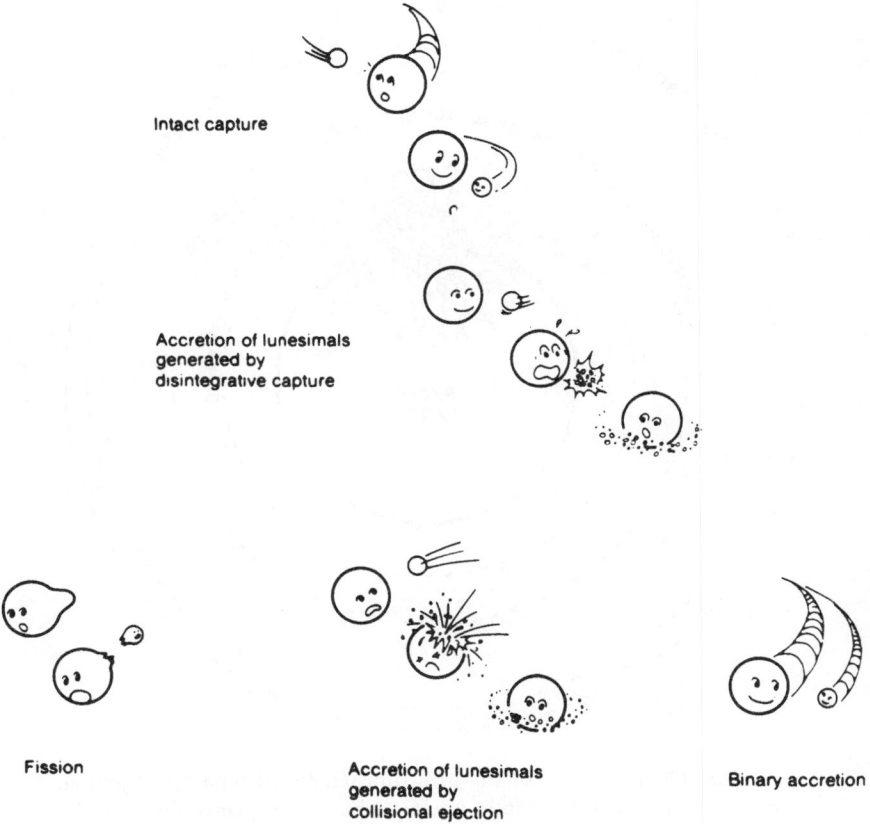

Intact capture

Accretion of lunesimals
generated by
disintegrative capture

Fission

Accretion of lunesimals
generated by
collisional ejection

Binary accretion

Figure 37. A cartoon of models for the origin of the Moon, by T. Matsui and Y. Abe, "Origin of the Moon and its early thermal evolution," in *Origin of the Moon*, ed. W. K. Hartmann et al., 1986, p. 454; copyright by Lunar and Planetary Institute.

The impact theory was endorsed by some scientists who had previously favored a terrestrial origin for the Moon, and was mentioned favorably by others.[24] Ringwood (D1985) favored "impacts by many large ($R > 100$ km), but not necessarily giant ($R > 1000$ km) planetesimals"; he regarded the Hartmann–Davis hypothesis as being closely related to his own earlier model but was skeptical of the single giant-impact version because it would lead to complete melting of the Earth. Ringwood also maintained that the lunar siderophile evidence contradicts Cameron's specific model, according to

24 Ringwood (1978; 1979; 244; 1986a; 1986b; 1986c). Whipple (1981). Wänke and Dreibus (1982, 1984). Brownlee (1985). Taylor (1986a, 1986b).

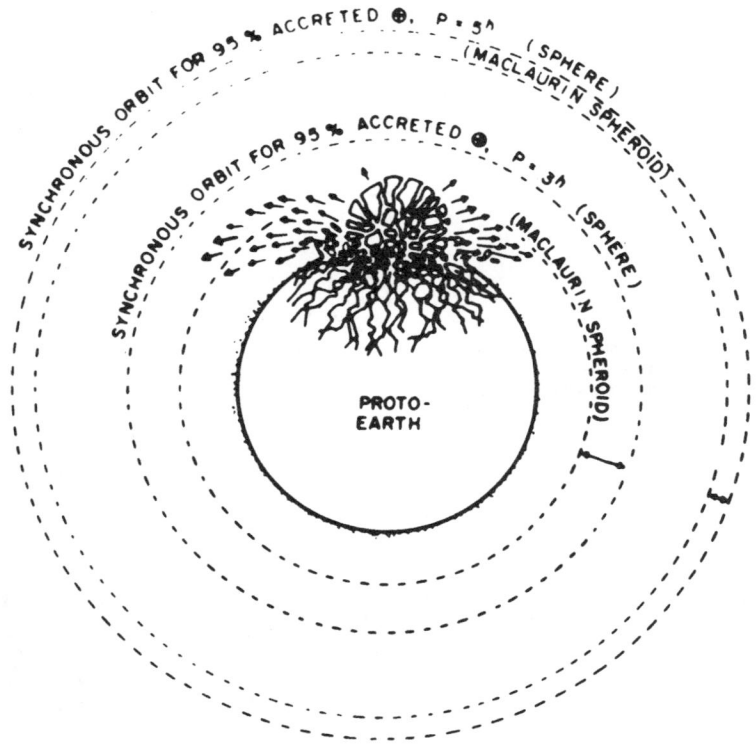

Figure 38. Circumstances of the giant impact. Initial high-speed jetting comes from the mantle–mantle interface, but if the planetesimal breaks up in tangential impact or before full interpenetration, much additional mass may be ejected from an effective launch point well outside Earth's atmosphere (schematically shown by stippling). Synchronous orbits are shown for Earth-rotation periods of three hours and five hours, both for a spherical model Earth and adjusted for the effects of Maclaurin spheroid shape appropriate to the period. (Latter curves after Weidenschilling, personal communication to Hartmann, 1984.) The latter curves are closer to Earth, facilitating lunar formation outside the synchronous orbit, where it must accrete in order to evolve outward tidally (W. K. Hartmann, "Moon origin: The impact-trigger hypothesis," in *Origin of the Moon*, ed. Hartmann et al., 1986, p. 601; copyright by Lunar and Planetary Institute).

which most of the material of the Moon was derived from the impactor, but is "marginally consistent with the Kipp/Melosh/Sonnett model in which about half of the protolunar material is derived from Earth's mantle" (letter to S.G.B., 11 November 1987).

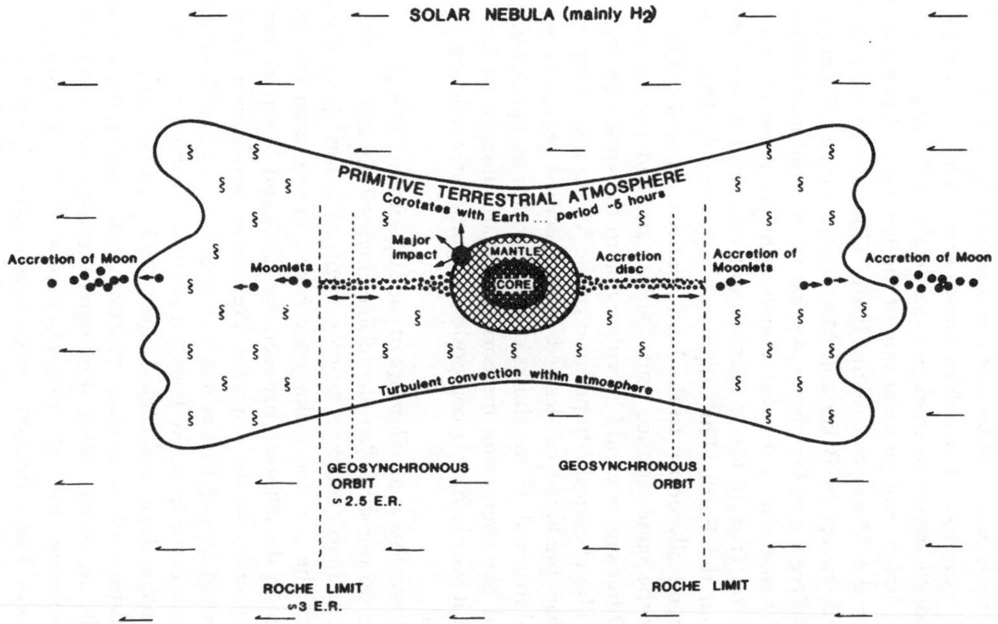

Figure 39. Ringwood's model showing the formation of the Moon via ejection of material from the Earth's mantle by impacts from late-accreting planetesimals in the presence of a corotating primitive terrestrial atmosphere (A. E. Ringwood, [1986b], "Terrestrial origin of the Moon," *Nature*, **322**, 327).

Durisen and Gingold (1986) suggested that their numerical calculations of fission, showing that the process produces a ring or disk rather than separate bodies (Durisen and Scott 1984), could be combined with impact in order to deal with the angular momentum problem.

A reporter for *Science* magazine wrote, "The idea that the impact of a Mars-size body on the young earth could have formed the moon has breathed new life into a long-stagnant field" (Kerr 1984). Similar language was used in an article in the Smithsonian's *Air and Space* magazine: The 1984 Hawaii conference "brought a fresh breeze of scientific thought to the problem of the origin of the moon – along with a sudden burst of research" (Frazier 1987: 89).

But when theorists started to work out the details of the impact hypothesis they found that it might not perform one of the functions that made it seem attractive: getting the Moon out of the Earth. Contrary to what had been generally assumed (except by Baker and Menzel; see earlier), Cameron stated at the Hawaii conference in October 1984 that in the collisions he had simulated, most of the material in the disk comes from the projectile rather than the Earth. Cameron (1985b) found that at most one-third of the lunar mass would come from the Earth. This might not make much difference if the projectile was chemically similar to the Earth, but that seemed unlikely unless it was formed at the same distance from the Sun and had a mass comparable to the Earth's. Otherwise it would be vulnerable to the same objection as the capture theory: The chemical composition of the Moon must simply be postulated rather than predicted or derived from specified processes acting on known terrestrial material. If one thinks of the impactor as a planet like Mars, made of material like shergottite meteorites that are thought to have come from Mars, then it would have a composition significantly different from the Earth and Moon.

Taylor (1987c) saw this as an *advantage* of the large-impact model. He argued that lunar samples do not show "an identifiable chemical signature of the terrestrial mantle, despite heroic attempts by proponents of fission" (e.g., Ringwood 1986a). If most of the material making the Moon comes from the impactor, then one can explain the *differences* between the Earth and Moon by attributing them to the composition of the impactor. Preliminary results from Sandia National Laboratories indicating that as much as 50 percent of the Moon-forming material might have come from the Earth's mantle were subsequently revised to only 10 or 20 percent (Kipp and Melosh 1987; Taylor 1987c: 1307).

Detailed calculations of the physical processes postulated in the impact theory were started only in the 1980s. Preliminary results were reported at the 1984 Hawaii conference by Cameron, Hartmann, Kaula and Ann Beachey, and D. J. Stevenson. The published proceedings volume (Hartmann, Phillips, and Taylor 1986) includes seven articles classified under the heading "Lunar Formation Triggered by Large Impact."[25] Three other articles related impact

---

25 Cameron (1986a). Benz, Slattery, and Cameron (1986). Hartmann (1986). Kipp and Melosh (1986). Melosh and Sonett (1986). Ringwood (1986a). Wänke and Dreibus (1986).

selenogony to general theories of impacts of large bodies in the early Solar System.[26] Three papers presented evidence *against* the giant-impact hypothesis: Paul Warren (1986) said the hypothesis has difficulty explaining his data on the MgO/FeO ratio; D. L. Turcotte and L. H. Kellogg (1986) argued that their isotopic data favor a cold origin rather than the hot origin implied by impact; and Solomon (1986) gave geophysical arguments against a hot origin. Stevenson (1987) reviewed the most recent research including contributions of Benz, Slattery, and Cameron; Wetherill; and his own work with A. C. Thompson. Alan Boss at the Carnegie Institution of Washington also published extensive studies of the impact model (1986a, 1986b). Several papers on aspects of the model were presented at the 18th Lunar and Planetary Science Conference.[27]

The calculations done in the mid-1980s indicated that the impact mechanism is capable of placing material into orbit around the Earth, but much more work would be needed to develop a well-defined model that is clearly distinguishable from other models, and even then this model might not appeal to scientists who think the Moon was made from terrestrial material.

4.4.6 Modifying the co-accretion theory

While the impact theory seemed to be more attractive than any other single hypothesis in the past 10 years, many lunar scientists favored some kind of modified co-accretion theory. Following the ideas of Ruskol (4.4.1) and Öpik (1961, 1962b, 1972), they assumed that a swarm of planetesimals began to collect around the Earth during its formation and later accumulated into a single satellite. Unlike the classical co-accretion theories, the newer theories assume that a large part of the selenogenic material is captured by the swarm after it has formed and processed by physical or chemical mechanisms in order to give the Moon a somewhat different composition than the Earth. No single incoming body contributes more than half of the Moon's substance – unlike the classical capture theory and Cameron's latest impact model. The same flux of planetesimals feeds the growing Earth with mass and angular momentum. The theories differ among each other in quantitative details, such as the time of arrival of the planetesimals during the Earth's formation, their sizes and extent of prior differentiation into iron cores and silicate mantles, the nature of their processing by the circumterrestrial swarm, and the explanation given for the failure of the same mechanism to produce similar satellites for other planets.

Ruskol (1971/1972a, 1971/1972b, 1972) argued that her model would provide more effective depletion of volatile elements at the edge of the swarm, where particles would be exposed to the solar wind, than at the center where

26 Hartmann and Vail (1986). Kaula and Beachey (1986). Wetherill (1986).
27 Benz, Cameron, and Slattery (1987). Kipp and Melosh (1987). Ringwood (1987). Taylor (1987b). Vickery and Melosh (1987).

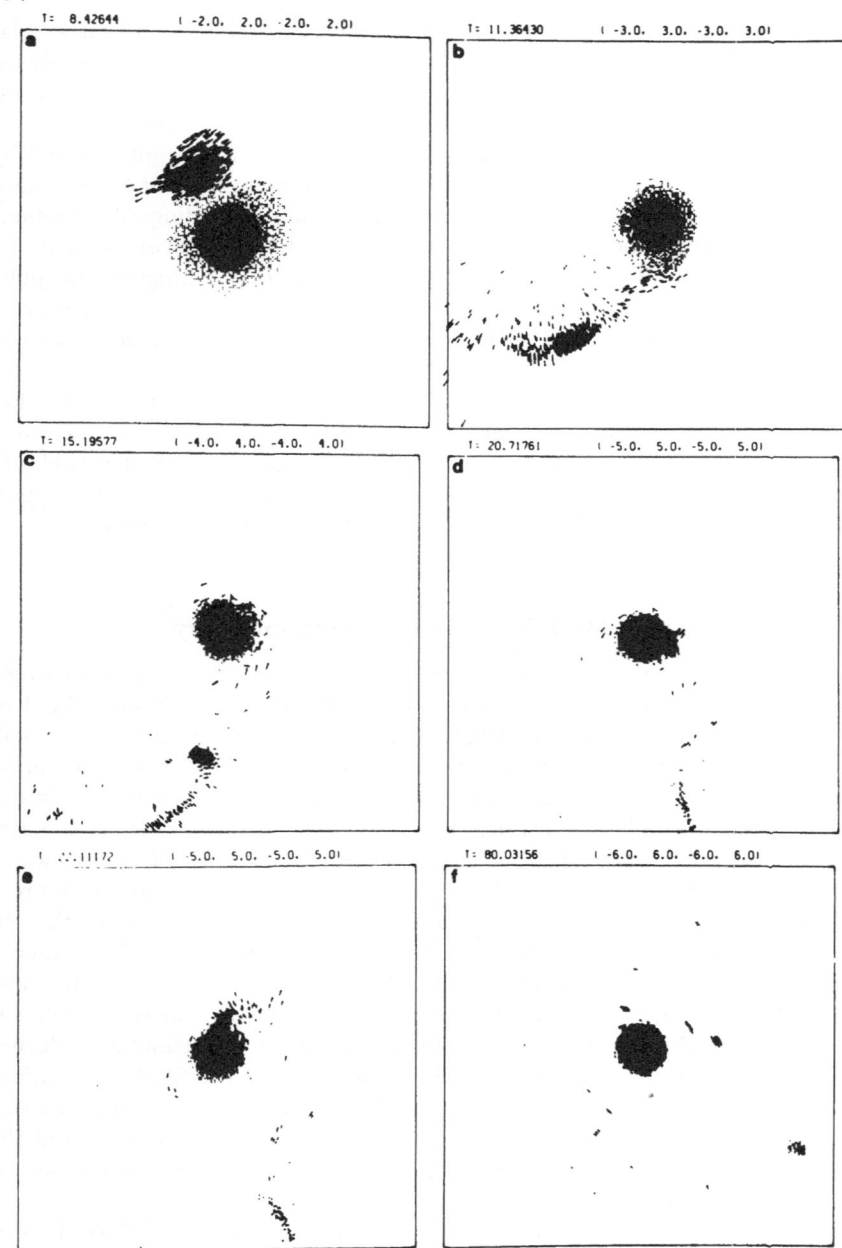

Figure 40. Snapshots of computer simulation of collision between proto-Earth and impactor, mass ratio 0.14. Velocity vectors are plotted at parti-cle locations. The velocity has been normalized to its maximum value in each frame. Time and coordinates of the four corners of the plotted field

the Earth itself formed, and that because of the long timescale for formation of the Earth and Moon (on the order of 10^8 years) such processes would produce the necessary effect better than in other models that are based on rapid or catastrophic processes.

Having abandoned the Lodchnikov–Ramsey hypothesis, Ruskol had to account for the much greater iron content of the Earth. She attributed the segregation of iron into the Earth rather than the Moon to the fact that silicate particles tend to break up into fine debris in collisions while iron-rich particles are more likely to survive and grow (cf. Orowan 1969). While the Earth accretes all the particles that strike it, the circumterrestrial swarm can more easily capture the finer silicate particles.

In 1975 Alan Harris, working with William Kaula at UCLA, showed how Ruskol's theory could be modified to explain how a large satellite such as the Earth's Moon could be formed by the same mechanism that produces small satellites around other planets.[28] According to Harris and Kaula, the Moon must have started growing in Earth orbit when the Earth was between 0.1 and 0.2 of its present mass. (This fraction was later revised to about 0.4 by Harris; see Harris 1978b.) Moreover, the planet's Q (specific dissipation function) must be less than about 1000:

The high Q of the outer planets does not allow a satellite embryo to survive a significant portion of the accretion process, thus only small bodies formed very late in the accumulation of the planet remain as satellites. The low Q of the terrestrial planets allows satellite embryos of these planets to survive during accretion, thus massive satellites such as Earth's Moon are expected. (Harris and Kaula 1975: 516)

The problem was then to explain why other terrestrial planets do not have massive satellites. Harris and Kaula concluded that they did originally have them but lost them because of tidal friction, as suggested by Burns (1973) and Ward and Reid (1973).

Another possibility was that a satellite of Venus was lost because of the rapid growth of its orbital eccentricity, an effect predicted by the theory if tidal dissipation in the satellite is not very large; the escape would have carried

Caption to Figure 40 *(cont.)*
are given in the upper line; time unit = 19.1 minutes; distance unit = 8.08×10^9 cm. For iron particles a solid black circle is plotted (W. Benz, W. L. Slattery, and A. G. W. Cameron, "The origin of the Moon and the single-impact hypothesis, II," *Icarus*, **71** [1987], 41; copyright by Academic Press).

28 Alan William Harris (b. 1944) received his Ph.D. in planetary and space science at the University of California, Los Angeles, in 1975. He has been a member of the technical staff of the Jet Propulsion Laboratory since 1974 and supervisor of the Earth and planetary physics group since 1983. He is currently studying asteroids and their possible danger to the Earth.

away most of the angular momentum of the system. Reimpact could possibly lead to the present retrograde rotation of Venus (Harris 1978b: 144; see also Kumar 1977; Donnison 1978). Alternatively, if the satellite orbit expanded while tidal dissipation slowed the rotation of the primary to the point where day and month are equal, and solar tidal forces then made the day longer than the month, the effect of tidal dissipation on the orbit must reverse and the satellite will be drawn in again to crash on the primary (Darwin 1905/1962: 288; Burns 1973). Ward and M. J. Reid (1973) suggested that solar tidal action could have produced this effect on satellites of Mercury and Venus. Thus, as argued by Harris and Kaula, the present lack of satellites of those planets does not refute a scenario that predicts the formation of satellites about other terrestrial planets.

Results of possible relevance to the compositional sorting of material in the circumterrestrial swarm were reported by the Anders group (Hertogen et al. 1977). They found that among highland samples from *Apollo 16* and *17*, the "most moonlike" (high refractories, low metals) bodies fell first, followed by bodies of progressively less Moon-like composition (as indicated by the Ir/Au ratio). This correlation suggested that the basin-forming objects were genetically related to the Moon in a way that would be expected from Ruskol's theory.

Another kind of evidence for co-accretion was brought forward by S. K. Runcorn: His theory of ancient lunar magnetism led to the conclusion that the rotational pole of the Moon has been changed by impacts of bodies that were once satellites of the Moon, and these satellites originated from the ring of bodies that formed the Moon (Runcorn 1982a, 1982b).

John Wasson and Paul Warren (1979, 1984) were skeptical about Ruskol's mechanism for separating iron from silicates in the circumterrestrial swarm, and proposed instead that the differentiation had already taken place in larger ($r \geqslant 50$ km) bodies. These would not be captured whole by the smaller particles in the swarm, but silicate fragments broken out of them could be captured; the iron cores would pass through the swarm and could be stopped only by a direct hit on the Earth. Making the previously differentiated body even larger would move this model into the category of "disintegrative capture" discussed by J. V. Smith (1974) and by J. Wood and H. Mitler (1974), except that the latter models proposed simultaneous capture and Roche-limit breakup.

A group calling itself the Tucson Lunar Origin Consortium (R. Greenberg, C. R. Chapman, D. R. Davis, M. J. Drake, W. K. Hartmann, F. L. Herbert, J. Jones, and S. J. Weidenschilling) proposed an integrated dynamical and geochemical approach, based on the following assumption:

As the Earth grew by planetesimal bombardment, a circumterrestrial cloud of particles was created from a combination of impact-ejected mantle material and planetesimals captured directly into orbit around the Earth. Such a swarm continued to capture planetesimals and to receive ejecta until the bombarding population thinned, the

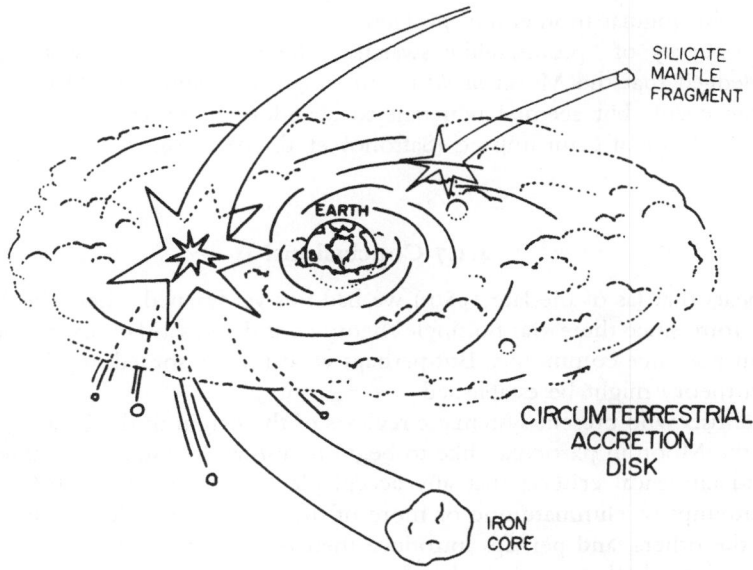

SILICATE
MANTLE
FRAGMENT

EARTH

CIRCUMTERRESTRIAL
ACCRETION
DISK

IRON
CORE

Figure 41. Schematic view of the circumterrestrial swarm. Compositional filtering occurs when the swarm is bombarded by heliocentric impacting planetesimals. Silicates are preferentially trapped, while the stronger iron cores pass through (S. J. Weidenschilling et al., "Origin of the Moon from a circumterrestrial disk," in *Origin of the Moon*, ed. W. K. Hartmann et al., 1986, p. 733; copyright by Lunar and Planetary Institute).

Earth stopped growing, and the Moon accreted in orbit. If Earth-mantle material dominates the swarm, the model resembles the fission hypothesis; if small planetesimals dominate, the model represents the "growth in Earth orbit" end-member; if the swarm were dominated by a single large planetesimal, we would essentially have a capture model. A model intermediate between these extremes appears most promising. (Greenberg et al. 1984b)

While recognizing the possible stochastic nature of a process dominated by large impacts, the Consortium treated the circumterrestrial swarm as a composition filter, "preferentially capturing small weak silicate bodies while passing large iron planetesimals (cores of broken parents)." The results were presented by Weidenschilling et al. (1986). The dynamical aspects of the model were treated by F. L. Herbert, D. R. Davis, and S. J. Weidenschilling (1986). Their preliminary results reported at the 1984 Hawaii meeting suggested that while capture of planetesimals could supply net prograde angular momentum to the system, it would not be adequate to maintain the swarm in orbit around a growing Earth without the help of a few very large impacts on the

Earth. But subsequent calculations indicated that the model might be able to resolve the angular momentum problem.

In a survey of "protosatellite swarms," the Ruskol–Safronov group acknowledged that the Moon *might* be an irregular satellite formed by a catastrophic event, but seemed reluctant to abandon or revise its co-accretion theory in favor of giant impacts (Safronov et al. 1986: 90).

4.4.7 Conclusions?

It appears that (as of the late 1980s) we had not yet learned where the Moon came from, since there was no single theory considered satisfactory by most of the lunar science community. But perhaps we can learn something about how such a theory might be established.

Scientists who present systematic reviews of the origin of the Solar System, or of the Moon in particular, like to begin by listing "constraints" – theoretical and empirical criteria that any acceptable hypothesis must satisfy. They then attempt to eliminate one or more of the existing theories on this basis, assess the others, and perhaps introduce their own new hypothesis or suggest further research that needs to be done in order to overcome difficulties in existing theories.

The structure of the 1984 Hawaii conference formalized this procedure. After introductory invited reviews (the first of which was a microcosm of the entire conference), the contributed papers were grouped under three categories of constraints – geophysical, chemical/petrological, and dynamical – followed by presentations of "my model of lunar origin" by several theorists. The closing session offered assessments of the models from the perspectives of the three constraints.

Unfortunately the algorithm did not converge to a unique solution. This could be seen rather dramatically when two of the invited reviewers, John Wood and Michael Drake, presented tables giving discrete judgments of the theories in terms of the constraints (a "report card" with letter grade, and a "truth table" with yes or no, respectively). The results were completely different, partly because Wood and Drake had divided the universes of theories and constraints in rather different ways, and partly because they made different judgments on how well the theories satisfied the constraints. Wood gave the highest grade to collisional ejection (impact), while Drake rated the double-planet over the fission hypothesis (including impact in the latter).

Later on, the discussion turned to the question of whether theories of lunar origin should be "testable." S. R. Taylor asserted that "the progress of science depends on the construction of testable hypotheses." He presented a quotation from Fred L. Whipple (1976: 16–17) to illustrate his philosophy:

I first note my faith in a basic general principle to be applied in exploring the frontiers of any field in science. One is attempting to find the way through a swampy morass of mutually conflicting observations, interpretations and theories. The principle: seize on a few "indisputable facts" that appear most relevant, momentarily ignore the remaining data. On these solid "facts" build conceptual structures or working hypotheses that are self-consistent, strongly rooted in the best evidence available. Then attempt to expand each working hypothesis by fitting in the successively weaker material. Rejection of material must be rigorously monitored. Is it irrelevant? Is it erroneous; is it misinterpreted because of incomplete theory or because of an unnecessary or erroneous assumption, either tacit or explicit? If none of these, another working hypothesis should be investigated.

Taylor claimed that the original fission hypothesis could be refuted by four differences between lunar and terrestrial compositions: (1) depletion of volatile element, (2) excess of FeO compared with the Earth's mantle, (3) depletion of siderophile elements, (4) excess of refractory elements. In order to account for these differences the fission hypothesis needs so many ad hoc adjustments that it becomes untestable. Hence, "separate terrestrial and lunar accretion from a population of fractionated precursor planetesimals provides a more reasonable explanation" (Taylor 1984).

Two objections were raised against Taylor's views. First, Ringwood and others disputed the statement that the Moon is enriched in refractories, based on a high estimate of uranium abundance from heat-flow measurements and estimates of aluminum in the crust. Second, Wasson and others disagreed that hypotheses must be testable. Kaula remarked that this philosophy reminded him of the drunk who lost his wallet and looked only near a street lamp because that's where he could see best. The basis of the second objection seemed to be that any *simple* testable theory was likely to be proved wrong (and already had been, in this case) so that the correct theory would probably be complicated and might not lead to testable predictions in the forseeable future. Its validity in the long run is not affected by our present inability to test it.

I suspect that the concern was not really about testability or "falsifiability," a term that the philosopher Karl Popper (1962) used as a criterion for whether hypotheses are *scientific* (not whether they are *true*). This criterion has been misused by creationists (see e.g., the law passed by the Arkansas legislature in 1981, subsequently overturned by a federal judge, requiring that public schools give equal time to creationism and evolution) and, until recently, by Popper himself (cf. Popper 1974, 1978, 1980) to argue that we can never obtain scientific knowledge about the past. Such misuse shows that it does not correspond to normal scientific practice. For, taken literally, it would imply that historical geology and much of astronomy can never be scientific since they deal with events so distant in time and space that they cannot be brought into terrestrial laboratories for controlled experiments. Obviously geologists and astronomers don't think that this limitation makes their fields unscientific, but they do worry about the tendency to adjust a theory to fit conflicting data rather than admitting error and looking for a new theory (Urey 1969a). John A. O'Keefe expressed this concern in a letter to Urey in 1974 because sociologist Ian Mitroff was about to publish a book claiming that lunar scientists "don't give up their ideas in the face of facts and don't think logically. He says we allow personal considerations to dominate our thinking" (quoted by Brush 1982b: 897). Mitroff's book (1974a) did indeed stress, as its title suggested, "the subjective side of science," but also presented evidence that lunar scientists themselves considered theorists to be dogmatic and oblivious to evidence that didn't fit their preconceived notions (1974a: 132, 162).

But Urey himself provided a counterexample when he gave up his own capture theory in the light of unfavorable evidence from the lunar samples (4.3.4), although he stuck to many of his other ideas about the Moon that most scientists consider to have been refuted. Other scientists (Cameron is a prime example) have also made radical switches in theory when they thought it appropriate, rather than attempting to fit new data into a rigid preconceived pattern.

The notion that theories of lunar origin are not testable probably accounts for the fact that half of the 42 lunar scientists in Mitroff's survey didn't really have much interest in such theories (Mitroff 1974a: 58). They presumably did not think that the origin question could be definitely answered by the means currently available, so they preferred to focus on other questions about the Moon that *could* be answered. In particular, one might argue that while the early thermal and dynamical history of the Moon can be established, and it is even possible to follow in some detail its formation from a circumterrestrial ring of gas, dust, and planetesimals, it is hopeless to extrapolate to an earlier epoch to determine where that ring came from. This is similar to the cosmological skepticism that maintains we cannot know what happened before the Big Bang – it's not a scientific question. But most of the scientists mentioned in this book do not share that kind of skepticism.

The history of other branches of physical science suggests that the greatest advances in selenogony are likely to come from exploring the consequences of models that are plausible but may represent an extreme situation, rather than attempting to reflect all the complexities of the real world. A good model "should not incorporate all the physical factors that are believed to influence the system; instead, it must be kept simple enough to permit the accurate mathematical calculation of its properties" (Brush 1983: 274). Darwin's model for the tidal evolution of the lunar orbit (*after* it was either captured or ejected from the Earth) and the Safronov–Wetherill model of planetesimal accretion (as applied to selenogony by Ruskol and by Harris and Kaula) are good examples. Calculations on capture (e.g., Malcuit and Winters 1987) may be useful even though that theory is now out of favor. Moreover, the availability of fast large-memory computers has significantly raised the level of expectation: Little heed would now be paid to the kind of qualitative arm-waving argument that was common in all areas of planetary science a couple of decades ago. The Hartmann–Davis and Cameron–Ward impact models have to meet these new standards; so far it appears that they do have the potential to yield progress, whether or not they satisfy the "constraints" that the current generation of lunar scientists wants to impose.

Abbreviations

Note: Titles of the more frequently cited journals in the Reference List are abbreviated.

Am. J. Phys.	American Journal of Physics
Am. J. Sci.	American Journal of Science
Ann. Physics	Annals of Physics
Ann. Rev. Astron. Astrophys.	Annual Review of Astronomy and Astrophysics
Ann. Rev. Earth Plan. Sci.	Annual Review of Earth and Planetary Science
Arch. Hist. Exact Sci.	Archive for History of Exact Sciences
Ark. Mat. Astr. Fysik	Arkiv for Matematik, Astronomi och Fysik
Astron. J.	Astronomical Journal
Astron. Nachr.	Astronomische Nachrichten
Astron. Zh.	Astronomicheskii Zhurnal
Astrophys. J.	Astrophysical Journal
Astrophys. Space Sci.	Astrophysics and Space Science
B. A. Report	Report of the . . . Meeting of the British Association for the Advancement of Science
Biog. Mem. N. A. S.	Biographical Memoirs of the National Academy of Sciences, U. S. A.
Bull. Am. Astron. Soc.	Bulletin of the American Astronomical Society
Bull. Geol. Soc. Am.	Bulletin of the Geological Society of America
Bull. Nat. Res. Counc.	Bulletin of the National Research Council (U. S. A.)
Carn. Inst. Yb.	Carnegie Institution of Washington Yearbook
C-H	Cameron papers at Harvard University Archives
CIW	Archives, Carnegie Institution of Washington. Washington, DC.
C. R. Acad. Sci. Paris	Comptes Rendus hebdomadaires des Seances de l'Academie des Sciences, Paris
C-UC	T. C. Chamberlin Papers, Department of Special Collections, Joseph Regenstein Library, University of Chicago
DSB	Dictionary of Scientific Biography, edited by C. C. Gillispie. New York: Scribner's.
Dokl. Akad. Nauk	Doklady Akademii Nauk SSSR
Earth Plan. Sci. Lett.	Earth and Planetary Science Letters

Edinburgh New Phil. J.	*Edinburgh New Philosophical Journal*
Eos	*Eos. Transactions of the American Geophysical Union*
Geoch. Cosmoch. Acta	*Geochimica et Cosmochimica Acta*
Geol. Mag.	*Geological Magazine*
Geophys. J.	*Geophysical Journal*
Geophys. J. Roy. Astron. Soc.	*Geophysical Journal of the Royal Astronomical Society*
Geophys. Res. Lett.	*Geophysical Research Letters*
Gerlands Beitr.	*Beitrage zur Geophysik, Zeitschrift für physikalische Erdkunde* (ed. G. Gerland).
Hist. Stud. Phys. Sci.	*Historical Studies in the Physical [and Biological] Sciences*
Icarus	*Icarus: International Journal of Solar System Studies*
Io Triumphe	*Io Triumphe*, Albion College Alumni Magazine
Irish Astron. J.	*Irish Astronomical Journal*
Isis	*Isis, An International Review Devoted to the History of Science and Its Cultural Influences*
Izv. Acad. Sci. USSR, Phys. Solid Earth	*Izvestiya, Academy of Science of the USSR, Physics of the Solid Earth*
Izv. A. N. Fiz. Zemli	*Izvestiya Akademii Nauk SSSR, Fizika Zemli*
J. Br. Astron. Ass.	*Journal of the British Astronomical Association*
J. Geol.	*Journal of Geology*
J. Geol. Ed.	*Journal of Geological Education*
J. Hist. Astron.	*Journal for the History of Astronomy*
J. Roy. Astron. Soc. Canada	*Journal of the Royal Astronomical Society of Canada*
LOA	Lick Observatory Archives
Lunar [& Plan.] Sci.	*Lunar [and Planetary] Science*, Abstracts of papers to be presented at the Lunar [and Planetary] Science Conference. Houston: Lunar and Planetary Institute.
Mém. Acad. Montpellier	*Mémoires de l'Académie des Sciences et Lettres, Montpellier: Section des Sciences*
Mem. Soc. Astron. Ital.	*Memorie della Società Astronomica Italiana*, Nuova Serie
M. N. Roy. Astron. Soc.	*Monthly Notices of the Royal Astronomical Society*
M. N. Roy. Astron. Soc. G. S.	*Monthly Notices of the Royal Astronomical Society, Geophysical Supplement*
NASA-A	Archives, National Aeronautics and Space Administration, NASA Headquarters, Washington, DC.
Notes Rec. Roy. Soc.	*Notes and Records of the Royal Society of London*
NTM	*NTM, Schriftenreihe für Geschichte der Naturwissenschaften, Technik und Medizin*

Origin of the Moon Conf. Abst.	[Abstracts of] *Papers presented to the Conference on the Origin of the Moon,* Kona, Hawaii, 13–16 October 1984. Houston: Lunar and Planetary Institute.
Phil. Mag.	*London, Edinburgh, and Dublin Philosophical Magazine and Journal of Science*
Phil. Trans.	*Philosophical Transactions of the Royal Society of London*
Phys. Rev.	*Physical Review*
Plan. Space Sci.	*Planetary and Space Science*
Pop. Astr.	*Popular Astronomy, A Critical Review of Astronomy and Allied Sciences.*
Prob. Cosmog.	*Problems of Cosmogony.* Translation of *Vopr. Kosmog.* Washington, DC: Joint Publications Research Service.
Proc. Am. Acad.	*Proceedings of the American Academy of Arts and Sciences,* Boston
Proc. Am. Assoc.	*Proceedings of the American Association for the Advancement of Science*
Proc. Am. Phil. Soc.	*Proceedings of the American Philosophical Society,* Philadelphia
Proc. . . . . Lunar [& Plan.] Sci. Conf.	*Proceedings of the . . . . . Lunar [and Planetary] Science Conference*
Proc. N. A. S.	*Proceedings of the National Academy of Sciences of the U. S. A.*
Proc. Roy. Soc.	*Proceedings of the Royal Society of London*
Prog. Th. Phys.	*Progress of Theoretical Physics*
Pub. Astron. Soc. Japan	*Publications of the Astronomical Society of Japan*
Pub. Astron. Soc. Pac.	*Publications of the Astronomical Society of the Pacific*
Q. J. Roy. Astron. Soc.	*Quarterly Journal of the Royal Astronomical Society*
Rep. Prog. Phys.	*Reports on Progress in Physics*
R–PU	Henry Norris Russell Papers, Princeton University Library
SCI	*Science Citation Index*
Sci. Am.	*Scientific American*
Sky & Tel.	*Sky and Telescope*
Smithsonian Institution Report	*Annual Report of the Board of Regents of the Smithsonian Institution.* Washington, DC.
Solar System Research	*Solar System Research.* Translation of *Astronomicheskii Vestnik*
Sov. Astron. AJ	*Soviet Astronomy AJ.* Translation of *Astron. Zh.*
Sov. Phys. JETP	*Soviet Physics JETP.* Translation of *Zhurnal eksperimentalnoi i teoreticheski fiziki*
Stud. Hist. Phil. Sci.	*Studies in History and Philosophy of Science*
Trans. [& Proc.] N. Z. Inst.	*Transactions [& Proceedings] of the New Zealand Institute*

U–UCSD   Harold C. Urey Papers, Library, Uni-
                                             versity of California, San Diego, at La
                                             Jolla.

Vopr. Kosmog.   *Voprosy Kosmogonii*
Z. f. Astrophysik   *Zeitschrift für Astrophysik*
Z. Physik   *Zeitschrift für Physik*
Zh. eksp. teor. fiz.   *Zhurnal eksperimentalnoi i teoreticheskoi
                                             fiziki*

Reference list and citation index

Note: When two dates are given, the first is the date of original presentation or publication, the second is the date of the actual publication consulted. "D" before the year means that the item is an unpublished document available in an archive (or in a few cases, a public lecture). Numbers at the end of journal citations are: [series number, if any], volume number (issue number, if needed), inclusive pages (year, if different from original publication). Boldface numbers at the end of each reference indicate sections where that reference is cited.

Abbot, C. G. (1925) *The Earth and the Stars*. New York: Van Nostrand. **1.3.3**

Abdel-Gawad, M. *See* Blander, M.

Abell, George (1985) The origin of the cosmos and the Earth. In *What Darwin Began*, ed. L. R. Godfrey, pp. 223–42. Boston: Allyn & Bacon. **4.4.4**

Abell, George O., Morrison, David, & Wolff, Sidney C. (1987) *Exploration of the Universe*. 5th edn. Philadelphia: Saunders College Publishing. **4.1.6**

Achates, Fidus (pseud.) (1947) The Washington Moultons, Forest Ray, '94, and Harold Glenn, 1907. *Io Triumphe* [Albion College Alumni Magazine], **12**, no. 2, 19–22. **1.1.1**

Acuña, Mario H., & Ness, Norman F. (1976) The main magnetic field of Jupiter. *J. Geophys. Res.*, **81**, 2917–22. **3.4.6**

Adachi, Isao. *See* Hayashi, C.

Adami, Lanford H. *See* Friedman, I.

Adams, John Couch (1853) On the secular variation of the Moon's mean motion. *Phil. Trans.*, **143**, 397–406. **4.2.2**

(1859) On the secular variation of the eccentricity and inclination of the Moon's orbit. *M. N. Roy. Astron. Soc.*, **19**, 206–8. **4.2.2**

Africano, John L. *See* Worden, S. P.

Ahrens, T. J. (1994) The origin of the Earth. *Physics Today*, **47** (8), 38–45. **1.1**

Ahrens, T. J. *See also* Boslough, M.; Lange, M.; O'Keefe, J. D.; Rigden, S. M.

Airy, George Biddell (1866) On the supposed effect of friction in the tides, influencing the apparent acceleration of the Moon's mean motion in longitude. *M. N. Roy. Astron. Soc.*, **26**, 221–35. **4.2.2**

Aitken, R. G. (1906) The nebular hypothesis. *Pub. Astron. Soc. Pac.*, **18**, 111–22. **1.2.6**

Albee, A. L., Burnett, D. S., Chodos, A. A., Eugster, O. J., Huneke, J. C., Papanastassiou, D. A., Podosek, F. A., Russ, G. Price II, Sanz, H. G., Tera, F., & Wasserburg, C. J. (1970) Ages, irradiation history, and chemical composition of lunar rocks from the Sea of Tranquillity. *Science*, **167**, 464–6. **4.4.4**

Alden, William C. *See* Leith, C. K.

Alexander, F. C., Jr. *See* Srinivasan, B.

Alexander, Tom (1964) *Project Apollo: Man to the Moon*. London: Muller. **4.3.2**

Alfvén, Hannes (1942a) Remarks on the rotation of a magnetized sphere, with application to solar rotation. *Ark. Mat. Astr. Fysik*, **28A**, no. 6. **3.4.2**

(1942b) On the cosmogony of the Solar System. *Stock Obs. Ann.*, **14**, no. 2. **1.1.3, 3.4.2, 3.4.4**

(1943a) On the cosmogony of the Solar System. *Stock. Obs. Ann.*, **14**, no. 5.
                                                                    3.4.2
(1943b) Non-solar planets and the origin of the Solar System. *Nature*, **152**, 721.
                                                                    3.4.8
(1946) On the cosmogony of the Solar System. *Stock. Obs. Ann.*, **14**, no. 9.
                                                                    1.1.3, 3.4.2
(1954) *On the origin of the Solar System*. Oxford: Clarendon Press.
                                                             1.1.3, 3.2.1, 3.4.3
(1960a) Cosmical electrodynamics. *Am. J. Phys.*, **28**, 613–18.   **3.4.3**
(1960b) On the origin of the Solar System. *New Scientist*, **7**, 1188–91.

                                                                    3.4.2, 3.4.3
(1965) Origin of the Moon. *Science*, **148**, 476–77.   **4.3.2**
(1966) *Worlds – Antiworlds*. San Francisco: Freeman.   **3.4.8**
(1967a) Partial corotation of a magnetized plasma. *Icarus*, **7**, 387–93.   **3.4.4**
(1967b) On the origin of the Solar System. *Q. J. Roy. Astron. Soc.*, **8**, 215–26.
                                                                    3.4.1, 3.4.4
(1968a) Asteroidal jet streams. *Astrophys. Space Sci.*, **4**, 84–102.   **3.4.7**
(1968b) On the structure of the Saturnian rings. *Icarus*, **8**, 75–81.   **3.4.4**
(1970) Jet streams in space. *Astrophys. Space Sci.*, **6**, 161–74.   **3.4.7**
(1976/1978) Origin of the Solar System. In *The origin of the Solar System* (NATO
    Advanced Study Institute, 1976), ed. S. F. Dermott, pp. 19–40. New York: Wiley
    (pub. 1978).   **2.2.3**
(1981a) The plasma phase of the evolution of the Solar System. In *The Solar System
    and its evolution*, ed. W. R. Burke, pp. 3–10.   **2.2.4**
(1981b) The Voyager 1/Saturn encounter and the cosmogonic shadow effect. *Astro-
    phys. Space Sci.*, **79**, 491–505.   **3.4.4**
(1983) Solar System history as recorded in the Saturnian ring structure. *Astrophys.
    Space Sci.*, **97**, 79–94.   **3.4.4, 3.4.7**
Alfvén, Hannes, & Arrhenius, Gustaf (1970) Mission to an asteroid. *Science*, **167**, 139–
    41.   **3.4.7**
(1973) Structure and evolutionary history of the Solar System. III. *Astrophys. Space
    Sci.*, **21**, 117–176.   **1.1.4, 3.4.4**
(1975) *Structure and evolutionary history of the Solar System*. Dordrecht: Reidel.
                                                                    1.1.4, 3.4.8
(1976b) *Evolution of the Solar System*. Washington, DC: National Aeronautics and
    Space Administration, report NASA-SP-345.   **1.1.4, 2.3.5**
Alfvén, Hannes, Axnäs, I., Brenning, N., & Lindqvist, P.-A. (1986) *Voyager* Saturnian
    ring measurements and the early history of the Solar System. *Plan. Space Sci.*, **34**,
    145–54.   **3.4.4**
Allen, J. *See* Taylor, S. R.
Aller, L. H., O'Mara, B. J., & Little, S. (1964) The abundances of iron and silicon in
    the Sun. *Proc. N. A. S.*, **51**, 1238–43.   **3.3.3**
Aller, Lawrence H. *See also* Goldberg, Leo; Ross, J. E.
Alter, Dinsmore (1960) Evolution of the Moon. In *Proceedings of Lunar and Planetary
    Exploration Colloquium*, vol. 2, no. 2, pp. 1–6. Downey, CA: Space Sciences Lab-
    oratory, North American Aviation Inc. (also published in *Pub. Astron. Soc. Pac.*,
    73 [1961], 5–14).   **4.3.2**
Al'tshuler, L. V., Krupnikov, K. K., Ledenev, B. N., Zhuchikhin, V. I., & Brazhnik,
    M. I. (1958) Dynamic compressibility and equation of state of iron under high
    pressure. *Sov. Phys. JETP*, **7**, 606–14. Translated from *Zh. Eksp. Teor. Fiz*, **34**,
    874–85.   **4.3.2**
Amaftunsky, A. (1909) Kosmogoniya. *Izvestiya Russkago Astronomischekaga Obshestva*,
    15, 135–44 (cited in *Astronomische Jahresbericht*, 1909).   **1.2.4**
[Ampère, A. M.] Roulin (1833) Theorie de la terre d'après M. Ampère. *Revue des
    deux mondes* [2], **3**, 96–107. English translation in *Edinburgh New Phil. J.*, **18**
    (1835), 339–47.   **3.3.4**

Anders, Edward (1962/1963) Meteorites and the early history of the Solar System. In *Origin of the Solar System* (proceedings of conference held in 1962), ed. R. Jastrow & A. G. W. Cameron, pp. 95–142. New York: Academic Press (pub. 1963).
                                                                                    3.3.3
(1964) *Chemical fractionation in meteorites.* NASA report COO-382-47. Russian translation in *Meteoritika*, **26**, 17ff.   **2.5.1, 3.3.3**
(1968) Chemical processes in the early Solar System, as inferred from meteorites. *Accounts of Chemical Research*, **1**, 289–98.   **3.3.3, 4.3.3**
(1971) Meteorites and the early Solar System. *Annual Review of Astronomy and Astrophysics*, **9**, 1–34.   **3.3.3, 3.3.5, 4.3.4**
(1977) Chemical compositions of the Moon, Earth and eucrite parent body. *Phil. Trans.*, A225, 23–40.   **4.4.4**
(1978) Procrustean science: Indigenous siderophiles in the lunar highlands, according to Delano and Ringwood. *Proc. 9th Lunar & Plan. Sci. Conf.*, 161–84.
                                                                                    4.4.4
Anders, Edward, Ganapathy, R., Keays, R. R., Laul, J. C., & Morgan, J. W. (1971) Volatile and siderophile elements in lunar rocks: Comparison with terrestrial and meteoritic basalts. *Proc. 2nd Lunar Sci. Conf.*, 1021–36.   **4.1.4, 4.4.3**
Anders, Edward, Ganapathy, R., Krähenbühl, Urs, & Morgan, John W. (1973) Meteoritic material on the Moon. *Moon*, **8**, 3–24.   **4.4.3**
Anders, Edward, & Heymann, Dieter (1969) Elements 112 to 119: Were they present in meteorites? *Science*, **164**, 821–3.   **2.5.1**
Anders, Edward. *See also* Barker, J. L., Jr.; DuFresne, E. R.; Fish, R. A.; Ganapathy, R.; Gros, Jacques; Hertogen, J.; Keays, R. R.; Larimer, J. W.; Lewis, R. S.; Morgan, J. W.; Srinivasan, B.; Vizgirda, J.; Wolf, R.
Anderson, Don L. (1972) The origin of the Moon. *Nature*, **239**, 263–5.   **4.4.4**
(1973) The composition and origin of the Moon. *Earth Plan. Sci. Lett.*, **18**, 301–16.
                                                                                    4.4.4
(1989) Composition of the Earth. *Science*, **243**, 367–70.   **3.3.3**
Anderson, Don L. *See* Ringwood, A. E.
Anderson, Herbert H. *See* Valley, G. E.
André, C. (1912) L'hypothèse nebulaire de Laplace et la théorie de la capture de M. T. J. J. See. *Scientia*, **11**, 153–69.   **1.2.4**
Andrews, D. J. (1982) Could the Earth's core and Moon have formed at the same time? *Geophys. Res. Lett.*, **9**, 1259–62.   **4.4.4**
Annell, C. S. *See* Cuttita, F.
Appenzeller, I., & Dearborn, D. S. P. (1984) Brightness variations caused by surface magnetic fields in pre-main-sequence stars. *Astrophys. J.*, **278**, 689–94.   **2.3.5**
Appenzeller, I., & Jordan, C., Editors (1987) *Circumstellar Matter* (Proceedings of International Astronomical Union Symposium, no. 122, Heidelberg, June 1986). Boston: Reidel.   **2.1.2**
Armellini, G. (1937) I problemi fondamentali della cosmogonia e la legge di Newton. *Rendiconti della R. Accademia dei Lincei, Classe di Scienze Fisiche, Matematiche a Naturali, Atti* [6], **25**, 209–15.   **1.3.5**
Arnett, W. David, & Wefel, John P. (1978) Aluminum-26 production from a stellar evolutionary sequence. *Astrophys. J. Lett.*, **224**, L139–42.   **2.5.3**
Arnold, J. R. (1969) Asteroid families and "jet streams." *Astron. J.*, **74**, 1235–42.
                                                                                    3.4.7
(1980) Condensation in the early Solar System. In *Early Solar System processes and the present Solar System*, ed. D. Lal, pp. 140–43. New York: North-Holland.
                                                                                    3.3.5
Arnold, J., Arrhenius, G., Eglinton, G., Frondel, C., Gast, P. W., MacGregor, I., Pepin, R., Strangway, D., Walker, R., Wasserburg G., & Zill, P. (1970) Summary of Apollo 11 Lunar Science Conference. *Science*, **167**, 449–51.   **4.4.3**

Arnould, M., Nørgaard, H., Thielemann, F.-K., & Hillebrandt, W. (1980) Synthesis of ^{26}Al in explosive hydrogen burning. *Astrophys. J.*, **237**, 931–50.   **2.5.3**

Arrhenius, Gustav O. S. (1962) Svante Arrhenius' contribution to Earth science and cosmology. *Levnadsteckningar över K. Svenska Vetenskapsakademiens ledamöter*, **157**, 345–59.   **1.2.4**

(1972/1974) Reconstruction of the history of the Solar System. In *On the origin of the Solar System* (proceedings of 1972 symposium), ed. H. Reeves, pp. 80–88. Paris: CNRS (pub. 1974).   **2.2.3**

Arrhenius, Gustav O. S., and De, B. R. (1973) Equilibrium condensation in a solar nebula. *Meteoritics*, **8**, 297–313.   **3.3.4**

Arrhenius, Gustaf *See also* Alfvén, H.; Arnold, J.; Brecher, A.

Arrhenius, Svante (1896) On the influence of carbonic acid in the air upon the temperature of the ground. *Phil. Mag.* [5], **41**, 231–76.   **1.2.4**

(1901) Zur Kosmogonie. *Archives Néerlandaises des Sciences Exactes et Naturelles* [2], **6**, 862–73.   **1.2.4**

(1903) *Lehrbuch der Kosmischen Physik*. Leipzig.   **1.2.4**

(1908) *Worlds in the making*. New York: Harper.   **1.2.4**

Artem'ev, A. V. (1969) Planetary rotation induced by elliptically orbiting particles. *Solar System Research*, **3**, 15–21. Translated from *Astronomicheskii Vestnik*, **3**, 18–25.   **1.2.3**

Artem'ev, A. V., & Radzievsky, V. V. (1965) The origin of the axial rotation of the planets. *Sov. Astron. AJ*, **9**, 96–99. Translated from *Astron. Zh.*, **42**, 124–8.

1.2.3, 3.1.1

Ashbrook, Joseph (1962) Astronomical scrapbook: The sage of Mare Island. *Sky & Tel.*, **24**, 193–202.   **1.2.4, 4.2.4**

Assousa, George E. *See* Herbst, W.

Audouze, J. *See* Reeves, H.; Vangioni-Flam, E.

Aumann, H. H., Gillespie, C. M., Jr., & Low, F. J. (1969) The internal powers and effective temperatures of Jupiter and Saturn. *Astrophys. J.*, **157**, L69–L72.

3.2.1

Aumann, H. H., Gillett, F. C., Beichman, C. A., De Jong, T., Houck, J. R., Low, F. J., Neugebauer, G., Walker, R. G., & Wesselius, P. R. (1984) Discovery of a shell around Alpha Lyrae. *Astrophys. J.*, **278**, L23–L27.   **2.1.2**

Axnäs, I. *See* Alfvén, H.; Petelski, E. F.

Babinet, Jacques (1861) Note sur un point de la cosmogonie de Laplace. *C. R. Acad. Sci. Paris*, **52**, 481–4.   **1.2.4**

Badash, Lawrence (1975) Rutherford, Ernest. *D. S. B.*, **12**, 12–36.   **1.2.1**

Baddenhausen, Hildegard *See* Wänke, H.

Baedecker, Philip A., Chou, Chen-Lin, Sundberg, L. L., & Wasson, John T. (1974) Volatile and siderophile trace elements in the soils and rocks of Taurus-Littrow. *Proc. 5th Lunar Sci. Conf.*, 1625–43.   **4.4.3**

Baedecker, Philip A., Chou, C.-L., & Wasson, J. T. (1972) The extra-lunar component in lunar soils and breccias. *Proc. 3rd Lunar Sci. Conf.*, 1343–59.   **4.4.3**

Baedecker, Philip A. *See* Wasson, J. T.

Baines, K. H., Schempp, W. V., & Smith, W. H. (1984) High-resolution observations of the 6815-Å band of methane in the major planets. *Icarus*, **56**, 534–42.

3.2.3

Baker, Howard B. (1954) The Earth participates in the evolution of the Solar System. *Occasional Papers of the Detroit Academy of Natural Sciences*, no. 3.   **4.1.5, 4.4.5**

Baker, James G. (1974) *On the origin of irregular satellites. I. Early events in the inner Solar System*. El Segundo, CA: Aerospace Corporation, Report ATR-74(8126). Foreword by D. H. Menzel.   **4.1.5, 4.4.5**

Balacesa, A. *See* Wänke, H.

Baldwin, Ralph B. (1965) *A fundamental survey of the Moon*. New York: McGraw-Hill.

4.3.3

Baldwin, Ralph B., & Wilhelms, Don E. (1992) Historical review of a long-over-looked paper by R. A. Daly concerning the origin and early history of the Moon. *J. Geophys. Res.*, **97**, 3837–43.   **4.4.5**

Ball, Robert Stawell (1881) A glimpse through the corridors of time: Origin and development of the Moon from the data of tidal evolution. *Nature*, **25**, 79–82, 103–107.   **4.2.3**

(1882a) Birth of the Moon by tidal evolution. *Knowledge*, **1**, 331–2, 352–3.   **4.2.3**

(1882b) On the occurrence of great tides since the commencement of the geological epoch. *Nature*, **27**, 201–203.   **4.2.3**

(1889) *Time and tide, a romance of the Moon.* London: Society for Promoting Christian Knowledge.   **4.2.3**

(1901) *The Earth's beginning.* London: Cassell.   **1.2.2**

Baluteau, J. P. *See* Gautier, D.

Banerjee, S. K. (1967) Fractionation of iron in the solar system. *Nature*, **216**, 781.   **3.3.3, 4.1.3**

Banerjee, S. K., & Hargraves, R. B. (1971) Natural remanent magnetization of carbonaceous Chondrites. *Earth Plan. Sci. Lett.*, **10**, 392–6.   **2.3.5**

(1972) Natural remanent magnetizations of carbonaceous chondrites and the magnetic field in the early solar system. *Earth Plan. Sci. Lett.*, **17**, 110–19.   **2.3.5**

Banerjee, S. K., & Mellema, J. P. (1976) A solar origin for the large magnetic field at 4.0 × 10^9 yr age. *Proc. 7th Lunar Sci. Conf.*, 3259–70.   **2.3.5**

Barker, J. L., Jr., & Anders, E. (1968) *Geoch. Cosmoch. Acta*, **32**, 627–45.   **4.4.3**

Bar-Matthews, Maryam. *See* MacPherson, G. J.

Barrell, Joseph (1907) Review of "The place of origin of the Moon – The volcanic problem" by W. H. Pickering (*J. Geol.*, 15, 22–38). *J. Geol.*, **15**, 503–7.   **4.2.4**

Barshay, Stephen S., & Lewis, John S. (1976) Chemistry of primitive solar material. *Ann. Rev. Astron. Astrophys.*, **14**, 81–94.   **4.4.4**

Barshay, Stephen S. *See also* Goettel, K. A.; Lewis, J. S.

Baschek, B. *See* Garz, T.

Bauer, H. H. (1985) *Beyond Velikovsky: The history of a public controversy.* Urbana: University of Illinois Press.   **4.1.5, 4.4.5**

Baxter, D. C., & Thompson, W. B. (1971). Jetstream formation through inelastic collisions. In *Physical Studies of Minor Planets* (IAU Colloquium No. 12), ed. T. Gehrels, pp. 319–26. Washington, DC: National Aeronautics and Space Administration, report NASA SP-267.   **3.4.7**

(1973) Elastic and inelastic scattering in orbital clustering. *Astrophys. J.*, **183**, 323–36.   **3.4.7**

Beachey, Ann E. *See* Kaula, W. M.

Beattie, E. H. (1912) Some thoughts on stellar impact. *J. Brit. Astron. Assoc.*, **22**, 378–81.   **1.2.1**

Beatty, J. Kelly (1979) The far-out worlds of Voyager 1, I. *Sky & Tel.*, **57**, 423–7.   **3.4.6**

Becher, Erich (1915) *Weltentwicklung: Ein Bild der unbelebten Natur.* Berlin: Reimer.   **1.2.4**

Becker, George F. (1898) Kant as a natural philosopher. *Am. J. Sci.* [4], **5**, 97–112.   **1.2.2, 1.2.5**

(1904) Present problems of geophysics. *Science*, **20**, 545–56.   **1.2.5**

Beebe, Reta F. *See* Smith, B. A.

Beesley, D. E. *See* De Groot, M.

Begemann, F. (1980) Isotopic anomalies in meteorites. *Rep. Prog. Phys.*, **43**, 1309–56.   **2.5**

Beichman, C. A. *See* Aumann, H. H.

Bennett, D. See Taylor, S. R.

Benz, W., Cameron, A. G. W. & Slattery, W. L. (1987) Planetary collision calculations: Origin of the Moon. *Lunar & Plan. Sci.*, **18**, 60–1.   **4.4.5**

Benz, W., Slattery, W. L., & Cameron, A. G. W. (1986) Short note: Snapshots from a three-dimensional modeling of a giant impact. In *Origin of the moon*, ed. W. K. Hartmann et al., pp. 617–20. Houston: Lunar and Planetary Institute.   **4.4.5**

(1987) The origin of the Moon and the single-impact hypothesis. II. *Icarus*, **71**, 30–45.   **4.4.5**

Berendzen, Richard, & Hart, Richard (1973) Adriaan van Maanen's influence on the island universe theory. *J. Hist. Astron.*, **4**, 45–56, 73–98.   **1.2.6**

Berendzen, Richard, Hart, Richard, & Seeley, Daniel (1976) *Man discovers the galaxies*. New York: Science History Publications.   **1.2.6**

Berget, A. (1904) *Physique du globe et météorologie*. Paris: Naud.   **1.2.2**

Bernas, R., Gradsztajn, E., Reeves, H., & Schatzman, E. (1967) On the nucleo-synthesis of lithium, beryllium, and boron. *Ann. Physics*, **44**, 426–78.   **2.3.4**

Bernas, R., Gradsztajn, E., Reeves, H., Yiou, F., & Schatzman, E. (1968) Sur la nucleosynthese du lithium, du beryllium et du bore et la formation du Système Solaire. In *Origin and distribution of the elements* (symposium, Paris, 1967), ed. L. H. Ahrens, pp. 111–16. New York: Pergamon Press (pub. 1968).   **2.3.4**

Berry, Arthur (1898/1961) *A short history of astronomy from earliest times through the nineteenth century*. London: Murray (1898). Reprint, New York: Dover Publications (1961).   **1.2.2, 4.2.2**

Berry, H. See Compston, W.

Bertout, Claude. See Bouvier, J.

Bethe, Hans A. (1983) The 1983 Nobel Prize in Physics. *Science*, **222**, 881–83.
                                                                                            2.3.2

Bézard, B. See Courtin, R.; Gautier, D.

Bickerton, Alexander William (1880) On the causes tending to alter the eccentricity of planetary orbits. *Trans. & Proc. N. Z. Inst.*, **13**, 149–54, 154–9.   **1.2.1**

(1894) Some recent evidence in favour of impact. *Trans. N. Z. Inst.*, **26**, 464–76 (iv) & plate LII.   **1.2.1, 1.2.4**

(1901) *The romance of the heavens*. London: Swan Sonnenschein.   **1.2.1**

(1904) Explosion of stars. *Knowledge* (n.s.) **1**, 261.   **1.2.4**

(1910) [Discussion of his theory] *J. Br. Astron. Assoc.*, **21**, 139–47.   **1.2.1**

(1911a) [Report of his remarks] *J. Br. Astron. Assoc.*, **21**, 139–42.   **1.2.4**

(1911b) The new astronomy. *Knowledge* (n.s.), **8**, 362–7, 379–83, 413–18, 453–9.
                                                                                            1.2.4

(1911c) The probability of solar grazes and the phenomena of suns plunging into nebulae. *J. Br. Astron. Assoc.* **21**, 195–6.   **1.2.1**

(1915) [Discussion remarks on Stromeyer's paper on Saturn's rings] *J. Br. Astron. Assoc.*, **25**, 114–15.   **1.2.1**

Binder, Alan B. (1974) On the origin of the Moon by rotational fission. *Moon*, **11**, 53–76.   **4.3.4**

(1975) On the petrology and structure of a gravitationally differentiated Moon of fission origin. *Moon*, **13**, 431–73.   **4.3.4**

(1977) Fission origin for the Moon: Accumulating evidence. *Lunar Sci.*, **8**, 118–20.
                                                                                            4.3.4

(1978a) The first few hundred years of evolution of a Moon of fission origin. *Proc. 11th Lunar Plan. Sci. Conf.*, 1931–9.   **4.4.4**

(1978b) On fission and the devolatilization of a Moon of fission origin. *Earth Plan. Sci. Lett.*, **41**, 381–5.   **4.4.3, 4.4.4**

(1980) On the internal structure of a Moon of fission origin. *J. Geophys. Res.*, **85**, 4872–80.   **4.4.4**

(1986) The binary fission origin of the Moon. In *Origin of the Moon*, ed. W. K. Hartmann et al., 499–516. Houston: Lunar and Planetary Institute.   **4.4.4**

Binder, Alan B., & Lange, Manfred (1977) On the thermal history of a Moon of fission origin. *Moon,* **17,** 29–45. 4.4.4

(1980) On the thermal history, thermal state, and related tectonism of a Moon of fission origin. *J. Geophys. Res.,* **85,** 3194–208. 4.4.4

Birch, P. V. *See* Millis, R. L.

Bird, R. B. See Hirschfelder, J. O.

Birkeland, K. (1912) Sur l'origine des planètes et de leurs satellites. *C. R. Acad. Sci. Paris,* **155,** 892–5. 1.2.4

(1913a) De l'origine des mondes. *Archives des Sciences* [4], **35,** 529–64. 1.2.4

(1913b) The origin of worlds. *Sci. Am. Supplement,* **76,** no. 1, 1, 8–9, 12, 20–1, 32. 1.2.4

Black, D. C. (1972) On the origins of trapped Helium, Neon and Argon isotopic variations in meteorites. *Geoch. Cosmoch. Acta,* **36,** 347–75, 377–94. 2.5.1

Black, D. C., & Bodenheimer, P. (1976) Evolution of rotating interstellar clouds. II. The collapse of protostars of 1, 2, and 5$M_\odot$. *Astrophys. J.,* **206,** 138–49. 2.4.1

Black, D. C., Levy, E. H., Cohen, M., McCarthy, D. W., Jr., Harrington, R. S., Stein, J. W., McMillan, R. S., & Russell, J. (1984) In search of other worlds. *Planetary Report,* **4,** no. 5. 2.1.2

Black, David C., & Matthews, M. S., Editors (1985) *Protostars and planets. II.* Tucson: University of Arizona Press. 2.2.5, 3.4.8

Black, David C. *See also* Bodenheimer, P.; Pollack, J. B.

Blackwell, D. E., Booth, A. J., & Petford, A. D. (1984) Is there an abundance anomaly for the 2.2 eV FeI lines in the solar spectrum? *Astron. Astrophys.,* **132,** 236–9. 3.3.3

Blamont, J. (1980/1982) A description of the atmosphere of Venus. In *Formation of planetary systems* (proceedings of 1980 conference), ed. A. Brahic, pp. 741–82. Toulouse: Cepadues-Editions (pub. 1982). 3.3.5

Blanchard, Doug (1985) [quoted in "Q & A column"] *New York Times,* 26 November, p. C9. 4.4.4

Blander, M., & Abdel-Gawad, M. (1969) The origin of meteorites and the constrained equilibrium condensation theory. *Geoch. Cosmoch. Acta,* **33,** 701–16. 3.3.4

Blander, M., & Katz, J. L. (1967) Condensation of primordial dust. *Geoch. Cosmoch. Acta,* **31,** 1025–34. 3.3.4

Bleakney, Walker *See* Manian, S. H.

Blum, K. *See* Wänke, H.

Bodenheimer, Peter (1974) Calculations of the early evolution of Jupiter. *Icarus,* **23,** 319–25. 3.2.1, 3.2.3

(1976) Contraction models for the evolution of Jupiter. *Icarus,* **29,** 165–71. 3.2.1

(1977) Calculations of the effect of angular momentum on the early evolution of Jupiter. *Icarus,* 31, 356–68. 3.2.1

(1982) Origin and evolution of the giant planets. In *The Comparative Study of the Planets,* ed. A. Coradini & M. Fulchignoni, pp. 25–48. Boston: Reidel. 3.2.2

Bodenheimer, Peter, & Black, David C. (1978) Numerical calculations of protostellar hydrodynamic collapse. In *Protostars and Planets,* ed. T. Gehrels, pp. 288–322. Tucson: University of Arizona Press. 2.4.1

Bodenheimer, Peter, Grossman, Allen S., DeCampli, William M., Marcy, Geoffrey, & Pollack, James. B. (1980) Calculations of the evolution of the giant planets. *Icarus,* 41, 293–308. 3.2.1

Bodenheimer, Peter. *See* Black, D. C.; Cameron, A. G. W.; Stewart, G. R.

Bogard, D. D., Nyquist, L. E., & Johnson, P. H. (1984) Noble gas contents of shergottites and implications for the Martian origin of meteorites. *Geoch. Cosmoch. Acta,* **48,** 1723–9. 4.4.4

Boischot, A. *See* Cuzzi, J. N.
Bonsack, Walter K., & Greenstein, Jesse L. (1960). The abundance of lithium in T Tauri stars and related objects. *Astrophys. J.,* **131,** 83–98.   2.3.1
Booth, A. J. *See* Blackwell, D. E.
Boslough, M., & Ahrens, T. (1983) Shock-melting and vaporization of anorthosite and implications for an impact-origin of the Moon. *Lunar & Plan. Sci.,* **14,** 63–9.
          4.4.5
Boss, Alan P. (1985) Protoearth mass shedding and the origin of the Moon. *Bull. Am. Astron. Soc.,* **17,** 713.   4.4.5
    (1986a) The origin of the Moon. *Science,* **231,** 341–5.   4.4.5
    (1986b) Protoearth mass shedding and the origin of the Moon. *Icarus,* **66,** 330–40.
          4.4.5
    (1988a) Terrestrial versus giant planet formation. In *Workshop on the origins of Solar Systems,* ed. J. A. Nuth & P. Sylvester, pp. 61–8. Houston: Lunar and Planetary Institute, Technical Report 88–04.   3.2.2
    (1988b) High temperatures in the early solar nebula. *Science,* **241,** 565–7.   3.3.5
    (1989) Low-mass star and planet formation. *Pub. Astron. Soc. Pac.,* **101,** 767–86.
          1.1.5, 3.2.2
Boss, Alan P., & Mizuno, H. (1985) Dynamic fission instability of dissipative proto-planets. *Icarus,* **63,** 134–52.   4.4.5
Boss, Alan P., & Peale, S. J. (1986) Dynamical constraints on the origin of the Moon. In *Origin of the Moon,* ed. W. K. Hartmann et al., pp. 59–101. Houston: Lunar and Planetary Institute.   4.4.5
Boss, Alan P. *See also* Mizuno, H.
Bouvier, Jerome, & Bertout, Claude (1986) Evidence for magnetic activity in T Tauri stars. In *Cool stars, stellar systems, and the Sun* (Proceedings of the Fourth Cambridge Workshop, at Santa Fe, NM, October 16–18, 1985), ed. M. Zeilik & D. M. Gibson, pp. 132–4. New York: Springer.   2.3.5
Bowen, Robert (1992) *Science and ideology in the Nazi state.* London: Belhaven.
          4.2.4
Bowie, William (1929) Possible origin of oceans and continents. *Gerlands Beitr.,* **21,** 178–82.   4.2.4
    (1930) Crustal changes due to Moon's formation. *Gerlands Beitr.,* **25,** 137–44.
          4.2.4
Boyce, Joseph. *See* Smith, B. A.
Boynton, W. V. (1975) Fractionation in the solar nebula: Condensation of yttrium and the rare earth elements. *Geoch. Cosmoch. Acta,* **39,** 569–84.   3.3.5
    (1985) Meteoritic evidence concerning conditions in the solar nebula. In *Protostars and planets.* II, ed. D. C. Black & M. S. Matthews, pp. 772–87. Tucson: University of Arizona Press.   2.2.3, 3.4.8
Brahic, Andre (1975) A numerical study of a gravitating system of colliding particles: Applications to the dynamics of Saturn's rings and to the formation of the Solar System. *Icarus,* **25,** 452–8.   3.4.7
    Editor (1982) *Formation of planetary systems.* Toulouse: Cepadues.   2.2.5
Brazhnik, M. I. *See* Al'tshuler, L. V.
Brecher, Aviva (1971) Early interplanetary magnetic fields and the remanent magnetization in meteorites. *Pub. Astron. Soc. Pac.,* **83,** 602–3.   2.3.5
    (1973) On the primordial condensation and accretion environment and the remanent magnetization of meteorites. In *Evolutionary and physical properties of meteoroids* (IAU Colloquium No. 13, Albany, June 1971), ed. C. L. Hemenway et al., pp. 311–29. Washington, DC: National Aeronautics and Space Administration, Report NASA SP-319.   2.3.5
Brecher, Aviva, & Arrhenius, Gustaf (1974) The paleomagnetic record in carbonaceous chondrites: Natural remanence and magnetic properties. *J. Geophys. Res.,* **79,** 2081–106.   2.3.5

Breneman, H. H., & Stone, E. C. (1985) Solar coronal and photospheric abundances from solar energetic particle measurements. *Astrophys. J.*, **299**, L57–L61.
                                                                                                    3.3.3
Brenning, N. *See* Alfvén, H.; Petelski, E. F.
Brett, Robin (1994) A. E. Ringwood: A remembrance. *Eos*, **75**, 323, 329.   3.3.2
Briggs, Geoffrey A. *See* Smith, B. A.
Brooks, Courtney G., Grimwood, James M., & Swenson, Loyd S., Jr. (1979) *Chariots for Apollo: A history of manned lunar spacecraft*. Washington, DC: National Aeronautics and Space Administration.   4.3.1
Brown, G. Malcolm (1976/1978) Chemical evidence for the origin, melting, and differentiation of the Moon. In *The origin of the solar system* (NATO Advanced Study Institute, 1976), ed. S. F. Dermott, pp. 597–609. New York: Wiley (pub. 1978).   4.3.4
Brown, Harrison (1949) A table of relative abundances of nuclear species. *Rev. Mod. Phys.*, **21**, 625–34.   1.1.3
        (1950a) The composition of our universe. *Physics Today*, **3**, no. 4 (April), 6–13.
                                                                                                    1.1.5
        (1950b) The composition and structure of the planets. *Astrophys. J.*, **111**, 641–53.
                                                                                                    1.1.5
Brown, Wilber K. (1970) *A model for formation of solar systems from massive supernova fragments*. Los Alamos: Los Alamos Scientific Laboratory, Report LA-4343.
                                                                                                    2.5.2
        (1971a) *Particle collisions in the opposing mass flow hypothesis: A facet of the supernova fragmentation model of solar system formation* Los Alamos: Los Alamos Scientific Laboratory, Report LA-4343, Supplement.   2.5.2
        (1971b) A Solar System formation model based on supernova shell fragmentation. *Icarus*, **15**, 120–34.   2.5.2
        (1974) A unified model of Solar System and galaxy formation based on explosive fragmentation. Los Alamos National Laboratory, report LA-5364.   2.5.2
Brownlee, Shannon (1985) A whacky theory of the Moon's birth. *Discover*, **6**, no. 3 (March), 64–9.   4.4.5
Bruno, Leonard C. (1979) *We have a sporting chance: The decision to go to the Moon*. Washington, DC: Library of Congress.   4.3.1
Brush, Stephen G. (1972) *Kinetic theory, III. The Chapman–Enskog solution of the transport equation for moderately dense gases*. New York: Pergamon Press.   3.1.1
        (1974) Should the history of science be rated X? *Science*, **183**, 1164–72.   2.2.4
        (1976) *The kind of motion we call heat: A history of the kinetic theory of gases in the 19th century*. Amsterdam: North Holland.   **1.1.3, 1.1.4, 1.2.3, 1.3.5, 3.1.1**
        (1979a) Scientific revolutionaries of 1905: Einstein, Rutherford, Chamberlin, Wilson, Stevens, Binet, Freud. In *Rutherford and physics at the turn of the century*, ed. Mario Bunge & William R. Shea, pp. 140–71. New York: Science History Publications.   **1.1.1, 1.2.2**
        (1979b) Looking up: The rise of astronomy in America. *American Studies*, **20**, no. 2, 41–67.   1.2.6
        (1982a) Finding the age of the Earth: By physics or by faith? *J. Geol. Ed.*, **30**, 34–58.   2.1.2
        (1982b) Nickel for your thoughts: Urey and the origin of the Moon. *Science*, **217**, 891–8.   **4.3.4, 4.4.7**
        (1983) *Statistical physics and the atomic theory of matter, from Boyle and Newton to Landau and Onsager*. Princeton: Princeton University Press.   **1.3.5, 4.4.7**
        (1990) Prediction and theory evaluation: Alfvén on space plasma phenomena. *Eos*, **71**, 19–33.   **2.2.2, 3.4.3, 3.4.5**
        (1992) Alfvén's programme in Solar System physics. *IEEE Transactions on Plasma Science*, **20**, 577–89.   3.4.5

(1993) Prediction and theory evaluation: Cosmic microwaves and the revival of the big bang. *Perspectives on Science*, **1**, 565–602.   **2.2.2**

(1995) Dynamics of theory change in science: The role of prediction. *PSA 1994*, **2**, 133–45.   **2.2.2**

Brush, Stephen G., & Landsberg, H. E. (1985) *The history of geophysics and meteorology: An annotated bibliography.* New York: Garland.   **2.1**

Brush, Stephen G. *See also* Levin, A. E.

Bryan, G. H. (1899) On the permanence of certain gases in the atmospheres of planets. *B. A. Report*, **69**, 634–63.   **1.2.3**

(1900) The kinetic theory of planetary atmospheres. *Nature*, **62**, 126.   **1.2.3**

Bryant, Walter W. (1907) *A history of astronomy.* London: Methuen.   **1.2.6**

Buchanan, Daniel (1916) The fallacy of the nebular hypothesis: Celestial mechanics and other facts that demand a new theory. *Sci. Am. Supplement*, **81**, 402–403. Reprinted from *Queens Quarterly*.   **1.2.6**

Bullen, K. E. (1951) Origin of the Moon. *Nature*, **167**, 29.   **4.3.2**

(1967) The origin of the Moon, IV. In *Mantles of the Earth and terrestrial planets*, ed. S. K. Runcorn, pp. 261–4. New York: Interscience.   **4.3.2**

(D1952a) Letter to H. C. Urey, 3 March 1952. Box 11, folder "Bullen, K. E.," U-UCSD.   **4.4.1**

(D1952b) Letter to H. C. Urey, 21 March 1952. Box 11, folder "Bullen, K. E.," U-UCSD.   **4.4.1**

(D1952c) Letter to H. C. Urey, 9 July 1952. Box 11, folder "Bullen, K. E.," U-UCSD.   **4.4.1**

Burbidge, E. M., Burbidge, G. R., Fowler, W. A., & Hoyle, F. (1957) Synthesis of the elements in stars. *Rev. Mod. Phys.*, **29**, 547–650.   **1.1.5, 1.3.4, 3.3.3**

Burbidge, G. R. *See* Burbidge, E. M.

Burnett, D. S., Fowler, W. A., & Hoyle, F. (1965) Nucleosynthesis in the early history of the solar system. *Geoch. Cosmoch. Acta*, **29**, 1209–41.   **2.3.4**

Burnett, D. S., Lippolt, H. J., & Wasserburg, G. J. (1966) The relative isotopic abundance of K^{40} in terrestrial and meteoritic samples. *Journal of Geophysical Research*, **71**, 1249–69.   **3.3.4**

Burns, G. J. (1911) The grazing impact theory. *J. Br. Astron. Ass.*, **21**, 160–1.
                                                                                        1.2.1

Burns, J. A. (1973) Where are the satellites of the inner planets? *Nature Physical Science*, **242**, 23–5.   **4.4.6**

(1975) The angular momentum of Solar System bodies: Implications for asteroid strengths. *Icarus*, **25**, 545–54.   **3.1.2**

(1977) Orbital evolution. In *Planetary Satellites*, ed. J. A. Burns, pp. 113–55. Tucson: University of Arizona Press.   **4.3.3, 4.4.4**

Burns, J. A., Showalter, M. R., & Cuzzi, J. N. (1980) Physical processes in Jupiter's ring: Clues to its origin by Jove! *Icarus*, **44**, 339–60.   **3.4.6**

Burns, J. A., Showalter, M. R., & Morfill, G. E. (1984) The ethereal rings of Jupiter and Saturn. In *Planetary Rings*, ed. R. Greenberg & A. Brahic, pp. 200–72. Tucson: University of Arizona Press.   **3.4.6**

Burstyn, Harold L. (1971) Ferrel, William. DSB, **4**, 590–3.   **4.2.2**

Busco, Pierre (1912) *L'evolution de l'astronomie au XIXe siècle.* Paris: Larousse.
                                                                                        1.2.4

Cabot, W., Canuto, V. M., Hubickyj, O., & Pollack, J. B. (1987) The role of turbulent convection in the primitive solar nebula. *Icarus*, **69**, 387–457.   **2.4.2**

Cadogan, Peter (1981) *The Moon – Our sister planet.* Cambridge University Press.
                                                                                        4.4.4

Caffee, M. W. *See* Swindle, T. D.

Cameron, A. G. W. (1960/1962) The origin of the solar system. (Presented at the National Academy of Science–National Research Council meeting, 20–2 June

1960) *Publications of Goddard Space Flight Center, 1959–1962*, **1**, 39–42.

2.4.1, 3.1

(1962a) The early chronology of the solar system. National Aeronautics and Space Administration, Technical Note D-1465. Reprinted in *Publications of Goddard Space Flight Center, 1959–1962*, 1. **2.4.1**

(1962b) The formation of the sun and planets. *Icarus*, **1**, 13–69.

1.1.5, 2.3.1, 2.3.4, 2.4.1, 2.5.1, 2.5.2, 3.1.2, 3.3.3

(1963b) The origin of the Solar System. In *Interstellar communication*, ed. A. G. W. Cameron, pp. 21–29. New York: Benjamin. **2.3.4, 2.3.5, 2.4.1**

(1963c) The origin of the atmospheres of Venus and the Earth. *Icarus*, **2**, 249–57.

4.1.3, 4.3.3

(1963d) The formation of the solar nebula. In *Origin of the Solar System*, ed. R. Jastrow & A. G. W. Cameron, pp. 85–94. New York: Academic Press.

1.1.5, 2.4.1, 3.3.3

(1964/1966) Planetary atmospheres and the origin of the moon. In *The Earth–Moon system* (conference, January 1964), ed. B. G. Marsden & A. G. W. Cameron, pp. 234–73. New York: Plenum Press (pub. 1966). **2.4.1, 4.3.3**

(1965) Origin of the Solar System. In *Introduction to space science*, edited by W. N. Hess, pp. 553–84. New York: Gordon & Breach. **1.3.4, 2.3.4, 2.4.1**

(1966) The early history of the sun. *Smithsonian Miscellaneous Collections*, **151**, no. 6.

1.1.5

(1969a) Physical conditions in the primitive solar nebula. In *Meteorite research* (proceedings of a symposium held in Vienna, 7–13 August 1968), ed. P. M. Millman, pp. 7–15. New York: Springer-Verlag. **1.1.5, 2.4.1, 3.3.3**

(1969b) The pre-Hayashi phase of stellar evolution. In *Low-luminosity stars*, ed. S. Kumar, pp. 423–31. New York: Gordon & Breach. **2.4.1**

(1970) Formation of the Earth–Moon system. *Eos*, **51**, 628–33. **4.4.4**

(1972) Orbital eccentricity of Mercury and the origin of the Moon. *Nature*, **240**, 299–300. **4.4.4**

(1973a) Properties of the solar nebula and the origin of the Moon. *Moon*, **7**, 377–83. **4.4.4**

(1973b) Interstellar grains in museums? In *Interstellar dust and related topics*, ed. J. M. Greenberg & H. C. van de Hulst, pp. 545–7. Dordrecht: Reidel.

2.5.1, 3.3.5

(1973c) Formation of the outer planets. *Space Science Reviews*, **14**, 383–91.

2.4.1

(1973d) Accumulation processes in the primitive solar nebula. *Icarus*, **18**, 407–50.

3.2.1

(1974) Models of the primitive solar nebula. In *On the origin of the Solar System*, ed. H. Reeves, pp. 56–70. Paris: CNRS. **2.4.1**

(1975a) The origin and evolution of the Solar System. *Sci. Am.*, **233**, no. 3 (September), 32–41. **3.3.5**

(1975b) Cosmogonical considerations regarding Uranus. *Icarus*, **24**, 280–4.

3.4.5

(1976/1978) The primitive solar accretion disk and the formation of the planets. In *The origin of the Solar System* (NATO Advanced Study Institute, 1976), ed. S. F. Dermott, pp. 49–74. New York: Wiley (pub. 1978). **2.4.2, .3.3.5**

(1978a) Physics of the primitive solar nebula and of giant gaseous protoplanets. In *Protostars and Planets*, ed. T. Gehrels, pp. 453–87. Tucson: University of Arizona Press. **3.1.2, 3.2.3, 3.3.5**

(1978b) Physics of the primitive solar accretion disk. *Moon and Planets*, **18**, 5–40.

2.4.2, 3.1.2, 3.2.3, 3.3.5

(1978c) The primitive solar accretion disk and the formation of the planets. In *The origin of the Solar System*, ed. S. F. Dermott, pp. 49–74. New York: Wiley.

2.4.2, 3.3.5

(1979a) On the origin of asteroids. In *Asteroids,* ed. T. Gehrels & M. S. Matthews, pp. 992–1007. Tucson: University of Arizona Press.   **3.3.5**

(1979b) The interaction between giant gaseous protoplanets and the primitive solar nebula. *Moon and Planets,* **21,** 173–83.   **3.2.1**

(1983) Origin of the atmospheres of the terrestrial planets. *Icarus,* **56,** 195–201.
4.4.5

(1984a) The rapid dissipation phase of the primitive solar nebula. *Lunar & Plan. Sci.,* **15,** 118–19.   **1.1.5, 2.4.2, 3.3.5**

(1984b) Star formation and extinct radioactivities. *Icarus,* **60,** 416–27.   **2.5.3**

(1985a) Formation and evolution of the primitive solar nebula. In *Protostars and planets.* II, ed. D. C. Black & M. S. Matthews, pp. 1073–99. Tucson: University of Arizona Press.   **2.4.2, 3.2.1, 3.2.3, 3.4.8**

(1985b) Formation of the prelunar accretion disk. *Icarus,* **62,** 319–27.
4.1.5, 4.1.6, 4.4.5

(1985c) The partial volatilization of Mercury. *Icarus,* **64,** 285–94.   **2.4.2, 3.3.5**

(1986a) The impact theory for origin of the Moon. In *Origin of the Moon,* ed. W. K. Hartmann et al., pp. 609–16. Houston: Lunar and Planetary Institute.   **4.4.5**

(1986b) Some autobiographical notes. In *Cosmogonical processes,* ed. W. D. Arnett et al., pp. 1–31. Utrecht: VNU Science.   **2.4, 2.4.1, 2.4.2**

(1988) Origin of the solar system. *Ann. Rev. Astron. Astrophys.,* **26,** 441–72.
2.2.4

(D1960) Letter to H. C. Urey, 8 December 1960. C-H.   **2.4.1**

Cameron, A. G. W., Colgate, S. A., & Grossman, L. (1973) Cosmic abundance of boron. *Nature,* **243,** 204–7.   **2.3.4**

Cameron, A. G. W., DeCampli, W. M., & Bodenheimer, P. (1982) Evolution of giant gaseous protoplanets embedded in the primitive solar nebula. *Icarus,* **49,** 298–312.
2.4.2, 3.2.1

Cameron, A. G. W., Höflich, P., Myers, P. C., & Clayton, D. D. (1995) Massive supernovae, Orion gamma rays, and the formation of the Solar System. *Astrophysical Journal,* **447,** L53–7.   **2.5.3**

Cameron, A. G. W., & Pine, M. R. (1973) Numerical models of the primitive solar nebula. *Icarus,* **18,** 377–406.   **2.4.1**

Cameron, A. G. W., & Pollack, J. B. (1976) On the origin of the Solar System and of Jupiter and its satellites. In *Jupiter,* ed. T. Gehrels, pp. 61–84. Tucson: University of Arizona Press.   **3.4.2**

Cameron, A. G. W., & Truran, J. W. (1977) The supernova trigger for the formation of the Solar System. *Icarus,* **30,** 447–61.   **1.1.5, 2.5.1, 2.5.2, 3.2.2**

Cameron, A. G W., & Ward, W. R. (1976) The origin of the moon. *Lunar Sci.,* **7,** 120–2.   **2.2.5, 2.4.1, 4.1.5, 4.4.5**

Cameron, A. G. W. *See also* Benz, W.; Consolmagno, G. J.; De Campli, W. M.; Ezer, D.; Haar, D. ter; Jastrow, R.; Marsden, B. G.; Mercer-Smith, J. A.; Perri, F.; Podolak, M.; Truran, J. W.; Ward, W. R.

Campbell, W. W. (1905) An address on astrophysics (at the St. Louis International Congress of Arts & Sciences). *Popular Science Monthly,* **66,** 297–318.   **1.2.4**

(1915) The evolution of stars and the formation of the Earth. *Scientific Monthly,* **1,** 1–17, 177–94, 238–55.   **1.2.6**

Canuto, V. M. *See* Cabot, W.

Carr, Michael H. *See* Smith, B. A.

Carron, M. K. *See* Cuttita, F.

Cendales, M. *See* Wänke, H.

Chamberlin, Rollin T. (1934) Thomas Chrowder Chamberlin. *Biog. Mem. N. A. S.,* **15,** 307–407.   **1.1.1, 1.2.3**

Chamberlin, Thomas Chrowder (1890/1965) The method of multiple working hypotheses. *Science* (o.s.), **15** (1890), 92–6. Reprinted, *Science,* **148** (1965), 754–9.
1.1.1

(1891) Present standing of the several hypotheses of the cause of the glacial period. *Sci. Am. Supplement*, **32**, 13075–6. **1.2.3**

(1897) A group of hypotheses bearing on climatic change. *J. Geol.*, **5**, 653–83. **1.2.3**

(1899a) On Lord Kelvin's address on the age of the Earth as an abode fitted for life. *Science*, **9**, 889–901; **10**, 11–18. **1.2.3**

(1899b) An attempt to frame a working hypothesis of the cause of glacial periods on an atmospheric basis. *J. Geol.*, **7**, 545–84, 667–85, 751–87. **1.2.3**, **1.2.4**

(1900) An attempt to test the nebular hypothesis by the relations of masses and momenta. *J. Geol.*, **8**, 58–73. **1.2.3**

(1901) On a possible function of disruptive approach in the formation of meteorites, comets, and nebulae. *Astrophy. J.*, **14**, 17–40. Reprinted in J. Geol., **9**, 369–93. **1.2.4**

(1903a) The origin of ocean basins on the planetesimal hypothesis. *Bull. Geol. Soc. Am.*, **14**, 548. **1.2.5**, **3.1.1**

(1903b) The origin of ocean basins on the planetessimal [*sic*] hypothesis. *Science*, **17**, 300–301. **1.2.5**

(1905) Fundamental problems of geology. *Carn. Inst. Yb.*, **3**, 195–228. **1.1.1**, **1.2.2**, **1.2.5**, **3.1.1**

(1906) Fundamental problems of geology. *Carn. Inst. Yb.*, **4**, 171–85. **1.1.1**

(1916) *The origin of the Earth*. Chicago: University of Chicago Press. **1.2.3**, **1.2.6**, **1.3.3**

(1920) Diastrophism and the formative processes. XIII. The bearings of the size and rate of infall of planetesimals on the molten or solid state of the Earth. *J. Geol.*, **28**, 665–701. **1.3.3**

(1922) Jones' criticism of Chamberlin's groundwork for the study of megadiastrophism. *Am. J. Sci.* [5], **4**, 253–73. **1.3.3**

(1924a) Review of Reid's "The planetesimal hypothesis and the Solar System" [*Am. J. Sci.*, **7** (1924), 37–64]. *J. Geol.*, **32**, 242–62. **1.2.6**, **1.3.4**

(1924b) Review of Jeffreys, *The Earth, its origin, history and physical constitution*. *J. Geol.*, **32**, 696–716. **1.3.3**

(1928) *The two solar families: The sun's children*. Chicago: University of Chicago Press. **1.2.6**, **1.3.3**

(D1895) Letter to G. C. Comstock, 11 December 1895. Letterbook II, pp. 165–7, C-UC. **1.2.3**

(D1896) Lecture, 27 January 1896. Typescript in box 3, folder 17, C-UC. **1.2.3**

(D1897a) Letter to G. J. Stoney, 6 September 1897. Letterbook IX, p. 428, C-UC. **1.2.3**

(D1897b) Letter to G. J. Stoney, 10 December 1897. Letterbook IX, p. 624, C-UC. **1.2.3**

(D1897c) Letter to C. R. Van Hise, 30 November 1897. Letterbook IX, p. 592, C-UC. **1.2.3**

(D1897d) Letter to F. R. Moulton, 10 September 1897. Letterbook IX, p. 434, C-UC. **1.2.3**

(1897e) Letter to Whitney, 9 October 1897. Letterbook IX, p. 476, C-UC. **1.2.3**

(D1897f) Letter to G. C. Comstock, 30 November 1897. University of Wisconsin Archives, Department of Astronomy, General Correspondence, series no. 7/4/2. Box 5. **1.2.3**

(D1897g) Letter to F. R. Moulton, 30 July 1897. Letterbook IX, p. 380, C-UC. **1.2.3**

(D1897h) Letter to F. R. Moulton, 24 November 1897. Letterbook IX, p. 589, C-UC. **1.2.3**

(D1897i) Letter to F. R. Moulton, 24 November 1897. Letterbook IX, p. 587, C-UC.   1.2.3

(D1898a) Letter to C. Van Hise, 22 January 1898. Letterbook IX, p. 698, C-UC.   1.2.3

(D1898b) Letter to H. Holt, 27 January 1898. Letterbook IX, p. 711, C-UC.   1.2.3

(D1898c) Letter to H. Holt, 13 August 1898. Letterbook X, p. 470, C-UC.   1.2.3

(D1898d) Letter to E. E. Barnard, 1 April 1898. Letterbook X, p. 172, C-UC.   1.2.3

(D1899) Letter to S. R. Cook, 14 January 1899. Letterbook IX, p. 181, C-UC.   1.2.3

(D1900a) Letter to J. Geikie, 13 March 1900. Letterbook XII, p. 235, C-UC.   1.2.3

(D1900b) Letter to F. R. Moulton, 19 January 1900. Letterbook XII, p. 161, C-UC.   1.2.3

(D1900c) Letter to J. Keeler, 30 January 1900. Copy from the tissue letter book, in box 1, folder 16, C-UC.   1.2.3

(D1900d) Letter to Keeler, 28 February 1900. Copy from the tissue letter book, in box I, folder 16, C-UC.   1.2.3

(D1900e) Letter to F. R. Moulton, 9 February 1900. Letterbook XII, p. 189, C-UC.   1.2.3

(D1901a) Memorandum dated 17 October 1901. Box 5, C-UC.   1.2.4

(D1901b) Letter to W. W. Campbell, 30 September 1901. Letterbook XIV, p. 213, C-UC.   1.2.4

(D1901c) Letter to W. W. Campbell, 18 October 1901. Letterbook XIV, p. 254.   1.2.4

(D1901d) Letters to W. W. Campbell and H. D. Curtis. Letterbook XVII, pp. 577, 690, C-UC.   1.2.4

(D1902a) Letter to C. D. Walcott, 23 January 1902. Copy in box 4, folder 20, C-UC.   1.2.5

(D1902b) Letter to C. D. Walcott, 28 January 1902. Chamberlin file at CIW.   1.2.5

(D1902c) Letter to C. D. Walcott, 14 June 1902. Chamberlin file at CIW.   1.2.5

(D1903a) Letter to G. C. Comstock, 6 March 1903. University of Wisconsin Archives, Astronomy Department General Correspondence series no. 7/4/1, box 7. Copy in Letterbook XVI, p. 680, C-UC.   1.2.3

(D1903b) Letter to the Librarian of Congress. Letterbook XVI, p. 681, C-UC.   1.2.3

(D1903c) Letter to H. L. Fairchild, 9 June 1903. Letterbook XVII, p. 269, C-UC.   1.2.5

(D1903d) An inquiry into the fundamental problems of geology as set forth in detail in a communication to Trustee C. D. Walcott under date of Jan. 23^{d}, 1902 & other letters. Application for Grant in Aid of Research, 2 January 1902 (i.e., 1903), received 6 January 1903. Chamberlin file at CIW.   1.2.5

(D1903e) Letter to C. D. Walcott, 10 January 1903. Chamberlin file at CIW.   1.2.5

(D1903f) Scheme for work under the Carnegie Institution, 7 January 1903. In folder of miscellaneous notes and calculations, box 5, C-UC.   1.2.5

(D1903g) Letter to C. D. Walcott, 27 February 1903. Chamberlin file at CIW.   1.2.5

(D1903h) Letter to F. R. Moulton, 12 January 1903. Letterbook XVI, pp. 404–406, C-UC.   1.2.5

(D1903i) Letter to F. R. Moulton, 28 November 1903. Letterbook XVII, p. 778.
                                                                                1.2.5
(D1904a) Letter to C. W. Hayes, 24 February 1904. Box 5, C-UC.   1.2.5
(D1904b) Letter to C. W. Hayes, 14 March 1904. Box 5, C-UC.   1.2.5
(D1904c) Letter to S. Newcomb, 17 March 1904. Newcomb Papers, Library of
Congress, container 19. Copy in box 5, C-UC.   1.2.5
(D1904d) Letter to G. F. Becker, 18 March 1904. Box 5, C-UC.   1.2.5
(D1904e) Letter to G. K. Gilbert, 18 March 1904. Letterbook XVIII, p. 102.
                                                                                1.2.5
(D1904f) Letter to S. Newcomb, 18 March 1904. Newcomb Papers, Library of
Congress, container no. 19. Copy in Letterbook XVIII, p. 105.   1.2.5
(D1904g) Letter to F. R. Moulton, 28 April 1904. Box 5, C-UC.   1.2.5
(D1904h) Letter to G. F. Becker, 11 April 1904. Box 5, C-UC.   1.2.5
(D1904i) Letter to G. K. Gilbert, 11 April 1904.   1.2.5
(D1904j) Letter to F. R. Moulton, 2 March 1904. Copy in Box 5, C-UC; also in
Letterbook XVIII, p. 55, C-UC.   **1.2.5, 1.2.6**
(D1904k) Letter to F. R. Moulton, 16 March 1904. Letterbook XVIII, p. 285.
                                                                                1.2.5
(D1904l) Letter to D. C. Gilman, 30 September 1904. Chamberlin file at CIW.
                                                                                1.2.5
(D1904m) Letter to E. C. Perisho, 29 February 1904. Letterbook XVIII, p. 49,
C-UC.   1.2.6
(D1904n) Letter to E. B. Frost, 2 March 1904. Frost correspondence at Yerkes
Observatory.   1.2.6
(D1904o) Letter to E. B. Frost, 17 March 1904. Frost correspondence at Yerkes
Observatory. Copy in Letterbook XVIII, p. 95, C-UC.   1.2.6
(D1904p) Letter to F. R. Moulton, 2 November 1904. Letterbook XVIII, p. 770,
C-UC.   1.2.6
(D1905a) Letter to E. B. Frost, 26 December 1905. Copy in box 1, folder 16,
C-UC.   1.2.2
(D1905b) Letter to R. S. Woodward, 28 October 1905b. Chamberlin file at CIW.
                                                                  1.2.5, 1.2.6
(D1906a) Letter to R. S. Woodward, 12 March 1906. Chamberlin file at CIW.
                                                                  1.2.5, 1.2.6
(D1906b) Letter to R. S. Woodward, 24 November 1906. Chamberlin file at CIW.
                                                                                1.2.5
(D1906c) Letter to Agnes Clerke, 14 March 1906. Box 4, folder 20, and box 5,
C-UC.   1.2.6
(D1906d) Letter to E. B. Frost, 13 December 1906. Letterbook XX, p. 401, C-UC.
                                                                                1.2.6
(D1906e) Letter to F. R. Moulton, 13 December 1906. Letterbook XX, p. 401,
C-UC.   1.2.6
(D1907a) Letter to R. S. Woodward, 7 October 1907. Chamberlin file at CIW.
                                                                                1.2.6
(D1907b) Letter to G. E. Hale, 20 April 1907. Letterbook XX, p. 789, C-UC.
                                                                                1.2.6
(D1908a) Letter to G. E. Hale, 27 May 1908. Letterbook XXI, p. 817, C-UC.
                                                                                1.2.6
(D1908b) Letter to R. S. Woodward, 19 August 1908. Letterbook XXI, p. 990,
C-UC.   1.2.6
(D1908c) Letter to R. S. Woodward, 13 October 1908. Chamberlin file at CIW.
                                                                                1.2.5
(D1908d) Letter to F. R. Moulton, 22 August 1908. Letterbook XXI, p. 994,
C-UC.   1.2.6

(D1908e) Letter to F. R. Moulton, 26 August 1908. Letterbook XXI, p. 1000, C-UC.   **1.2.6**

(D1909a) Letter to R. S. Woodward, 6 September 1909. Chamberlin file at CIW.
                                                                               1.2.5

(D1909b) Letter to R. S. Woodward, 19 October 1909. Chamberlin file at CIW.
                                                                               1.2.5

(D1910a) Letter to R. S. Woodward, 14 January 1910. Chamberlin file at CIW.
                                                                               1.2.5

(D1910b) Letter to R. S. Woodward, 11 February 1910. Chamberlin file at CIW.
                                                                               1.2.5

(D1913) Letter to C. G. Abbot, 3 May 1913. Secretarial Records of C. D. Walcott, record number 45, box 31, Smithsonian Institution Archives.   **1.2.4**

(D1920a) Notes on Jeans's Problems of cosmogony and stellar dynamics. 11 February 1920. Typed copy in box 3, C-UC.   **1.3.3**

(D1920b) Editorial. Our new responsibility. 17 February 1920. Typed copy in box 3, C-UC.   **1.3.3**

(D1928a) Memorandum on original announcement of the planetesimal hypothesis, 4 October 1928. Box 5, C-UC.   **1.2.5**

(D1928b) Reminiscences of Prof. T. C. Chamberlin (given orally to Mr. J. V. Nash). Chicago, 5 October 1928. Typescript in box 1, folder 4, C-UC.
                                                                        1.2.6, 1.3.3

(D19XX) Preliminary notes on the consequences of collisions between larger and smaller nebulae as a possible mode of origin of the solar system. Undated manuscript, box 5, C-UC.   **1.2.3**

(D19XY) Memorandum of Chamberlin's studies on the fundamental doctrines of geology. Undated, submitted to Carnegie Institution of Washington to support application for research grant. Chamberlin file at CIW.   **1.2.4, 1.2.5**

(D19XZ) Proposition relative to the grant of the Carnegie Institution for investigations in the fundamental doctrines of geology. Undated; apparently a memorandum to President Harper, University of Chicago. Copy in box 4, folder 20, C-UC.   **1.2.5**

Chamberlin, T. C., & Moulton, F. R. (1900) Certain recent attempts to test the nebular hypothesis. *Science,* **12,** 201–208.   **1.2.3, 1.3.5**

(1909) The development of the planetesimal hypothesis. *Science,* **30,** 642–5.
                                                                        1.2.3, 1.2.6

Chamberlin, T. C., & Salisbury, R. C. (1904) *Geology.* I, *Geologic processes and their results.* New York: Holt.   **1.2.3**

(1906) *Geology.* II. *Earth history, genesis – Paleozoic.* 2nd edn. New York: Holt.
                                                                  1.2.2, 1.2.5, 1.2.6

Champagne, A. E., Howard, A. J., & Parker, P. D. (1983a) Threshold states in [26]Al. I. Experimental investigations. *Nuclear Physics,* **402,** 159–78.   **2.5.3**

(1983b) Threshold states in [26]Al. II, Extraction of resonance strengths. *Nuclear Physics,* **402,** 179–88.   **2.5.3**

(1983c) Nucleosynthesis of [26]Al at low stellar temperatures. *Astrophys. J.,* **269,** 686–9.   **2.5.3**

Chandrasekhar, S. (1943) Stochastic problems in physics and astronomy. *Rev. Mod. Phys.,* **15,** 1–89.   **3.1.1**

(1946) On a new theory of Weizsäcker on the origin of the Solar System. *Rev. Mod. Phys.,* **18,** 94–102.   **1.1.4**

(1949) On Heisenberg's elementary theory of turbulence. *Proc. Roy. Soc.,* **A200,** 20–33.   **1.1.4**

Chandrasekhar, S., & Haar, D. ter (1950) The scale of turbulence in a differentially rotating gaseous medium. *Astrophys. J.,* **111,** 187–90.   **1.1.4**

Chang, Sherwood. *See* Kerridge, J. F.

Chapman, Clark R. (1982) *Planets of rock and ice: From Mercury to the moons of Saturn.* New York: Scribner.   **2.5.1, 3.4.8, 4.1.5**

Chapman, Clark R., McCord, Thomas B., & Johnson, Torrence V. (1973) Asteroid spectral reflectivities. *Astron. J., 78,* 126–40.   **3.4.7**

Chapman, Clark R. *See also* Gradie, J. C.; Greenberg, R.; Weidenschilling, S. J.

Chappell, B. W. *See* Compston, W.

Chase, S. C. *See* Ingersoll, A. P.

Chedin, A. *See* Gautier, D.

Chen, Dao-Han *See* Tai, W.-S.

Chevalier, Roger A. (1974) The evolution of supernova remnants. I. Spherically symmetric models. *Astrophys. J., 188,* 501–16.   **2.5.2**

Chevalier, Roger A., & Theys, John C. (1975) Optically thin radiating shock waves and the formation of density inhomogeneities. *Astrophys. J., 195,* 53–60.

                                                                              2.5.2

Chièze, J.-P. *See* Vangioni-Flam, E.

Chodos, A. A. *See* Albee, A. L.

Chou, Chen-Lin. *See* Baedecker, P. A.; Wasson, J. T.

Christian, R. P. *See* Cuttita, F.

Cisne, J. L. *See* Turcotte, D. L.

Clark, Sydney P., Jr., Turekian, Karl K., & Grossman, Lawrence (1972) Model for the early history of the Earth. In *The nature of the solid Earth,* ed. E. C. Robertson, pp. 3–18. New York: McGraw-Hill.   **3.3.4**

Clark, Sydney P., Jr. *See also* Grossman, L.; Turekian, K. K.

Clarke, W. B., De Laeter, J. R., Schwarcz, H. P., & Shane, K. C. (1970) Aluminum 26–magnesium 26 dating of feldspar in meteorites. *J. Geophys. Res., 75,* 448–62.
                                                                              2.5.1

Clayton, Donald D. (1975a) *The dark night sky: A personal adventure in cosmology.* New York: Quadrangle.   **2.5.1**

(1975b) Extinct radioactivities. *Astrophys. J., 199,* 765–9.   **2.5.1**

(1977) Solar system isotopic anomalies: Supernova neighbor or presolar carriers? *Icarus, 32,* 255–69.   **2.5.2**

(1978a) Precondensed matter: Key to the early Solar System. *Moon and Planets, 19,* 109–37.   **2.5.2, 3.3.5**

(1978b) An integration of special and general isotopic anomalies in r-process nuclei. *Astrophys. J., 224,* 1007–1012.   **2.5.3**

(1978c) On strontium isotopic anomalies and odd-A p-process abundances. *Astrophys. J., 224,* L93–L95.   **2.5.3**

(1979a) Supernovae and the origin of the Solar System. *Space Science Reviews, 24,* 147–226.   **2.5.1, 3.3.5**

(1979b) On isotopic anomalies in samarium. *Earth Plan. Sci. Lett., 42,* 7–12.
                                                                              2.5.3

(1980a) Internal chemical energy: Heating of parent bodies and oxygen isotopic anomalies. *Nukleonika, 25,* 1477–1490.   **3.3.5**

(1980b) Chemical energy in cold-cloud aggregates: The origin of meteoritic chondrules. *Astrophys. J. Letters, 239,* L37–L41.   **3.3.5**

(1981a) Origin of Ca-Al-rich inclusions. II. Sputtering and collisions in the three-phase interstellar medium. *Astrophys. J., 251,* 374–86.   **2.5.1, 2.5.2**

(1981b) Some key issues in isotopic anomalies: Astrophysical history and aggregation. *Proc. Lunar & Plan. Sci. Conf., 12B,* 1781–1802.   **3.3.5**

(1982) Cosmic chemical memory: A new astronomy. *Q. J. Roy. Astron. Soc., 23,* 174–212.   **2.5.2, 3.3.5**

(1984) ^{26}Al in the interstellar medium. *Astrophys. J., 280,* 144–9.   **2.5.3**

(1986) Isotopic anomalies and SUNOCON survival during galactic evolution. In *Cosmogonical processes,* ed. W. D. Arnett et al., pp. 101–22. Utrecht: VNU Science Press.   **2.5.2**

Clayton, Donald D., Dwek, E., & Woosley, S. E. (1977) Isotopic anomalies and proton irradiation in the early Solar System. *Astrophys. J.,* **214,** 300–15.
                 2.3.4, 2.5.1

Clayton, D. D. *See* Cameron et al. (1995).

Clayton, Robert, Grossman, Lawrence, & Mayeda, Toshiko (1973) A component of primitive nuclear composition in carbonaceous meteorites. *Science,* **182,** 485–8.
                 2.5.1

Clayton, Robert, & Mayeda, Toshiko K. (1975) Genetic relations between the Moon and meteorites. *Proc. 6th Lunar Sci. Conf.,* 1751–69.
                 4.3.4

Clayton, Robert N., Mayeda, Toshiko K., & Molini-Velsko, Carol A. (1985) Isotopic variations in Solar System material: Evaporation and condensation of silicates. In *Protostars and planets.* II, ed. D. C. Black & M. S. Matthews, pp. 755–71. Tucson: University of Arizona Press.
                 3.3.5, 3.4.8

Clayton, Robert N., Onuma, N., & Mayeda, T. K. (1976) A classification of meteorites based on oxygen isotopes. *Earth Plan. Sci. Lett.,* **30,** 10–18.
                 4.4.4

Clayton, Robert N. *See also* Grossman, L.; Onuma, N.

Clerke, Agnes (1902) *A popular history of astronomy during the nineteenth century.* 4th edn. London: Black.
                 1.2.4

(1903) *Problems in astrophysics.* London: Black.
                 1.2.6

(1905) *Modern cosmogonies.* London: Black.
              1.2.4, 1.2.6

Clifford, A. C. (1925) New stars. *New Zealand Journal of Science and Technology,* **8,** 1–7.
                 1.2.4

Cloud, Preston E., Jr. (1968) Atmospheric and hydrospheric evolution in the primitive Earth. *Science,* **160,** 729–36.
                 4.3.3

Cohen, M. *See* Black, D. C.

Colburn, D. S. *See* Sonett, C. P.

Coleman, P. J., Jr. *See* Russell, C. T.

Colgate, Stirling A. *See* Cameron, A. G. W.

Collins, Stewart A. *See* Smith, B. A.

Collinson, D. W. *See* Runcorn, S. K.

Colombo, G. *See* Franklin, F. A.

Compston, W., Berry, H., Vernon, M. J., Chappell, B. W., & Kaye, M. J. (1971) Rubidium–strontium chronology and chemistry of lunar material from the Ocean of Storms. *Proc. 2nd Lunar Sci. Conf.,* 1471–86.
                 4.4.4

Compston, W. *See also* Gray, C. M.

Comstock, G. C. (D1895) Letter to T. C. Chamberlin, 13 December 1895. Box 1, folder 15, C-UC.
                 1.2.3

Condie, Kent C. (1982) *Plate tectonics and crustal evolution.* 2nd edn. New York: Pergamon Press.
                 4.4.4

Conrath, B. J. *See* Hanel, R. A.

Consolmagno, Guy J., & Cameron, A. G. W. (1980) The origin of the "FUN" anomalies and the high temperature inclusions in the Allende meteorite. *Moon and Planets,* **23,** 3–25.
                2.5.2, 2.5.3

Cook, Allan F., II. *See* Smith, B. A.

Cook, S. R. (1899) On the escape of gases from the planets according to the kinetic theory. *Proc. Am. Assoc.,* 120–1.
                 1.2.3

(1900) On the escape of gases from planetary atmospheres according to the kinetic theory. *Astrophys. J.,* **11,** 36–43.
                 1.2.3

Cooper, Henry S. F., Jr. (1970) *The Moon rocks.* New York: Dial Press.
                 4.3.4

(1983) *Imaging Saturn: The Voyager flights to Saturn.* New York: Holt, Rinehart, & Winston.
                 3.4.4

(1987) Letter from the Space Center. *New Yorker,* **63,** no. 16 (June 8), 71–81
                4.1.6, 4.4.5

Cooper, J. A., Richards, J. R., & Stacey, F. D. (1967) Possible new evidence bearing on the lunar capture hypothesis. *Nature,* **215,** 1256.
                 4.3.3

Cortie, A. L. (1899) The origin of the Solar System. *American Catholic Quarterly Review*, **24**, 19–27.   1.2.2

Couper, Heather, & Henbest, Nigel (1985) *The planets.* London. Pan Books.
                                                                                 2.5.3, 4.4.4

Courtin, R., Gautier, D., Marten, A., Bézard, B., & Hanel, R. A. (1984) The composition of Saturn's atmosphere at northern temperate latitudes from *Voyager* IRIS spectra: NH_3, PH_3, C_2H_2, C_2H_6, CH_3D, CH_4, and the Saturnian D/H isotopic ratio. *Astrophys. J.*, **287**, 899–916.   3.2.3

Cravens, Hamilton (1971) The abandonment of evolutionary social theory in America: The impact of academic professionalization upon American sociological theory, 1890–1920. *American Studies*, **12**, 5–20.   1.2.2

Cravens, Hamilton, & Burnham, John C. (1971) Psychology and evolutionary naturalism in American thought, 1890–1940. *American Quarterly*, **23**, 635–57.
                                                                                      1.2.2

Cremin, A. W. *See* Williams, I. P.

Croll, James (1889) *Stellar evolution and its relations to geological time.* New York: Appleton.   1.2.1

Crommelin, A. C. D. (1904) Phoebe: Saturn's ninth satellite. *J. Br. Astron. Ass.*, **15**, 32–5.   1.2.4

      (1923) The solar system. In *Hutchinson's Splendour of the heavens*, ed. T. E. R. Phillips & W. H. Steavenson, chap. 3. London: Hutchinson.   1.3.3

      (1938) Recent progress in astronomy. *Reports of the Physical Society of London*, **5**, 82–99.   1.3.5

Cruikshank, Dale P. (1983) The development of studies of Venus. In *Venus*, ed. D. M. Hunten et al., pp. 1–9. Tucson: University of Arizona Press.   3.3.5

Curtiss, C. F. *See* Hirschfelder, J. O.

Cuttita, Frank, Rose, H. J., Jr., Annell, C. S., Carron, M. K., Christian, R. P., Ligon, D. T., Jr., Dwornik, E. J., Wright, T. L., & Greenland, L. P. (1973) Chemistry of twenty-one igneous rocks and soils returned by the Apollo 15 mission. *Proc. 4th Lunar Sci. Conf.*, 1081–96.   4.4.3

Cuzzi, Jeffrey (1978) The rings of Saturn: State of current knowledge and some suggestions for future studies. In *The Saturn system*, ed. D. M. Hunten et al., pp. 73–102. Washington, DC: National Aeronautics and Space Administration, NASA Conference Publication no. 2068.   3.4.4

Cuzzi, J. N., Lissauer, J. J., Esposito, L. W., Holberg, J. B., Marouf, E. A., Tyler, G L., & Boischot, A. (1984) Saturn's rings: Properties and processes. In *Planetary rings*, ed. R. Greenberg & A. Brahic, pp. 73–199. Tucson: University of Arizona Press.
                                                                                      3.4.4

Cuzzi, J. N. *See also* Burns, J. A.; Lissauer, J. J.; Pollack, J. B.

Dai, W., & Hu, Z. (1980) On the origin of the Solar System. *Scientia Sinica*, **23**, 862–79.   3.4.2

Dakowsky, M. (1969) The possibility of extinct superheavy elements occurring in meteorites. *Earth Plan. Sci. Lett.*, **6**, 152–4.   2.5.1

Dallaporta, N., & Secco, L. (1973) Sulla teoria di Hoyle della formazione del sistema solare. *Mem. Soc. Astron. Ital.* (n.s.), **43**, 705–13.   2.3.5

      (1975) On Hoyle's theory of the origin of the Solar System. *Astrophys. Space Sci.*, **37**, 335–64.   2.3.5

Daly, R. A. (1933) *Igneous rocks and the depths of the Earth.* Rev. edn. New York: McGraw-Hill.   4.4.5

      (1946) Origin of the Moon and its topography. *Proc. Am. Phil. Soc.*, **90**, 104–19.
                                                                                      4.4.5

Danielson, G. Edward. *See* Smith, B. A.

Danielsson, Lars (1969) Statistical arguments for asteroidal jet streams. *Astrophys. Space Sci.*, **5**, 53–8.   3.4.7

Darden, Lindley, & Maull, Nancy (1977) Interfield theories. *Philosophy of Science*, **44**, 43–64.   **1.3.5**

Darwin, Erasmus (1790/1825) *Botanic Garden*. I. London: Johnson (1825 edn., first pub. 1790).   **4.2.1**

Darwin, George Howard (1878) On the precession of a viscous spheroid. *Nature*, **18**, 580–2.   **4.1.1, 4.2.3**

(1879) On the precession of a viscous spheroid, and on the remote history of the Earth. *Phil. Trans.*, **170**, 447–538.   **1.2.3, 4.1.1, 4.2.3**

(1880) On the secular changes in the elements of the orbit of a satellite revolving about a tidally distorted planet. *Phil. Trans.*, **171A**, 713–891.   **4.2.3**

(1886) Tidal friction and the evolution of a satellite. *Nature*, **33**, 367–8; **34**, 287–8.
                                                                                         4.2.4

(1888) On the mechanical condition of a swarm of meteorites, and on theories of cosmogony. *Phil. Trans.*, **180**, 1–69.   **1.2.3, 1.2.6, 1.3.1**

(1898) *The tides and kindred phenomena in the solar system*. Boston: Houghton Mifflin.
                                                                                         1.2.2

(1905) Cosmical evolution. (Address by the president). *B. A. Report*, **75**, 3–32.
                                                                                         1.2.6

(1905/1962) *The tides and kindred phenomena in the solar system*. Boston: Houghton Mifflin (1905). Reprint, San Francisco: Freeman (1962).   **4.2.3, 4.4.6**

(1907–1916) *Scientific Papers*. 5 vols. Cambridge University Press.
                                                                                 4.2.3, 4.2.4

(1909) A theory of the evolution of the Solar System. *Internationale Wochenschrift*, **3**, 921–32.   **1.2.3, 1.2.6**

Davidson, J. A. *See* Harper, D. A.

Davidson, M. (1911) Theory of grazing collisions. Is it necessary to suppose that the impact actually occurs? *J. Br. Astron. Assoc.*, **21**, 371–7.   **1.2.1**

Davies, Merton E. *See* Smith, B. A.

Davis, Donald R. *See* Greenberg, R.; Hartmann, W. K.; Herbert, F.; Weidenschilling, S. J.

De, B. R. (1972/1978) A 1972 prediction of Uranian rings based on the Alfvén critical velocity effect. *Moon and Planets*, **18**, 339–42.   **3.4.5**

De, B. R. *See also* Arrhenius, G.

Dean, John Candee (1913) From whence came the Earth? *Westminster Review*, **180**, 635–44.   **1.2.4**

Dearborn, D. S. P. *See* Appenzeller, I.

DeCampli, W. M., & Cameron, A. G. W. (1979) Structure and evolution of isolated giant gaseous protoplanets. *Icarus*, **38**, 367–91.   **2.4.2, 3.2.1**

DeCampli, W. M. *See also* Bodenheimer, P.; Cameron, A. G. W.

Decker, Deborah A. *See* Goettel, K. A.

Degeweij, J., Gradie, J. C., & Zellner, B. (1978) Minor planets and related objects. XXV. *UBV* photometry of 145 faint asteroids. *Astron. J.*, **83**, 643–50.   **3.4.7**

De Groot, Mart, McCrea, W., Wilson, A. H., Jr., Singer, S. F., Mullan, D. J., Helin, E. F., Beesley, D. E., & McFarland, J. (1986) Ernst Julius Öpik, 1893–1985, the man and the scientist. *Irish Astron. J.*, **17**, 411–42.   **3.4.8**

De Jong, T. *See* Aumann, H. H.

De Laeter, J. R. *See* Clarke, W. B.

Delano, J. W., & Ringwood, A. E. (1978) Indigenous abundances of siderophile elements in the lunar highlands: Implications for the origin of the Moon. *Moon and Planets*, **18**, 385–425.   **4.4.4**

Delano, J. W. *See also* Ringwood, A. E.

De La Rue, Warren (1866) Address delivered by the president, Warren de la Rue, Esq., on presenting the Gold Medal of the Society to Professor J. C. Adams, Director of the Cambridge Observatory. *M. N. Roy. Astron. Soc.*, **26**, 157–89.
                                                                                         4.2.2

Delaunay, C. E. (1865) Sur l'existence d'une cause nouvelle ayant un action sensible sur la valeur de l'equation seculaire de la Lune. *C. R. Acad. Sci. Paris*, **61**, 1023–32.   **4.2.2**

(1866a) Remarques de M. Delaunay à l'occasion de cette comunication. *C. R. Acad. Sci. Paris*, **62**, 165–6.   **4.2.2**

(1866b) Sur l'accélération apparente du moyen mouvement de la Lune due aux actions du Soleil et de la Lune sur les eaux de la mer. *C. R. Acad. Sci. Paris*, **62**, 197–200.   **4.2.2**

(1866c) Réponse a la note de M. Allégret inséreé au Compte Rendu de la Séance du 26 février. *C. R. Acad. Sci. Paris*, **62**, 575–9.   **4.2.2**

(1866d) Sur la controverse relative a l'equation seculaire de la Lune. *C. R. Acad. Sci. Paris*, **62**, 704–707.   **4.2.2**

(1866e) Note sur la question du ralentissement de la rotation de la Terre. *C. R. Acad. Sci. Paris*, **62**, 1107.   **4.2.2**

(1866f) Conference sur l'astronomie et en particulier sur le ralentissment du mouvement de rotation de la Terre. *Revue des Cours Scientifique*, **3**, 321–31.
                                                                                   4.2.2

Delaunay, L. (1905) *La science geologique.* Paris: Colin.   **1.2.2**

DeMarcus, Wendell C. (1958) The constitution of Jupiter and Saturn. *Astron. J.*, **63**, 2–28.   **3.2.1**

(D1956) Letter to H. C. Urey, 14 March 1956. Box 12, folder "DeMarcus, W. C.," U-UCSD.   **4.4.1**

Descartes, Rene (1664/1824–26) *Le Monde, ou traite de la Lumière.* Paris: Jacques le Gras (1664). Reprinted in *Oeuvres des Descartes*, ed. V. Cousin, Vol. 4, pp. 215–332. Paris: Levrault (1824–1826). Also in *Oeuvres des Descartes*, ed. C. Adam & P. Tannery, Vol. 9, pp. 1–118. Paris: Cerf (1909).   **2.1, 4.2.1**

Deutsch, S. *See* Hanappe, F.

De Vaucouleurs, Gerard (1962) The Moon and planets. In *The challenges of space*, ed. H. Odishaw, pp. 142–51. Chicago: University of Chicago Press.   **4.3.1**

Dicke, R. H. (1957) Principle of equivalence and the weak interactions. *Rev. Mod. Phys.*, **29**, 355–62.   **4.2.4**

(1964/1966) The secular acceleration of the Earth's rotation and cosmology. In *The Earth–Moon system* (proceedings of conference held in 1964), ed. B. G. Marsden & A. G. W. Cameron, pp. 98–164. New York: Plenum Press (pub. 1966).
                                                                                   4.4.2

Dickey, J. O. *See* Yoder, C. F.

Dickey, J. S., Jr. *See* Marvin, U. B.

D[ingle], H[erbert] (1929) The planetesimal hypothesis. *Nature*, **123**, 555–7. Signed "D.H."   **1.3.3**

Dingle, Herbert (1932) The origin of the Solar System. *Nature*, **129**, 333–5.
                                                                                   1.3.3

Dodd, R. J., & Napier, W. McD. (1974) Direct simulation of collision processes. II: The growth of planetesimals. *Astrophys. Space Sci.*, **29**, 51–9.   **3.1.2**

Dodd, R. J. *See also* Napier, W. McD.

Doel, Ronald Edmund (1990) *Unpacking a myth: Interdisciplinary research and the growth of Solar System astronomy.* Ph.D. dissertation, Princeton University.   **1.1.4**

Dole, Stephen H. (1970) Computer simulation of the formation of planetary systems. *Icarus*, **13**, 494–508.   **3.1.1**

Dollfus, Audouin (1967) Un nouveau satellite de Saturne. *C. R. Acad. Sci. Paris*, **264B**, 822–4.   **3.4.4**

(1968) La découverte du 10ᵉ satellite de Saturne. *L'Astronomie*, **82**, 253–62.
                                                                                   3.4.4

Dolmage, Cecil G. (1909) *Astronomy to-day.* 2nd edn. London: Seeley.   **1.2.6**

Donahue, T. M., & Pollack, J. B. (1983) Origin and evolution of the atmosphere of Venus. In *Venus*, ed. D. M. Hunten et al., pp. 1003–36. Tucson: University of Arizona Press.   **3.3.5**

Donahue, T. M. *See also* Hoffman, J. H.

Donn, Bertram (1975) Problems of cosmic chemistry. II. Condensation processes in high temperature clouds. *Mémoires de la Société Royale Scientifique de Liège* [6], **9**, 499–509. Also published as *Astrophysique et Spectroscopie,* Proceedings of the 20th Astrophysics Colloquium.   3.3.4

Donnison, J. R. (1978) The escape of natural satellites from Mercury and Venus. *Astrophys. Space Sci.,* **59**, 499–501.   4.4.6

Donnison, J. R., & Williams, I. P. (1974) The evolution of Jupiter from a protoplanet. *Astrophys. Space Sci.,* **29**, 387–96.   3.2.1

(1978) On the origin and evolution of Jupiter and Saturn. *Astrophys. Space Sci.,* **53**, 241–3.   3.2.1

(1985) Orbital stability during the early stages of planetary formation. *M. N. Roy. Astron. Soc.,* **216**, 521–7.   3.2.1

Donovan, Arthur. *See* Laudan, R.

Dorling, Jon (1979) Bayesian personalism, the methodology of scientific programmes, and Duhem's problem. *Stud. Hist. Phil. Sci.,* **10**, 177–87.   4.2.2

Dorman, H. J. *See* Nakamura, Y.

Dormand, J. R. (1973) The Solar System and its origin. *Physics Education,* **8**, 475–81.
                                                                                                                  2.2.1, 2.3.5

Dormand, J. R., & Woolfson, M. M. (1971) The capture theory and planetary condensation. *M. N. Roy. Astron. Soc.,* **151**, 307–31.   1.1.4, 3.2.1

(1974) The evolution of planetary orbits. *Proc. Roy. Soc.,* **A340**, 349–65.
                                                                                                                  1.1.4, 3.2.1

(1977) Interactions in the early Solar System. *M. N. Roy. Astron. Soc.,* **180**, 243–79.
                                                                                                                  1.1.4, 3.2.1

(1989) *The origin of the Solar System: The capture theory.* New York: Halsted (Wiley).
                                                                                                                  1.1.4, 3.2.1

Dorschner, J. (1974) Heutige astronomische Vorstellungen über die Entstehung des Planetensystems. *Die Sterne,* **50**, 91–100.   2.3.5, 2.4

(1979) Apollo und der Mond. Eine Betrachtung zum 10. Jahrestag der Mondlandung. *Die Sterne,* **55**, 129–43.   4.4.4

Dr. Lowell's Cosmogony (1910) Review of *The Evolution of Worlds* [by P. Lowell]. *Living Age,* **265**, 699–700.   1.2.4

Drake, Michael J. (1986) Is lunar bulk material similar to Earth's mantle? In *Origin of the Moon,* ed. W. K. Hartmann et al., pp. 105–124. Houston: Lunar and Planetary Institute.   4.4.4, 4.4.5

Drake, M. J. *See also* Greenberg, R.; Jones, J. H.; Kaula, W. M.; Kreutzberger, M.; Newsom, Horton E.; Weidenschilling, S. J.

Dreibus, Gerlind, Kruse, H., Spettel, B., & Wänke, H. (1977) The bulk composition of the Moon and the eucrite parent body. *Proc. 8th Lunar Science Conf.,* 211–27.
                                                                                                                  4.4.4

Dreibus, Gerlind, & Wänke, Heinrich (1979) On the chemical composition of the Moon and the eucrite parent body and a comparison with the composition of the Earth, the case of Mn, Cr, and V. *Lunar & Plan. Sci.,* **10**, 315–17.   4.4.4

Dreibus, Gerlind. *See also* Wänke, H.

Drobyshevskii, E. M. (1978) The origin of the Solar System. Implications for trans-Neptunian planets and the nature of long-period comets. *Moon and Planets,* **18**, 145–94.   4.4.4

Duennebier, F. *See* Nakamura, Y.

Du Fresne, E. R., & Anders, E. (1961) The record in the meteorites, V. A thermometer mineral in the Mighei carbonaceous chondrite. *Geoch. Cosmoch. Acta,* **23**, 200–208.   3.3.2

(1962) On the chemical evolution of the carbonaceous chondrites. *Geoch. Cosmoch. Acta,* **26**, 1085–1114.   3.3.2

Dugan, Raymond Smith. *See* Russell, H. N.

Duncan, John Charles (1926) *Astronomy.* New York: Harper.   **1.2.6**

Dunham, E. *See* Elliot, James L.

Dunthorne, Richard (1749) A letter . . . concerning the acceleration of the Moon. *Phil. Trans.,* **46,** 162–72.   **4.2.2**

Durisen, Richard H., & Gingold, Robert A. (1986) Numerical simulations of fission. In *Origin of the Moon,* ed. W. K. Hartmann et al., pp. 487–98. Houston: Lunar and Planetary Institute.   **4.4.5**

Durisen, Richard H., & Scott, E. H. (1984) Implications of recent numerical calculations for the fission theory of the origin of the Moon. *Icarus,* **58,** 153–8.   **4.4.4, 4.4.5**

Dwek, E. *See* Clayton, D. D.

Dwornik, E. J. *See* Cuttita, F.

E[ddington], A. S. (1906) A criticism of Sir George Darwin's theories. *Observatory,* **29,** 179–81 [signed "A. S. E."].   **4.2.4**

Eddington, A. S. (1909/1970) Some recent results of astronomical research (lecture at Royal Institution, London, 26 March 1909). In *The Royal Institution library of science – astronomy,* **2,** pp. 97–112. New York: American Elsevier (1970).   **1.2.5**

     (1926) *The internal constitution of the stars.* Cambridge University Press.   **1.3.4**

Edgeworth, K. E. (1949) The origin and evolution of the solar system. *M. N. Roy. Astron. Soc.,* **109,** 600–609.   **3.1.1**

Edmunds, M. G. (1977) An induced birth for the solar system? *Nature,* **267,** 393–4.   **2.5.2**

Edwards, A. C. (1982) The age problem. In *Progress in cosmology,* ed. A. W. Wolfendale, pp. 291–303. Boston: Reidel.   **2.1.2**

Eglinton, G. *See* Arnold, J.

Eichhorn, Heinrich *See* Gatewood, G.

Elliot, James L., Dunham, E., & Mink, D. (1977) The rings of Uranus. *Nature,* **267,** 328–30.   **3.4.5, 3.4.6**

Elliot, James L., & Kerr, Richard A. (1984) *Rings: Discoveries from Galileo to Voyager.* Cambridge, MA: MIT Press.   **3.4.5**

Elliot, James L., & Nicholson, P. D. (1984) The rings of Uranus. In *Planetary rings,* ed. R. Greenberg & M. S. Matthews, pp. 25–72. Tucson: University of Arizona Press.   **3.4.5**

Elsasser, Walter M. (1963) Early history of the Earth. In *Earth science and meteoritics,* ed. J. Geiss & E. D. Goldberg, pp. 1–30. Amsterdam: North-Holland.   **4.3.3**

Elsasser, Walter M. *See also* Urey, H. C.

Emme, Eugene M. (1968) Historical perspectives on Apollo. *Journal of Spacecraft and Rockets,* **5,** 369–82.   **4.3.1**

Emme, Eugene M., & Hall, R. Cargill (D1976) Interview with Harold C. Urey, 18 October 1976. Transcript in NASA-A.   **4.3.2**

Ennis, Jacob (1867) *The origin of the stars.* New York: Appleton.   **4.2.1**

Epstein, R. I. *See* Mercer-Smith, J. A.

Epstein, Samuel, & Taylor, Hugh P., Jr. (1970) $^{18}O/^{16}O$, $^{30}Si/^{28}Si$, D/H, and $^{13}C/^{12}C$ studies of lunar rocks and materials. *Science,* **167,** 533–5.   **4.4.3**

Esat, T. M. *See* Papanastassiou, D. A.

Esposito, L. W. *See* Cuzzi, J. N.

Essene, E. *See* Ringwood, A. E.

Eugster, O. J. *See* Albee, A. L.

Ewing, M. *See* Nakamura, Y.

Ezell, Edward Clinton, & Ezell, Linda Neuman (1984) *On Mars: Exploration of the red planet, 1958–1978.* Washington, DC: National Aeronautics and Space Administration, report NASA SP-4212.   **3.3.1**

Ezell, Linda Neuman. *See* Ezell, E. C.

Ezer, Dilhan, & Cameron, A. G. W. (1962) Early solar evolution. *Sky & Tel.*, **24**, 328–30.   **2.4.1**
    (1963) The early evolution of the sun. *Icarus*, **1**, 422–41.   **1.1.5, 2.4.1, 3.3.3**
    (1965) A study of solar evolution. *Canadian Journal of Physics*, **43**, 1497–1517.   **3.3.3**

Fahr, H. J. *See* Petelski, E. F.

Falk, Sydney W., & Schramm, David N. (1979) Did the Solar System start with a bang? *Sky & Tel.*, **58**, 18–22.   **2.5.2**

Fanale, Fraser P. (1976) Martian volatiles: Their degassing history and geochemical fate. *Icarus*, **28**, 179–202.   **3.3.5**

Farinella, Paolo, & Paolicchi, P. (1977) Conservation laws and mass distribution in the planet formation process. *Moon*, **17**, 401–408.   **3.1.2**

Faulkner, J., Griffiths, K., & Hoyle, F. (1963) On the Hayashi effect in the early phases of gravitational contraction of the sun. *M. N. Roy. Astron. Soc.*, **126**, 1–10.   **2.3.3**

Feigelson, Eric D. (1982) X-ray emission from young stars and implications for the early Solar System. *Icarus*, **51**, 155–63.   **2.5.3**

Fenton, Carroll, & Fenton, Mildred (1945) *Giants of geology.* Garden City, NY: Doubleday.   **1.1.1, 1.2.3**

Fenton, Mildren. *See* Fenton, C.

Fernie, J. D. (1970) The historical quest for the nature of the spiral nebulae. *Pub. Astron. Soc. Pac.*, **82**, 1189–1230.   **1.2.6**

Ferraro, V. C. A. (1937) The non-uniform rotation of the sun and its magnetic field. *M. N. Roy. Astron. Soc.*, **97**, 458–72.   **1.1.3, 3.4.2**

Ferrel, William (1853) On the effect of the Sun and Moon upon the rotatory motion of the Earth. *Astron. J.*, **3**, 138–41.   **4.2.2**
    (1866) Note on the influence of the tides in causing an apparent secular acceleration of the Moon's mean motion. *Proc. Amer. Acad.*, **6**, 379–83, 390–93.   **4.2.2**
    (1895) Autobiographical sketch. *Biog. Mem. N. A. S.*, **3**, 287–99. (Written in 1888.)   **4.2.2**

Ferrin, Ignacio (1978) Azimuthal brightness variation of Saturn's ring A and size of particles. *Nature*, **271**, 528–9.   **3.4.7**

Fesharaki, Fereidun. *See* Smith, K. R.

Feyerabend, P. K. (1962) Explanation, reduction, and empiricism. In *Minnesota Studies in the Philosophy of Science*, vol. **3**, ed. H. Feigl & G. Maxwell, pp. 28–97. Minneapolis: University of Minnesota Press.   **1.3.5**
    (1965) Problems of empiricism. In *Beyond the edge of certainty*, ed. R. G. Colodny, pp. 145–260. Englewood Cliffs, NJ: Prentice-Hall.   **1.3.5**

Field, George B. (1963) The origin of the Moon. *American Scientist*, **51**, 349–54.   **4.3.3**

Fillius, W. (1976) The trapped radiation belts of Jupiter. In *Jupiter*, ed. T. Gehrels, pp. 896–927. Tucson: University of Arizona Press.   **3.4.6**

Finch, D. G. (1982) The evolution of the Earth–Moon system. *Moon and Planets*, **26**, 109–14.   **4.3.3**

Fish, R. A., Goles, G. G., & Anders, E. (1960) The record in the meteorites. III. On the development of meteorites in asteroidal bodies. *Astrophys. J.*, **132**, 243–58.   **2.3.2, 2.5.1, 3.3.2**

Fisher, David E. (1987) *The birth of the Earth: A wanderlied through space, time, and the human imagination.* New York: Columbia University Press.   **3.1.1, 3.2.2, 3.2.3**

Fisher, Osmond (1882) On the physical cause of the ocean basins. *Nature*, **25**, 243–4.   **4.1.1, 4.2.3**

Fiske, John (1876) *The unseen world and other essays*. Boston: Osgood.   **1.2.4**

Fison, Alfred H. (1898) *Recent advances in astronomy*. London: Blackie.   **1.2.2**

Foldvari-Vogl, M. *See* Sztrokay, K. I.

Forbes, Eric G. (1972). *The Euler–Mayer correspondence (1751–1755): A new perspective on eighteenth century advances in the lunar theory*. New York: American Elsevier.

    4.2.2

Fowler, William A. (1984) Experimental and theoretical nuclear astrophysics: The quest for the origin of the elements. *Rev. Mod. Phys.*, **56**, 149–79.   **2.5.3**

    (1987) The age of the observable universe. *Q. J. Roy. Astron. Soc.*, **28**, 87–108.

    2.1.2

    (1992) From steam to stars to the early universe. *Ann. Rev. Astron. Astrophys.*, **30**, 1–9.   **2.3.2**

Fowler, William A., Greenstein, Jesse L., & Hoyle, Fred (1961) Deuteronomy: Synthesis of deuterons and the light nuclei during the early history of the Solar System. *Am. J. Phys.*, **29**, 393–403.   **1.1.5, 1.3.4, 2.3.2, 2.5, 2.5.1**

    (1962) Nucleosynthesis during the early history of the solar system. *Geophys. J.*, **6**, 148–220.   **2.3.2, 2.5, 2.5.1**

Fowler, William A., & Hoyle, Fred (1960) Nuclear cosmochronology. *Ann. Physics*, **10**, 280–302.   **2.3.1**

Fowler, William A. *See also* Burbidge, E. M.; Burnett, D. S.; Reeves, H.; Ward, R. A.

Frankel, Henry (1976) Alfred Wegener and the specialists. *Centaurus*, **20**, 305–24.

    1.3.5

    (1979) The career of continental drift theory: An application of Imre Lakotos' analysis of scientific growth to the rise of drift theory. *Stud. Hist. Phil. Sci.*, **10**, 21–66.   **1.3.5**

Franklin, F. A., & Colombo, G. (1970) A dynamical model for the radial structure of Saturn's rings. *Icarus*, **12**, 338–47.   **3.4.4**

Frazier, Kendrick (1987) Moon? Boom! *Air & Space/Smithsonian*, **1**, no. 5. (January), 84–91.   **4.1.6, 4.4.5**

Fredriksson, K. *See* Kurat, G.

Freeman, John W. (1976/1978) The primordial solar magnetic field. In *The origin of the Solar System* (NATO Advanced Study Institute, 1976), ed. S. F. Dermott, pp. 635–640. New York: Wiley (pub. 1978).   **3.4.2**

French, Bevan M. (1977) *The Moon book*. New York: Penguin Books.

    4.3.1, 4.4.4

    (1978) *What's new on the Moon?* Washington, DC: U.S. Government Printing Office. See also NASA report EP-131, reprinted in *Sky & Tel.*, **53** (1977), 164–9.

    4.4.4

    (1981) The Moon. In *The New Solar System*, ed. J. K. Beatty et al., pp. 71–82. Cambridge University Press/Cambridge, MA: Sky Publishing.   **4.4.4**

Friedlander, Michael W. (1985) *Astronomy: From Stonehenge to quasars*. Englewood Cliffs, NJ: Prentice-Hall.   **2.3.5, 4.4.4**

Friedman, Irving, O'Neill, James R., Adami, Lanford H., Gleason, Jim D., & Hardcastle, Kenneth (1970) Water, hydrogen, deuterium, carbon, carbon-13, and oxygen-18 content of selected lunar material. *Science*, **170**, 538–40.   **4.4.3**

"From an Oxford note-book" (1929) *Observatory*, **52**, 126–8.   **1.3.3**

Frondel, C. *See* Arnold, J.

Frost, Edwin B. (1911) The contribution of astronomy to general culture. *Pop. Astron.*, **19**, 463–71.   **1.2.6**

    (1933) *An astronomer's life*. Boston: Houghton Mifflin.   **1.2.6**

    (D1905) Letter to T. C. Chamberlin, 15 March 1905. Box 5, C-UC.   **1.2.6**

Fukuoka, Takaaki. *See* Gooding, James L.

Gale, Henry G. *See* Michelson, A. A.

Gamow, George, & Hynek, J. A. (1945) A new theory by C. F. von Weizsäcker on the origin of the planetary system. *Astrophys. J.*, **101**, 249–54.   **1.1.4**

Ganapathy, R., & Anders, E. (1974) Bulk compositions of the Moon and Earth, estimated from meteorites. *Proc. 5th Lunar Sci. Conf.*, 1181–1206.
                                                                        4.1.4, 4.4.4
Ganapathy, R., & Grossman, L. (1976) The case for an unfractionated $^{244}Pu/^{238}U$ ratio in high-temperature condensates. *Earth Plan. Sci. Lett.*, **31**, 386–92.
                                                                        3.3.3
Ganapathy, R., Keays, R. R., Laul, J. C., & Anders, E. (1970) Trace elements in Apollo 11 lunar rocks: Implications for meteorites influx and origin of Moon. *Proceedings of the Apollo 11 Lunar Science Conference*, ed. A. A. Levinson, vol. 2, pp. 1117–42. New York: Pergamon Press.   **4.1.4, 4.4.3, 4.4.4**
Ganapathy, R., Laul, J. C., Morgan, J. W., & Anders, E. (1972) Moon: Possible nature of the body that produced the Imbrian Basin, from the composition of Apollo 14 samples. *Science*, **175**, 55–9.   **4.4.3**
Ganapathy, R., Morgan, J. W., Krähenbühl, U., & Anders, E. (1973) Ancient meteoritic components in lunar highland rocks: Clues from trace elements in Apollo 15 and 16 samples. *Proc. 4th Lunar Sci. Conf.*, 1239–61.   **4.4.3**
Ganapathy, R. *See also* Keays, R. R.; Morgan, J. W.
Gareis, A. (1901) Beiträge zur Kosmogonie. *Mittheilungen aus dem Gebiete des Seewesens*, **29**, 877–918.   **1.2.4**
Garwin, Laura (1989) Tales of a lost magma ocean. *Nature*, **338**, 19–20.   **4.1.6**
Garz, T., & Kock, M. (1969) Experimentelle Oszillatorenstärken von Fe I-Linien. *Astronomy and Astrophysics*, **2**, 274–9.   **3.3.3**
Garz, T., Kock, M., Richer, J., Baschek, B., Holweger, H., & Unsöld, A. (1969) Abundances of iron and some other elements in the sun and meteorites. *Nature*, **223**, 1254–5.   **3.3.3, 4.1.4, 4.3.3, 4.4.3**
Gast, P. W. (1960) Limitations on the composition of the upper mantle. *J. Geophys. Res.*, **65**, 1287–97.   **3.3.3**
Gast, P. W. *See also* Arnold, J.
Gasteyer, C. E. (1970) Forest Ray Moulton, April 29, 1872–December 7, 1952. *Biog. Mem. N. A. S.*, **41**, 341–55.   **1.1.1**
Gatewood, George, & Eichhorn, Heinrich (1973) An unsuccessful search for a planetary companion of Barnard's star (BD + 4°3561). *Astron. J.*, **78**, 769–76.
                                                                        2.1.2
Gault, D. E. *See* Murray, B. C.
Gautier, D., Bezard, B., Marten, A., Baluteau, J. P., Scott, N., Chedin, A., Kunde, V., & Hanel, R. A. (1982) The C/H ratio in Jupiter from the *Voyager* infrared investigation. *Astrophys. J.*, **257**, 901–12.   **3.2.3**
Gautier, D., & Owen, Tobias (1983) Cosmogonical implications of helium and deuterium abundances on Jupiter and Saturn. *Nature*, **302**, 215–18.   **3.2.3**
  (1985) Observational constraints on models for giant planet formation. In *Protostars and Planets*. II, ed. D. C. Black & M. S. Matthews, pp. 832–46. Tucson: University of Arizona Press.   **1.1.5, 2.4.2, 3.2.3**
Gautier, D. *See* Courtin, R.
Gehrels, Tom, Editor (1978) *Protostars and planets: Studies of star formation and of the origin of the Solar System*. Tucson: University of Arizona Press.   **2.2.5**
Geikie, Archibald (1894) *Class-book of geology*. London: Macmillan.   **1.2.2**
Geikie, J. (1900) A white-hot liquid earth and geological time. *Scottish Geographical Magazine*, **16**, 60–7.   **1.2.3**
Gershberg, R. E. (1982) A hypothesis on the hydromagnetic activity of T Tau-type stars and related objects. *Astron. Nachr.*, **303**, 251–61.   **2.3.5**
Gerstenkorn, Horst (1955) Uber Gezeitenreibung beim Zweikörperproblem. *Z. f. Astrophysik*, **36**, 245–74.   **4.1.2, 4.3.2**
  (1957) Veränderungen des Erde-Monde-Systems durch Gezeitenreibung. *Z. f. Astrophysik*, **42**, 137–55.   **4.3.2**

(1967a) The importance of tidal friction for the early history of the Moon. *Proc. Roy. Soc.*, **A296**, 293–7.   **4.3.2, 4.3.3**

(1967b) On the controversy over the effect of tidal friction upon the history of the Earth–Moon system. *Icarus*, **7**, 160–7.   **4.3.2, 4.3.3, 4.4.2**

(1968) A reply to Goldreich. *Icarus*, **9**, 394–5.   **4.3.2, 4.4.2**

(1969) The earliest past of the Earth–Moon system. *Icarus*, **11**, 189–207.   **4.1.2**

Gheury de Bray, Maurice (1925) L'hypothèse cosmogonique et la pluralité des mondes habités. *Ciel et Terre*, **41**, 8–14.   **1.3.3**

Gifford, A. C. (1925) New stars. *New Zealand Journal of Science and Technology*, **8**, 1–7.   **1.2.4**

Gilbert, Grove Karl (1893) The Moon's face: A study of the origin of its features. *Bulletin of the Washington Philosophical Society*, **12**, 241–92.   **4.2.4**

Gillespie, C. M., Jr. *See* Aumann, H. H.

Gillett, F. C. *See* Aumann, H. H.

Gillispie, Charles Coulston, Fox, Robert W., & Grattan-Guinness, Ivor (1978) Laplace, Pierre-Simon, Marquis de. *D.S.B.*, **15**, 273–356.   **4.2.2**

Gingerich, Owen *See also* Lang, K. R.

Gingold, Robert A., & Monaghan, J. J. (1980) The Roche problem for polytropes in central orbits. *M. N. Roy. Astron. Soc.*, **191**, 897–924.   **1.1.4**

Gingold, Robert A. *See also* Durisen, R.

Giuli, R. T. (1968a) On the rotation of the Earth produced by gravitational accretion of particles. *Icarus*, **8**, 301–23.   **1.2.3, 3.1.1**

(1968b) Gravitational accretion of small masses attracted from large distances as a mechanism for planetary rotation. *Icarus*, **9**, 186–90.   **1.2.3, 3.1.1**

Gleason, Jim D. *See* Friedman, Irving

Gnedin, Yu. N., Pogodin, M. A., & Red'kina, N. P. (1986) A determination of magnetic fields of T Tau and Ae/Be Herbig stars using the parameters of their linear polarization. In *Upper main sequence stars with anomalous abundances* (90th I. A. U. Colloquium, Crimea, USSR, May 1985), ed. C. R. Cowley et al., pp. 87–90. Boston: Reidel.   **2.3.5**

Gnedin, Yu. N., & Red'kina, N. P. (1984) The magnetic field of T Tauri. *Soviet Astronomy Letters*, **10**, 255–7. Translated from *Pis'ma Astronomicheski Zhurnal*, **10**, 613–19.   **2.3.5**

Goettel, Kenneth A. (1976) Models for the origin and composition of the Earth, and the hypothesis of potassium in the Earth's core. *Geophysical Surveys*, **2**, 369–97.
                                                                                  3.3.4

Goettel, Kenneth A., & Barshay, S. S. (1976/1978) The chemical equilibrium model for condensation in the solar nebula: Assumptions, implications, and limitations. In *The origin of the Solar System* (NATO Advanced Study Institute, 1976), ed. S. F. Dermott, pp. 611–27. New York: Wiley (pub. 1978).   **3.3.4, 3.4.3**

Goettel, Kenneth A., Shields, Janet A., & Decker, Deborah A. (1981) Density constraints on the composition of Venus. *Proc. 12th Lunar Plan. Sci. Conf.*, 1507–16.
                                                                                  3.3.5

Gold, Thomas (1955) The lunar surface. *M. N. Roy. Astron. Soc.*, **115**, 585–604.
                                                                                  4.2.4

(1960) Abundance of lithium and origin of the Solar System. *Astrophys. J.*, **132**, 274–5.   **2.3.1**

(1964) Fields and particles in interplanetary space. In *Space exploration and the Solar System*, pp. 181–93. New York: Academic Press.   **3.4.5**

Gold, Thomas, & Hoyle, Fred (1960) On the origin of solar flares. *M. N. Roy. Astron. Soc.*, **120**, 89–105.   **2.3.1, 2.3.2**

Gold, Thomas *See also* Bondi, H.

Goldberg, Leo, Müller, Edith A., & Aller, Lawrence H. (1960) The abundances of the elements in the solar atmosphere. *Astrophys. J. Supplement*, **5**, 1–137.   **3.3.3**

Goldreich, Peter (1963) On the eccentricity of satellite orbits in the Solar System. *M. N. Roy. Astron. Soc.*, **126**, 257–68.   **4.4.2**
(1966) History of the lunar orbit. *Reviews of Geophysics*, **4**, 411–39.   **4.4.2, 4.4.4**
(1968) On the controversy over the effect of tidal friction upon the history of the Earth–Moon system: A reply to comments by H. Gerstenkorn. *Icarus*, **9**, 391–3.   **4.3.2, 4.4.2**
(1972) Tides and the Earth–Moon system. *Sci. Am.*, **226**, no. 4, 42–52.   **4.4.2**
Goldreich, Peter, & Ward, William R. (1973) The formation of planetesimals. *Astrophys. J.*, **183**, 1051–61.   **1.1.5, 3.1.2**
Goldschmidt, V. M. (1937) The principles of distribution of chemical elements in minerals and rocks. *Journal of the Chemical Society* (London), 655–73.   **1.1.3**
(1958) *Geochemistry*. Oxford: Clarendon Press.   **4.3.4**
Goldsmith, Donald (1985) *The evolving universe*. 2nd edn. Menlo Park, CA: Benjamin/Cummings.   **2.5.3**
Goldsmith, Donald, & Owen, Tobias (1980) *The search for life in the universe*. Menlo Park, CA: Benjamin/Cummings.   **4.4.4**
Goldstein, B. E. *See* Russell, C. T.
Goles, G. G. (1969) Cosmic abundances. *Handbook of geochemistry*, ed. K. H. Wedepohl, vol. I, pp. 116–33. Berlin: Springer Verlag.   **3.3.3**
Goles, G. G. *See also* Fish, R. A.
Gooding, James L., Keil, Klaus, Fukuoka, Takaaki, & Schmitt, Roman (1980) *Earth Plan. Sci. Lett.*, **50**, 171–80.   **3.3.5**
Gopalan, K., Kaushal, S., Lee-Hu, C., & Wetherill, G. W. (1970) Rubidium-strontium, uranium, and thorium-lead dating of lunar material. *Science*, **167**, 471–3.   **4.4.4**
Gore, J. Ellard (1893) *The visible universe: Chapters on the origin and construction of the universe*. London: Lockwood.   **1.2.3**
(1902) The nebular hypothesis. *Gentlemen's Magazine*, **293**, 178–85.   **1.2.4**
(1906) The new ccsmogony. *Knowledge* (n.s.), **3**, 523–26; *Pop. Astron.*, **14**, 515–22.   **1.2.6**
(1907) *Astronomical essays historical and descriptive*. London: Chatto & Windus.
                                                                                1.2.6
Goswami, J. N., & Lal, D. (1979) Formation of the parent bodies of the carbonaceous chondrites. *Icarus*, **40**, 510–21.   **3.1.2**
Graboske, Harold C., Jr., Pollack, James B., Grossman, Allen S., & Olness, Robert J. (1975) The structure and evolution of Jupiter: The fluid contraction stage. *Astrophys. J.*, **199**, 265–81.   **3.2.1**
Graboske, Harold C., Jr. *See also* Grossman, A. S.; Pollack J. B.
Gradie, Jonathan C., Chapman, Clark R., & Williams, James G. (1979) Families of minor planets. In *Asteroids*, ed. T. Gehrels, pp. 359–90. Tucson: University of Arizona Press.   **3.4.7**
Gradie, Jonathan C., & Zellner, B. (1977) Asteroid families: Observational evidence for common origins. *Science*, **197**, 254–5.   **3.4.7**
Gradie, Jonathan C. *See also* Degewij, J.
Gradsztajn, E. *See* Bernas, R.
Graham, A. *See* Taylor, S. R.
Grant, Robert (1852) *History of physical astronomy*. London: Bohn.   **4.2.2**
Gray, C. M., & Compston, W. (1974) Excess ^{26}Mg in the Allende meteorite. *Nature*, **251**, 495–7.   **2.5.1**
Greeley, Ronald. *See* Murray, B.
Greenberg, Richard (1979) Growth of large, late-stage planetesimals. *Icarus*, **39**, 141–50.   **3.1.2**
(1980) Collisional growth of planetesimals. *Moon and Planets*, **22**, 63–6.   **3.1.2**
Greenberg, Richard, Chapman, C. R., Davis, D. R., Drake, M. J., Hartmann, W. K., Herbert, F., Jones, J. H., & Weidenschilling, S. J. (1984b) An integrated dynami-

cal and geochemical approach to lunar origin modelling. In *Origin of the Moon Conf. Abst.* **4.4.6**

Greenberg, Richard, Hartmann, W. K., Chapman, C. R., & Wacker, J. F. (1978a) The accretion of planets from planetesimals. In *Protostars and planets,* ed. T. Gehrels, pp. 599–622. Tucson: University of Arizona Press. **3.1.2**

Greenberg, Richard, Wacker, J., Chapman, C. R., & Hartmann, W. K. (1978b) Planetesimals to planets: A simulation of collisional evolution. *Icarus,* **35,** 1–26. **3.1.2**

Greenberg, Richard, Weidenschilling, Stuart J., Chapman, Clark R., & Davis, Donald R. (1984a) From icy planetesimals to outer planets and comets. *Icarus,* **59,** 87–113. **3.1.2**

Greenberg, Richard. *See also* Torbett, M.; Weidenschilling, S. J.

Greenland, L. P. *See* Cuttita, Frank

Greenstein, Jesse L. *See* Bonsack, W. K.; Fowler, W. A.

Gregory, J. W. (1912) *The making of the Earth.* London: Williams & Norgate. **1.2.4**

Grevesse, N. (1984) Accurate atomic data and solar photospheric abundances. *Physica Scripta,* **T8,** 49–58. **3.3.3**

Griffiths, K. *See* Faulkner, J.

Grimwood, James M. *See* Brooks, C. G.

Gros, Jacques, Takahashi, Hiroshi, Hertogen, Jan, Morgan, John W., & Anders, Edward (1976) Composition of the projectiles that bombarded the lunar highlands. *Proc. 7th Lunar Sci. Conf.,* 2403–425. **4.4.4**

Gros, Jacques. *See also* Morgan, J. W.

Grossman, Allen S., Pollack, James B., Reynolds, Ray T., Summers, Audrey L., & Graboske, H. C., Jr. (1980) The effect of dense cores on the structure and evolution of Jupiter and Saturn. *Icarus,* **42,** 358–79. **3.2.1**

Grossman, Allen S. *See also* Bodenheimer, P.; Graboske, H. C., Jr.; Pollack, J. B.

Grossman, Lawrence (1972) Condensation in the primitive solar nebula. *Geoch. Cosmoch. Acta,* **36,** 597–619. **3.3.3**

(1973) Refractory trace elements in Ca-Al-rich inclusions in the Allende meteorite. *Geoch. Cosmoch. Acta,* **37,** 1119–40. **3.3.3**

(1975) Petrography and mineral chemistry of Ca-rich inclusions in the Allende meteorite. *Geoch. Cosmoch. Acta,* **39,** 433–54. **3.3.3**

(1980) Refractory inclusions in the Allende meteorite. *Ann. Rev. Earth Plan. Sci.,* **8,** 559–608. **2.5**

(1981) Telltale inclusions. *Natural History,* **90,** no. 4 (April), 68–71. **2.5**

Grossman, L. A., & Clark S. P., Jr. (1973) High temperature condensation in chondrites and the environment in which they formed. *Geoch. Cosmoch. Acta,* **37,** 635–50. **3.3.3**

Grossman, L., Clayton, R. N., & Mayeda, T. K. (1974) Oxygen isotopic compositions of lunar soils and Allende inclusions and the origin of the Moon. *Lunar Sci.,* **5,** 298–300. **4.3.4**

Grossman, L., & Larimer, J. W. (1974) Early chemical history of the solar system. *Reviews of Geophysics and Space Physics,* **12,** 71–101. **3.3.3, 3.3.4**

Grossman, L., & Olsen, E. (1974) Origin of the high-temperature fraction of C2 chondrites. *Geoch. Cosmoch. Acta,* **38,** 173–87. **3.3.3**

Grossman, Lawrence, & Steele, Ian M. (1976) Amoeboid olivine aggregates in the Allende meteorite. *Geoch. Cosmoch. Acta,* **40,** 149–55. **3.3.3**

Grossman, Lawrence. *See also* Cameron, A. G. W.; Clark, S. P., Jr.; Clayton, R. N.; Ganapathy, R.; MacPherson, G. J.; Olsen, E. J.

Groves, Gordon W. (1960) On tidal torque and eccentricity of a satellite's orbit. *M. N. Roy. Astron. Soc.,* **121,** 497–502. **4.4.2**

Gurevich, L. E., & Lebedinskii, A. I. (1950) Ob obrazovanii planet (On the formation of planets). *Izvestiya Akademii Nauk SSSR, Seriya Fizicheskii,* **14,** 764–99. English

translation in *Origin of the Solar System: Soviet research, 1925–1991*, ed. A. E. Levin & S. G. Brush. New York: American Institute of Physics (1995).
                                                                    1.1.5, 3.1.1
Gutenberg, Beno (1930) Hypotheses on the development of the Earth. *Journal of the Washington Academy of Science,* **20,** 17–25.   **4.2.4**
Haar, D. ter (1944) On the origin of smoke particles in the interstellar gas. *Astrophys. J.,* **100,** 288–99.   **1.1.3**
  (1948) Studies on the origin of the Solar System. *Kongelige Danske Videnskabernes Selskab, Matematisk-Fysiske Meddelelser,* **25,** no. 3.   **1.1.3, 1.1.4**
  (1967) On the origin of the Solar System. *Ann. Rev. Astron. Astrophys.,* **5,** 267–78.
                                                                            2.2.1
Haar, D. ter, & Cameron, A. G. W. (1963) Historical review of theories of the origin of the Solar System. In *Origin of the Solar System,* ed. R. Jastrow & A. G. W. Cameron, pp. 1–37. New York: Academic Press.   **2.4.1**
Haar, D. ter. *See also* Chandrasekhar. S.
Hämeen-Antilla, K. A. (1977) Statistical mechanics of Keplerian orbits. IV. Concluding remarks. *Astrophys. Space Sci.,* **51,** 429–37.   **3.4.7**
  (1983) Collisions in self-gravitating clouds of planetesimals. *Moon and Planets,* **28,** 267–303.   **3.4.7**
Hale, George Ellery (1902) Stellar evolution in the light of recent research. *Popular Science Monthly,* **60,** 291–313.   **1.2.4, 1.2.5**
  (1908) *The study of stellar evolution: An account of some recent methods of astrophysical research.* Chicago: University of Chicago Press.   **1.2.6**
  (D1899a) Letter to F. R. Moulton, 13 June 1899. Letterbook 7, p. 153, Yerkes Observatory.   **1.2.3**
  (D1899b) Letter to F. R. Moulton, 22 June 1899. Letterbook 7, p. 187, Yerkes Observatory.   **1.2.3**
  (D1900) Letter to F. R. Moulton, 20 February 1900. Letterbook 8, p. 71, Yerkes Observatory.   **1.2.3**
Hall, R. Cargill (1977) *Lunar impact: A history of Project Ranger.* Washington, DC: National Aeronautics and Space Administration, report NASA SP-4210.
                                                                            4.3.1
Hall, R. Cargill. *See also* Emme, E. M.
Halley, Edmond (1695) Some account of the ancient state of the City of Palmyra, with short remarks upon the inscriptions found there. *Phil. Trans.,* **19,** 160–75.
                                                                            4.2.2
Halm, Jacob E. (1905a) On Professor Seeliger's theory of temporary stars. *Proc. Roy. Soc. Edinb.,* **25,** 513–52.   **1.2.4**
  (1905b) Some suggestions on the nebular hypothesis. *Proc. Roy. Soc. Edinb.,* **25,** 553–61.   **1.2.4**
Halstead, B. (1980) Popper: Good philosophy, bad science? *New Scientist,* **87,** 215–17.
                                                                            2.2.2
Hammond, Allen L. (1974) Exploring the Solar System. III. Whence the Moon? *Science,* **186,** 911–13.   **4.4.5**
  (1976) Presolar grains: Isotopic clues to solar system origin. *Science,* **192,** 772–3.
                                                                            4.3.4
Hanappe, F., Vosters, M., Picciotto, E., & Deutsch, S. (1968) Chimie des nieges antarctiques et taux de deposition de matiere extraterrestre – deuxieme article. *Earth Plan. Sci. Lett.,* **4,** 487–96.   **4.4.3**
Hanel, R. A., Conrath, B. J., Herath, L. W., Kunde, V. G., & Pirraglia, J. A. (1981) Albedo, internal heat, and energy balance of Jupiter: Preliminary results of the Voyager infrared investigation. *J. Geophys. Res.,* **86,** 8705–12.   **3.2.1**
Hanel, R. A., Conrath, B. J., Kunde, V. G., Pearl, J. C., & Pirraglia, J. A. (1983) Albedo, internal heat flux, and energy balance of Saturn. *Icarus,* **53,** 262–85.
                                                                            3.2.1

Hanel, R. A. *See also* Courtin, R.; Gautier, D.

Hansen, Kirk S. (1982) Secular effects of oceanic tidal dissipation on the Moon's orbit and the Earth's rotation. *Reviews of Geophysics and Space Physics*, **20**, 457–80.
                                                                      4.3.3, 4.4.2

Hardcastle, Kenneth. *See* Friedman, I.

Hargraves, R. B. *See* Banerjee, S. K.

Harper, D. A., Loewenstein, R. F., & Davidson, J. A. (1984) On the nature of the material surrounding Vega. *Astrophys. J.*, **285**, 808–12.   **2.1.2**

Harrington, R. S. *See* Black, D. C.

Harris, A. W. (1977) An analytic theory of planetary rotation rates. *Icarus*, **31**, 168–74.
                                                                      1.2.3, 3.1.1

(1978a) The formation of the outer planets. *Lunar & Plan. Sci.*, **9**, 459–61.
                                                                      1.1.5, 3.2.2

(1978b) Satellite formation. II. *Icarus*, **34**, 128–45.   **4.4.4, 4.4.6**

Harris, A. W., & Kaula, W. M. (1975) A co-accretional model of satellite formation. *Icarus*, **24**, 516–24.   **4.1.4, 4.4.4, 4.4.6**

Harris, A. W., *See also* Kaula, W. M.; Stevenson, D. J.

Harris, Marvin (1968) *The rise of anthropological theory: A history of theories of culture.* New York: Crowell.   **1.2.2**

Harris, P. G., & Tozer, D. C. (1967) Fractionation of iron in the Solar System. *Nature*, **215**, 1449–51.   **3.3.3, 4.1.3**

Hart, Richard. *See* Berendzen, R.

Hartmann, William K. (1965) Terrestrial and lunar flux of large meteorites in the last two billion years. *Icarus*, **4**, 157–65.   **4.1.3**

(1966) Early lunar cratering. *Icarus*, **5**, 406–18.   **4.1.5, 4.4.5**

(1973). *Moons and planets.* Rev. edn. Belmont, CA: Wadsworth Publishing.
                                                                      4.4.4

(1976) The Moon's early history. *Astronomy*, **4**, no. 9. (September), 7–16.   **4.4.5**

(1978) The early history of planet Earth. *Astronomy*, **6**, no. 8 (August), 6–19.
                                                                      4.4.5

(1983) *Moons and planets.* Belmont, CA: Wadsworth Publishing Co.   **4.4.5**

(1986) Moon origin: The impact-trigger hypothesis. In *The origin of the Moon*, ed. W. K. Hartmann et al., pp. 579–608. Houston: Lunar & Planetary Institute.
                                                                      4.4.5

(1987) *Astronomy: The cosmic journey.* Rev. edn. Belmont, CA: Wadsworth Publishing.   **2.2.1**

(1989) Piecing together Earth's early history. *Astronomy*, **17**, no. 6 (June), 24–34.
                                                                      4.1.6

Hartmann, William K., & Davis, D. R. (1975) Satellite-sized planetesimals and lunar origin. *Icarus*, **24**, 504–15.   **3.1.2, 4.1.5, 4.4.5**

Hartmann, William K., Phillips, R. J., & Taylor, G. J., Editors (1986) *Origin of the Moon.* Houston: Lunar and Planetary Institute.   **2.2.2, 4.4.5**

Hartmann, William K., & Vail, S. M. (1986) Giant impactors: Plausible size and populations. In *Origin of the Moon*, ed. W. K. Hartmann et al., pp. 551–6. Houston: Lunar and Planetary Institute.   **4.4.5**

Hartmann, William K. *See also* Greenberg, R.; Weidenschilling, S. J.

Hashimoto, Akihito. *See* MacPherson, G. J.

Hawkins, J. W. *See* Urey, H. C.

Hayashi, Chushiro (1961) Stellar evolution in early phases of gravitational contraction. *Pub. Astron. Soc. Japan*, **13**, 450–2.   **1.1.5, 3.3.3**

Hayashi, Chushiro, Adachi, I., & Nakazawa, K. (1976) Formation of the planets. *Prog. Th. Phys.*, **55**, 945–6.   **3.2.2**

Hayashi, Chushiro, Nakazawa, Kiyoshi, & Adachi, Isao (1977) Long-term behavior of planetesimals and the formation of the planets. *Pub. Astron. Soc. Japan*, **29**, 163–96.   **1.1.5, 3.2.2**

Hayashi, Chushiro, Nakazawa, Kiyoshi, & Nakagawa, Y. (1985) Formation of the Solar System. In *Protostars and planets*. II, ed. D. C. Black & M. S. Matthews, pp. 1100–53. Tucson: University of Arizona Press.   3.4.8

Hayashi, Chushiro. *See also* Mizuno, H.; Nakazawa, K.

Head, J. W. *See* Kaula, W. M.

Heintz, W. D. (1976) Systematic trends in the motions of suspected stellar companions. *M. N. Roy. Astron. Soc.*, **175**, 533–5.   2.1.2

(1978) Reexamination of suspected unresolved binaries. *Astrophys. J.*, **220**, 931–4.   2.1.2

Heisenberg, Werner (1948a) Zur statistischen Theorie der Turbulenz. *Zeitschrift für Physik*, **124**, 628–57.   1.1.4

(1948b) On the theory of statistical and isotropic turbulence. *Proc. Roy. Soc.*, **A195**, 402–6.   1.1.4

Helin, E. F. *See* De Groot, Mart

Helmholtz, Hermann von (1854/1962) *Ueber die Wechselwirkung der Naturkräfte und die darauf bezüglichen neuesten Ermittelungen der Physik*. Königsberg: Gräfe & Unzer. English translation in Helmholtz's *Popular Scientific Lectures*, pp. 59–92. New York: Dover Publications (1962).   4.2.2

Henbest, Nigel. *See* Couper, Heather

Henderson-Sellers, A. (1983). *The origin and evolution of planetary atmospheres*. Bristol: Adam Hilger.   3.3.4

Henkel, F. W. (1909) The birth and death of worlds. *Knowledge* (n.s.), **6**, 6–8.   1.2.4

(1910a) New theories of the evolution of stellar system. *Science Progress in the Twentieth Century*, **5**, 82–90.   1.2.4

(1910b) New methods in astronomy (and their application to the evolution of our system). *Knowledge* (n.s.), **7**, 350–4.   1.2.4

Hennecke, E. W. *See* Manuel, O. K.

Henon, Michel (1978) Motion of enclosed particles around a central mass point: Errors in the "apples in a spacecraft" model. *Science*, **199**, 692–3.   3.4.7

Henyey, L. G., LeLevier, Robert, & Levee, R. D. (1955) The early phases of stellar evolution. *Pub. Astron. Soc. Pac.*, **67**, 154–60.   2.3.2

Herbert, Floyd, Davis, Donald R., & Weidenschilling, S. J. (1986) Formation and evolution of a circumterrestrial disk: Constraints on the origin of the Moon in geocentric orbit. In *Origin of the Moon*, ed. W. K. Hartmann et al., pp. 701–30. Houston: Lunar and Planetary Institute.   4.4.6

Herbert, Floyd. *See also* Greenberg, R.; Weidenschilling, S. J.

Herbig, G. H. (1962) The properties and problems of T Tauri stars and related objects. In *Advances in astronomy and astrophysics*, ed. Z. Kopal, pp. 47–103. New York: Academic Press.   3.3.3

Herbst, William, & Assousa, George E. (1977) Observational evidence for supernova-induced star formation: Canis Major R1. *Astrophys. J.*, **217**, 473–87.   2.5.2

(1978) The role of supernovae in star formation and spiral structure. In *Protostars and planets*, ed. T. Gehrels, pp. 368–83. Tucson: University of Arizona Press.   2.5.2

Herbst, William, & Rajan, R. Sundar (1980) On the role of a supernova in the formation of the Solar System. *Icarus*, **42**, 35–42.   2.5.2

Herczeg, Tibor (1967/1969) Cosmogonical aspects in the theory of planetary interiors. In *The application of modern physics to the Earth and planetary interiors* (NATO Advanced Study Institute, 1967), ed. S. K. Runcorn, pp. 301–9. New York: Wiley (pub. 1969).   2.4

(1968) Planetary cosmogonies. *Vistas in Astronomy*, **10**, 175–206.   2.3.5

Herndon, J. M. (1978) Re-evaporation of condensed matter during the formation of the Solar System. *Proc. Roy. Soc.*, **A363**, 283–8.   3.3.5

Hertogen, Jan, Janssens, Marie-Josee, Takahashi, H., Palme, H., & Anders, E. (1977) Lunar basins and craters: evidence for systematic compositional changes of bombarding population. *Proc. 8th Lunar Sci. Conf.*, 17–45.   **4.4.4, 4.4.6**

Hertogen, Jan. *See* Gros, J.; Morgan, J. W.

Herzog, G. F. *See* Keays, R. R.

Hetherington, Norriss S. (1972) Adriaan van Maanen and internal motions in spiral nebulae: A historical review. *Q. J. Roy. Astron. Soc.*, **13**, 25–39.   **1.2.6**

(1974) Edwin Hubble on Adriaan van Maanen's internal motions in spiral nebulae. *Isis*, **65**, 390–3.   **1.2.6**

(1975) The simultaneous "discovery" of internal motions in spiral nebulae. *J. Hist. Astron.*, **6**, 115–25.   **1.2.4, 1.2.6**

(1988) *Science and objectivity.* Ames: Iowa State University Press.   **1.2.6**

(1994) Converting an hypothesis into a research program: T. C. Chamberlin, his planetesimal hypothesis, and its effect on research at the Mt. Wilson Observatory. In *The Earth, the heavens and the Carnegie Institution of Washington*, ed. G. Good, pp. 113–23. Washington, DC: American Geophysical Union.   **1.2.6**

Heymann, Dieter. *See* Anders, E.

Hibberson, W. *See* Ringwood, A. E.

Higuchi, H. *See* Morgan, J. W.

Hill, E. L. *See* Luyten, W. J.

Hillebrandt, W., & Thielemann, F.-K. (1982) Nucleosynthesis in novae: A source of Ne-E and ^{26}Al? *Astrophys. J.*, **255**, 617–23.   **2.5.3**

Hillebrandt, W. *See also* Arnould, M.

Hinners, Noel W. (D1972) Memorandum, 15 August 1972. In Urey file at NASA-A.   **4.3.2**

Hipkin, R. G. *See* Scrutton, C. T.

Hirschfelder, J. O., Curtiss, C. F., & Bird, R. B. (1954) *Molecular theory of gases and liquids.* New York: Wiley.   **3.1.1**

Hobbs, W. E. (1922) *Earth evolution and its facial expression.* London: Macmillan.   **1.3.3**

Hobson, E. W. (1921–22/1968) *The domain of natural science* (Gifford Lectures at University of Aberdeen, 1921–22). Repr., New York: Dover Publications (1968).   **1.3.3**

Hodge, Paul (1984) The cosmic distance scale. *American Scientist*, **72**, 474–82.   **2.1.2**

Hodges, R. R. *See* Hoffman, J. H.

Höflich, P. *See* Cameron, A. G. W.

Hörbiger, Hanns (1913) *Hörbigers Glacial-Kosmogonie. Eine neue Entwickelungsgeschichte des Weltalls und des Sonnenysystems . . .* , ed. P. Fauth. Kaiserslautern: Kayser.   **4.2.4**

Hoffman, J. H., Hodges, R. R., Donahue, T. M., & McElroy, M. B. (1980) Composition of the Venus lower atmosphere from the Pioneer Venus mass spectrometer. *J. Geophys. Res.*, **85**, 7882–90.   **3.3.5**

Hofmeister, H. *See* Wänke, H.

Hohenberg, C. M. *See* Swindle, T. D.

Hoinkes, G. *See* Kurat, G.

Holberg, J. B. *See* Cuzzi, J. N.

Holdren, John P. *See* Smith, K. R.

Holmberg, Erik. *See* Reuyl, D.

Holmes, Arthur (1913) *The age of the Earth.* New York: Harper & Brothers.   **1.2.6, 1.3.2**

(1915) Radioactivity and the Earth's thermal history. *Geol. Mag.* (decade VI), **2**, 60–71, 102–12; **3** (1916), 265–74.   **1.2.6, 1.3.2**

(1931) Radioactivity and Earth movements. *Transactions of the Geological Society of Glasgow*, **18**, 559–606.   **1.3.2**

Holweger, H. *See* Garz, T.

Horn, H. S. (1986) Notes on empirical ecology. *American Scientist,* **74,** 572–3.
                                                                        2.2.2

Houck, J. R. *See* Aumann, H. H.

Houpis, Harry L. F., & Mendis, D. A. (1983) The fine structure of the Saturnian ring
    system. *Moon and Planets,* **29,** 39–46.   **3.4.7**

Hovey, E. O. (1905) Geology and geography at the Denver meeting of the American
    Association for the Advancement of Science. *Sci. Am. Supplement,* **52,** 21504–5.
                                                                        1.2.4

Howard, A. J. *See* Champagne, A. E.

Howe, Herbert A. (1896) *A study of the sky.* Meadville, PA: Flood & Vincent.
                                                                        1.2.2

Hoyle, Fred (1944) On the origin of the Solar System. *Proceedings of the Cambridge
    Philosophical Society,* **40,** 256–8.   **2.3.1**

    (1945) Note on the origin of the Solar System. *M. N. Roy. Astron. Soc.,* **105,** 175–
    8.   **2.3.1, 2.5.2**

    (1955) *Frontiers of astronomy.* New York: Harper.   **2.3.1**

    (1960) On the origin of the solar nebula. *Q. J. Roy. Astron. Soc.,* **1,** 28–55.
                                                                 1.1.5, 2.3.1, 3.3.3

    (1965) *Galaxies, nuclei, and quasars.* New York: Harper & Row.   **2.3.1**

    (1969) Planetary formation and lunar material. *Science,* **166,** 401.   **2.3.5**

    (1971) Origin of the solar nebula. In *Highlights of astronomy,* ed. C. de Jaeger, vol. 2,
    pp. 195–203. Dordrecht: Reidel.   **2.3.5**

    (1972) The history of the Earth. *Q. J. Roy. Astron. Soc.,* **13,** 328–45.   **2.3.5**

    (1979) *The cosmogony of the solar system.* Swansea: University College Cardiff Press/
    Christopher Davies Publishers.   **2.3.5**

    (1986) The abundances of the elements and some reflections on the cosmogony of
    the solar system. In *Earth and the human future,* ed. K. R. Smith et al., pp. 28–44.
    Boulder: Westview Press.   **2.3.1, 3.4.2**

Hoyle, Fred, & Ireland, J. G. (1960) Note on the transference of angular momentum
    within the galaxy through the agency of a magnetic field. *M. N. Roy. Astron.
    Soc.,* **121,** 253–9.   **2.3.2**

Hoyle, Fred, & Wickramasinghe, N. C. (1968) Condensation of the planets. *Nature,*
    217, 415–18.   **2.3.3**

Hoyle, Fred. *See also* Burbidge, E. M.; Burnett, D. S.; Faulkner, J.; Fowler, W. A.;
    Gold, T.; Reeves, H.

Hoyt, William Graves (1980) *Planets X and Pluto.* Tucson: University of Arizona Press.
                                                                        3.4.6

    (1982) G. K. Gilbert's contribution to selenology. *J. Hist. Astron.,* **13,** 155–67.
                                                                        4.2.4

    (1987) *Coon Mountain controversies: Meteor Crater and the development of impact theory.*
    Tucson: University of Arizona Press.   **4.1.5**

Hu, Z. *See* Dai, W.

Huang, Su-Shu (1973) Extrasolar planetary systems. *Icarus,* **18,** 339–76.   **1.1.3**

Hubbard, N. J. *See* Pomeroy, J. H.

Hubbard, W. B. (1968) Thermal structure of Jupiter. *Astrophys. J.,* **152,** 745–54.
                                                                        3.2.1

    (1969) Thermal models of Jupiter and Saturn. *Astrophys. J.,* **155,** 333–44.   **3.2.1**

    (1981) Constraints on the origin and interior structure of the major planets. *Phil.
    Trans.,* **A303,** 315–26.   **3.2.2, 3.2.3**

Hubbard, W. B., & MacFarlane, J. J. (1980) Structure and evolution of Uranus and
    Neptune. *J. Geophys. Res.,* **85,** 225–34.   **3.2.2**

Hubickyj, O. *See* Cabot, W.

Huggins, William (1892) The new star in Auriga (lecture at the Royal Institution, 13
    May 1892). *Proceedings of the Royal Institution,* **13,** 615–24.   **1.2.4**

Huneke, J. C. *See* Albee, A. L.

Hunt, G. E. *See* Smith, B. A.

Hwaung, Golden. *See* Manuel, O. K.

Hynek, J. A. *See* Gamow, G.

Imbrie, John (1982) Astronomical theory of the Pleistocene ice ages: A brief historical review. *Icarus,* **50,** 408–22. 4.2.2

Imbrie, John, & Imbrie, Katherine Palmer (1979) *Ice ages: Solving the mystery.* Short Hills, NJ: Enslow Publishers. 1.2.1, 4.2.2

Imbrie, Katherine Palmer. *See* Imbrie, J.

Ingersoll, Andrew P., Orton, G. S., Münch, G., Neugebauer, G., & Chase, S. C. (1980) Pioneer Saturn infrared radiometer: Preliminary results. *Science,* **207,** 439–43. 3.2.1

Ingersoll, Andrew P. *See also* Smith, B. A.

Ip, Wing-Huen (1975) Application of the theory of jet stream to the asteroid belt. *Astrophys. Space Sci.,* **38,** 475–81. 3.4.7

(1976a) Dynamical study of the Hilda asteroids. I. Resonant orbital motion of the PLS objects from the Palomar–Leiden survey. *Astrophys. Space Sci.,* **44,** 373–83. 3.4.7

(1976b) A note on the cosmogony of the asteroidal belt. *Astrophys. Space Sci.,* **45,** 235–41. 3.4.7

(1978) On the origin of Uranus' rings. *Nature,* **272,** 802–3. 3.4.7

Ip, W.-H., & Mendis, D. A. (1974) On the effects of accretion and fragmentation in interplanetary matter streams. *Astrophys. Space Sci.,* **30,** 233–41. 3.4.7

Ireland, J. G. *See* Hoyle, F.

Irifune, T. *See* Kato, T.

Irons, James C. (1896) *Autobiographical sketch of James Croll with memoir of his life and work.* London: Stanford. 1.2.1

Jacobson, A. S. *See* Mahoney, W. A.

Jacoby, Harold (1915) *Astronomy: A popular handbook.* New York: Macmillan. 1.2.6

Jagoutz, E. *See* Wänke, H.

Jakeŝ, P. *See* Taylor, S. R.

Jaki, Stanley L. (1978) *Planets and planetarians: A history of theories of the origin of planetary systems.* New York: Wiley. 2.1

Janssens, Marie-Josee. *See* Hertogen, Jan

Jastrow, Robert (1959) *Red giants and white dwarfs.* Rev. edn. New York: New American Library. 4.3.1

(1960) The exploration of the Moon. *Sci. Am.,* **202,** no. 5 (May), 61–9. 4.3.1

(1961) Statement. In *NASA scientific and technical programs.* Hearings before the Committee on Aeronautical and Space Sciences, U. S. Senate, 87th Congress, 1st Session, 28 February and 1 March 1961, pp. 111–27. Washington, DC: U.S. Government Printing Office. 4.3.2

(1981a) A red Solar System. *Science Digest,* **89,** no. 4 (May), 14, 116. 4.3.1

(1981b) Exploring the Moon. In *Space science comes of age,* ed. P. A. Hanle & V. D. Chamberlain, pp. 45–50. Washington, DC: National Air & Space Museum/ Smithsonian Institution Press. 4.3.1, 4.3.5

Jastrow, Robert, & Cameron, A. G. W., Editors (1962/1963) *Origin of the Solar System.* Proceedings of a Conference (1962). New York: Academic Press (pub. 1963). 2.2.5

Jastrow, Robert, & Newell, Homer (1963) Why land on the Moon? *Atlantic,* **212,** no. 2 (August), 41–50. 4.3.2

Jeans, J. H. (1901) The stability of a spherical nebula. *Proc. Roy. Soc.,* **68,** 454–5. 1.3.1

(1902) The stability of a spherical nebula. *Phil. Trans.*, **199**, 1–58.   **1.3.1**
(1903) On the vibrations and stability of a gravitating planet. *Phil. Trans.*, **201**, 157–
  84.   **1.3.1**
(1905) On the density of *Algol* variables. *Astrophys. J.*, **22**, 93–102.   **1.3.1**
(1917) The motion of tidally distorted masses, with special reference to theories of
  cosmogony. *Memoirs of the Royal Astronomical Society*, **62**, 1–48.   **1.1.1**
(1919) *Problems of cosmogony and stellar dynamics* (Adams Prize Essay for 1917). Cambridge University Press.   **1.1.1, 1.2.4, 1.2.6, 1.3.1, 1.3.3, 1.3.4, 1.3.5, 2.2.3**
(1929a) *Astronomy and cosmogony*. 2nd edn. Cambridge University Press.   **1.2.6**
(1929b) The planetesimal hypothesis. *Observatory*, **52**, 172–3.   **1.3.3**
(1932) *The mysterious universe.* Rev. edn. Cambridge Univerity Press.

          1.1.1, 1.3.4
(1938) The origin of the planets. *Science and Culture*, **4**, 73–5.   **1.3.4**
(1942) Origin of the Solar System. *Nature*, **149**, 695.   **1.1.1, 1.3.4**
(1943) The evolution of the solar system. *Endeavour*, **2**, 3–11.   **2.2.5**
Jeffery, P. M. *See* Keays, R. R.
Jeffreys, Harold (1916a) On certain possible distributions of meteoric bodies in the
  Solar System. *M N. Roy. Astron. Soc.*, **77**, 84–112.   **1.1.1, 1.3.2**
(1916b) The compression of the Earth's crust in cooling. *Phil. Mag.*, **32**, no. 6, 575–
  91.   **1.3.2**
(1917a) Theories regarding the origin of the Solar System. *Science Progress*, **12**, 52–
  62.   **1.1.1, 1.3.2**
(1917b) The resonance theory of the origin of the Moon. *M. N. Roy. Astron. Soc.*,
  78, 116–31.   **4.2.4**
(1918) On the early history of the Solar System. *M. N. Roy Astron. Soc.*, **78**, 424–
  42.   **1.1.1**
(1920a) The chief cause of the lunar secular acceleration. *M. N. Roy. Astron. Soc.*,
  80, 309–17.   **4.2.4**
(1920b) Tidal friction in shallow seas. *Phil. Trans.*, **A221**, 239–64.   **4.2.4, 4.4.2**
(1921) On certain geological effects of the cooling of the Earth. *Proc. Roy. Soc.*,
  A100, 122–49.   **1.3.2**
(1924) *The Earth: Its origin, history and physical constitution.* Cambridge University
  Press.   **1.3.2, 1.3.3**
(1925a) The evolution of the Earth. In *Evolution in the light of modern knowledge*, pp.
  31–58. London: Blackie.   **4.2.4**
(1925b) Origin of the Solar System: A Reply to "T. C. C." *Am. J. Sci.* [5], **9**, 395–
  405.   **1.3.2, 1.3.3**
(1929a) *The Earth.* 2nd edn. New York: Macmillan/Cambridge University Press.

          1.3.2
(1929b) The planetesimal hypothesis. *Science*, **69**, 245–6.   **1.3.3**
(1929c) (no title) *Observatory*, **52**, 173–7.   **1.3.3**
(1929d) Collision and the origin of rotation in the Solar System. *M. N. Roy. Astron.
  Soc.*, **89**, 636–41.   **1.3.4**
(1929e) The early history of the Solar System on the collision theory. *M. N. Roy.
  Astron. Soc., Supplement*, **89**, 731–8.   **1.3.4**
(1930) The resonance theory of the origin of the Moon (second paper). *M. N. Roy.
  Astron. Soc.*, **91**, 169–73.   **4.1.1, 4.2.4**
(1931a) Origin of Solar System explained in a new theory. *New York Times*, 3 May
  1931, Sect. IX, p. 4.   **1.3.4**
(1935) Origin of the Solar System. *Nature*, **136**, 932–3.   **1.3.4, 1.3.5**
(1944) Origin of the Solar System. *Nature*, **153**, 140.   **1.1.3**
(1948) The origin of the Solar System. *M. N. Roy. Astron. Soc.*, **108**, 94–103.

          1.1.3, 1.3.5
(1952) The origin of the Solar System. *Proc. Roy. Soc.*, **A214**, 281–91.   **1.3.5**
(1973) Developments in geophysics. *Ann. Rev. Earth Plan. Sci.*, **1**, 1–13.   **1.3.2**

(1974) *Collected papers of Sir Harold Jeffreys on geophysics and other sciences.* III. New York: Gordon & Breach.   **1.3.2**

(D1934) Letter to H. N. Russell, 2 April [1934]. Box 32, R-PU.   **1.3.4**

Jewitt, D. C. (1982) The rings of Jupiter. In *Satellites of Jupiter,* ed. D. Morrison & M. S. Matthews, pp. 44–64. Tucson: University of Arizona Press.   **3.4.6**

Johnson, P. H. *See* Bogard, D. D.; Taylor, S. R.

Johnson, Torrence V. *See* Chapman, C. R.; Smith, B. A.

Jones, H. Spencer (1922) *General astronomy.* London: Arnold/New York: Longmans, Green.   **1.3.3**

Jones, John H., & Drake, M. J. (1986) Constraints on the origin of the Moon. *Geoch. Cosmoch. Acta,* **50,** 1827.   **4.4.4**

Jones, John H. *See* Greenberg, R.; Kreutzberger, M. E.; Weidenschilling, S. J.

Jones, William F. (1922) A critical review of Chamberlin's groundwork for the study of megadiastrophism. *Am. J. Sci.* **3,** no. 5, 393–413.   **1.3.3**

Jordan, C. *See* Appenzeller, I.

Kaempffert, Waldemar (1909) The life of a star. *Outlook,* **92,** 707–16.   **1.2.4**

Käppeler, F. *See* Mathews, G. J.

Kant, Immanuel (1754a) Untersuchung der Frage, ob die Erde in ihrer Umdrehung um die Achse, wodurch sie die Abwechselung des Tages und der Nacht hervorbringt, einige Veränderung seit den ersten Zeiten ihres Ursprungs erlitten habe, und woraus man sich ihrer Versichern könne, welche von der königl. Akademie der Wissenschaften zu Berlin zum Preise für das jetztlaufende Jahraufgegeben worden. In *Kant's gesammelte Schriften,* vol. I, pp. 183–91. Berlin: Reimer (1910).   **4.2.2**

(1754b) Die Frage: Ob die Erde veralte? physikalisch erwogen. In *Kant's gesammelte Schriften,* vol. I, pp. 193–213. Berlin: Reimer (1910).   **4.2.2**

(1755/1981) *Universal natural history and theory of the heavens.* Trans. with introduction and notes by S. L. Jaki. Edinburgh: Scottish Academic Press (1981).   **4.2.2**

(1910) *Kant's gesammelte Schriften.* Berlin: Reimer.   **4.2.2**

Kato, T., Ringwood, A. E., & Irifune, T. (1988) Experimental determination of element partitioning between silicate perovskites, garnets and liquids: Constraints on early differentiation of the mantle. *Earth Plan. Sci. Lett.,* **89,** 123–45.   **4.1.6**

Katz, J. L. *See* Blander, M.

Kaula, William M. (1964) Tidal dissipation by solid friction and the resulting orbital evolution. *Reviews of Geophysyics,* **2,** 661–85.   **4.4.2**

(1968) *An introduction to planetary physics: The terrestrial planets.* New York: Wiley.   **2.4**

(1971) Dynamical aspects of lunar origin. *Reviews of Geophysics and Space Physics,* **9,** 217–38.   **4.4.2, 4.4.4**

(1974/1977) Mechanical processes affecting differentiation of protolunar material. In *The Soviet–American Conference on Cosmochemistry of the Moon and Planets (Moscow, June 1974),* ed. J. H. Pomeroy & N. J. Hubbard, pp. 805–13. Washington, DC: National Aeronautics and Space Administration, report NASA SP 370 (pub. 1977).   **4.4.3**

(1977a) Review of *Structure and evolutionary history of the solar system* by Alfvén & Arrhenius. *Icarus,* **31,** 181–2.   **3.4.7, 3.4.8**

(1977b) On the origin of the Moon, with emphasis on bulk composition. *Proc. 8th Lunar Sci. Conf.,* 321–31.   **3.1.2, 4.1.6, 4.4.4**

(1979) Thermal evolution of Earth and Moon growing by planetesimal impacts. *J. Geophys. Res.,* **84,** 999–1008.   **3.1.2, 4.1.6**

(1981) Inferences from other bodies for the Earth's composition and evolution. In *Evolution of the Earth,* ed. R. J. O'Connell, pp. 141–6. Washington, DC: American Geophysical Union/Boulder: Geological Society of America.   **4.4.5**

(1986) Formation of the sun and its planets. In *Physics of the Sun,* ed. P. A. Sturrock, vol. 3, pp. 1–32. Boston: Reidel.   **3.3.5**

(D1987) The Earth as a planet (Presented at "Quo Vadimus" Symposium, XIX General Assembly, International Union of Geodesy and Geophysics, Vancouver, 14 August 1987).   4.4.5

Kaula, William M., & Beachey, A. E. (1986) Mechanical models of close approaches and collisions of large protoplanets. In *Origin of the Moon*, ed. W. K. Hartmann et al., pp. 567–76. Houston: Lunar and Planetary Institute.   4.4.5

Kaula, William M., Drake, M. J., & Head, J. W. (1986) The Moon. In *Satellites*, ed. J. A. Burns & M. S. Matthews, pp. 581–628. Tucson: University of Arizona Press.   4.4.4

Kaula, William M., & Harris, A. W. (1973) Dynamically plausible hypothesis of lunar origin. *Nature*, **245**, 367–9.   4.4.4

(1975) Dynamics of lunar origin and orbital evolution. *Reviews of Geophysics and Space Physics*, **13**, 363–71.   4.4.4, 4.4.5

Kaula, William M. *See also* Harris, A. W.; Kobrick, M.

Kaushal, S. *See* Gopalan, K.

Kavanagh, R. W. *See* Skelton, R. T.

Kaye, Maureen J. *See* Compston, W.; Taylor, S. R.

Keays, R. R., Ganapathy, R., Laul, J. C., Anders, E., Herzog, G. F., & Jeffery, P. M. (1970) Trace elements and radioactivity in lunar rocks: Implications for meteorite infall, solar-wind flux, and formation conditions of the Moon. *Science*, **167**, 490–3.   4.4.3

Keays, R. R. *See* Ganapathy, R.

Keil, Klaus. *See* Gooding, James L.

Keller, James E. (1900) The Crossley Reflector of the Lick Observatory. *Astrophys. J.*, **11**, 325–49.   1.2.3

Kellogg, L. H. *See* Turcotte, D. L.

Kelly, W. R., & Wasserburg, G. J. (1978) Evidence for the existence of ^{107}Pd in the early Solar System. *Geophys. Res. Lett.*, **5**, 1079–1082.   2.5.2

(1980) [Letter to the editor] *Sky & Tel.*, **59**, 14–15.   2.5.2

Kelvin, Lord [William Thomson] (1897/1899) The age of the Earth as an abode fitted for life. *Journal of the Victoria Institute for 1897*, **31**, 11–35 (pub. 1899); *Phil. Mag.* [5], **47** (1899), 66–90; *Science*, **9** (1899), 704–11.   1.2.3

(1901) On ether and gravitational matter through infinite space. *Phil. Mag.* [6], **2**, 161–77.   1.3.1

(1908) On the formation of concrete matter from atomic origins. *Phil. Mag.* [6], **15**, 397–413.   4.2.4

Kerfoot, J. B. (1912) The new nebular hypothesis. Prof. See's book on the evolution of the stellar system, condensed and explained for the layman. *World To-day* [*Hearst's Magazine*], **21**, 1665–76.   1.2.4

Kerr, Richard A. (1983) Where was the Moon eons ago? *Science*, **221**, 1166.
                                                                                  4.4.2

(1984) Making the Moon from a big splash. *Science*, **226**, 1060–1.   4.1.6, 4.4.5

(1989) Making the Moon, remaking Earth. *Science*, **243**, 1433–5.   4.1.6

Kerr, Richard A. *See also* Elliot, J.

Kerridge, John F. (1977) Iron: Whence it came, where it went. *Space Science Reviews*, **20**, 3–68.   3.3.4, 3.3.5

(1979) Fractionation of refractory lithophilic elements among chondritic meteorites. *Proc. 10th Lunar & Plan. Sci. Conf.*, 989–96.   3.3.5

Kerridge, John F., & Chang, Sherwood (1985) Survival of interstellar material in meteorites: Evidence from carbonaceous material. In *Protostars and planets*. II, ed. D. C. Black & M. S. Matthews, pp. 738–54. Tucson: University of Arizona Press.
                                                                                  3.4.8

Kerz, Ferdinand (1884) *Erinnerungen an Sätze aus der Physik und der Mechanik des Himmels*. Leipzig: Veit.   1.2.3

Kesson, S. E. *See* Ringwood, A. E.

Khitarov, N. I. *See* Kuskov, O. L.

Kiang, T. (1966) Bias-free statistics of orbital elements of asteroids. *Icarus*, **5**, 437–49.
                                                                                    3.4.7

Kigoshi, K. *See* Reed, G. W.

Kipp, M. E., & Melosh, H. J. (1986) Short note: A preliminary numerical study of colliding planets. In *Origin of the Moon*, ed. W. K. Hartmann et al., pp. 643–7. Houston: Lunar and Planetary Institute.   **4.4.5**
   (1987) A numerical study of the giant impact origin of the Moon: The first half hour. *Lunar & Plan. Sci.*, **18**, 491–2.   **4.4.5**

Kippax, John R. (1914) *The call of the stars*. New York: Putnam.   **1.2.6**

Kirkwood, Daniel (1869) On the nebular hypothesis, and the approximate commensurability of the planetary periods. *M. N. Roy. Astron. Soc.*, **29**, 96–102.
                                                                                    1.2.5

Klein, Hermann J. (1903) Neue Untersuchungen über die frühesten Zustände der Erde und des Mondes. *Sirius*, **36**, 1–15.   **1.2.2**

Knott, Cargill Gilston (1919) The propagation of earthquake waves through the Earth, and connected problems. *Proceedings of the Royal Society of Edinburgh*, **39**, 157–208.   **1.3.3**

Kobrick, Michael, & Kaula, William M. (1979) A tidal theory for the origin of the solar nebula. *Moon and Planets*, **20**, 61–101.   **1.1.4**

Kock, M. *See* Garz, T.

Kohman, Truman P. *See* Simanton, J. R.

Komuro, T. *See* Nakazawa, K.

Kopal, Zdenek (1970) Bickerton, Alexander William. *DSB*, **2**, 123.   **1.2.1**

Kozlovskaya, S. V. *See* Safronov, V. S.

Krähenbühl, Urs. *See* Anders, E.; Ganapathy, R.; Morgan, J. W.

Kraft, Robert P. (1967) Studies of stellar rotation. V. The dependence of rotation on age among solar-type stars. *Astrophys. J.*, **150**, 551–70.   **3.4.2**

Kreimendahl, Frank M. *See* Lewis, J. S.

Kreutzberger, Melanie, Drake, M. J., & Jones, John H. (1986) Origin of the Earth's Moon: Constraints from alkali volatile trace elements. *Geoch. Cosmoch. Acta*, **50**, 91–8.   **4.4.4**

Krupnikov, K. K. *See* Al'tshuler, L. V.

Kruse, H. *See* Dreibus, G., Wänke, H.

Kuhn, Jeffrey R. *See* Worden, S. P.

Kuhn, Thomas S. (1970) *The structure of scientific revolutions*. 2nd edn. Chicago: University of Chicago Press.   **1.3.5, 2.2.4, 4.1.5**
   (1974) Second thoughts on paradigms. In *The structure of scientific theories*, ed. F. Suppe, pp. 459–82. Urbana: University of Illinois Press.   **1.3.5**

Kuiper, G. P. (1951a) On the origin of the Solar System. In *Astrophysics*, ed. J. A. Hynek, Chap. 8. New York: McGraw-Hill.   **1.1.4**
   (1951b) On the origin of the Solar System. *Proc. N. A. S.*, **37**, 1–14.   **1.1.4**
   (1951c) On the evolution of the protoplanets. *Proc. N. A. S.*, **37**, 383–93.
                                                                                    1.1.4
   (1956a) The origin of Earth and planets. *J. Geophys. Res.*, **61**, 398–405.   **1.1.4**
   (1956b) The formation of the planets. *J. Roy. Astron. Soc. Canada*, **50**, 57–68, 105–21, 158–76.   **1.2.6, 1.3.4**
   (1959) The Moon. *J. Geophys. Res.*, **64**, 1713–19.   **4.3.2**
   (1974) On the origin of the Solar System. I. *Celestial Mechanics*, **9**, 321–48.
                                                                                    1.1.4

Kumar, S. S. (1977) The escape of natural satellites from Mercury and Venus. *Astrophys. Space Sci.*, **51**, 235–8.   **4.4.6**

Kunde, V. *See* Gautier, D., Hanel, R. A.

Kurat, G., Hoinkes, G., & Fredriksson, K. (1975) Zoned Ca-Al rich chondrule in Bali: New evidence against the primordial condensation model. *Earth Plan. Sci. Lett.*, **26**, 140–4.   **3.3.5**

Kuroda, P. K. (1961) The time interval between nucleosynthesis and the formation of the Earth. *Geoch. Cosmoch. Acta*, **24**, 40–7.   **2.3.1, 2.4.1**

Kushner, David (1989) The controversy surrounding the secular acceleration of the Moon's mean motion. *Arch. Hist. Ex. Sci.*, **39**, 291–316.   **4.2.2**

(1993) Sir George Darwin and a British school of geophysics. *Osiris*, **8**, 196–223.
                                                                             4.1.1

Kuskov, O. L., & Khitarov, N. I. (1982) Initial stage of evolution of the Earth: Problems of geochemistry. *Izv. Acad. Sci. USSR, Phys. Solid Earth*, **18**, 447–56. Translated from *Izv. A. N. Fiz. Zemli*.   **3.3.4**

Lago, M. T. V. T. (1984) A new investigation of the T Tauri star RU Lupi. III. The wind model. *M. N. Roy. Astron. Soc.*, **210**, 323–40.   **2.3.5**

Lakatos, Imre (1970) Falsification and the methodology of scientific research programmes. In *Criticism and the growth of knowledge*, ed. I. Lakatos & A. Musgrave, pp. 91–195. Cambridge University Press.   **1.3.5, 2.2.4**

Lal, D. *See* Goswami, J. N.

Lambeck, Kurt (1975) Effects of tidal dissipation in the oceans on the Moon's orbit and the Earth's rotation. *J. Geophys. Res.*, **80**, 2917–25.   **4.4.2**

(1977) Tidal dissipation in the oceans – Astronomical, geophysical and oceanographic consequences. *Phil. Trans.*, **A287**, 545–94.   **4.4.2**

Lammlein, D. *See* Nakamura, Y.

Landsberg, H. E. *See* Brush, S. G.

Lang, Kenneth R., & Gingerich, Owen (1979) *Source book in astronomy and astrophysics, 1900–1975*. Cambridge, MA: Harvard University Press.   **1.3.1**

Lange, Manfred, & Ahrens, T. J. (1982) The evolution of an impact-generated atmosphere. *Icarus*, **51**, 96–120.   **3.3.5**

Lange, Manfred *See also* Binder, A. B.

Lankford, John (1980) A note on T. J. J. See's observations of craters on Mercury. *J. Hist. Astron.*, **11**, 129–32.   **4.2.4**

Lanoix, M., Strangway, D. W., & Pearce, G. W. (1978) The primordial magnetic field preserved in chondrules of the Allende meteorite. *Geophys. Res. Lett.*, **5**, 73–6.
                                                                             2.3.5

Lanoix, M. *See* Sugiura, N.

Larimer, J. W. (1967) Chemical fractionation in meteorites. I. Condensation of the elements. *Geoch. Cosmoch. Acta*, **31**, 1215–38.   **3.3.3, 4.3.4**

Larimer, J. W., & Anders, E. (1967) Chemical fractionation in meteorites. II. Abundance patterns and their interpretation. *Geoch. Cosmoch. Acta*, **31**, 1239–70.
                                                                             3.3.3, 4.4.4

(1970) Chemical fractionations in meteorites. III. Major element fractionations in chondrites. *Geoch. Cosmoch. Acta*, **34**, 367–87.   **3.3.3**

Larimer, J. W. *See* Grossman, L.

Larson, Richard B. (1969) Numerical calculations of the dynamics of a collapsing proto-star. *M. N. Roy. Astron. Soc.*, **145**, 271–95.   **1.1.5, 3.3.5**

(1972a) The collapse of a rotating cloud. *M. N. Roy. Astron. Soc.*, **156**, 437–58.
                                                                             2.4.1

(1972b) The evolution of spherical protostars with masses $0.25M_O$ to $10M_O$. *M. N. Roy. Astron. Soc.*, **157**, 121–45.   **3.3.5**

(1974) Collapse calculations and their implications for the formation of the Solar System. In *On the origin of the solar system*, ed. H. Reeves, pp. 142–153. Paris: CNRS.   **2.4.1**

(1988) This week's Citation Classic. *Current Contents*, no. 5 (February 1), 18.
                                                                             3.3.5

Larsson, Ulf (1993) Physics in a stronghold of engineering: Professorial appointments at the Royal Institute of Technology, 1922–1925. In *Center on the periphery*, ed. S. Lindqvist, pp. 58–75. Canton, MA: Science History Pubs. **1.1.3**

Latham, G. See Nakamura, Y.

Latimer, W. M. (1950) Astrochemical problems in the formation of the Earth. *Science*, **112**, 101–4. **4.4.3**

Laudan, Larry (1976) Two dogmas of methodology. *Philosophy of Science*, **43**, 585–97. **1.3.5**

(1977) *Progress and its problems*. Berkeley: University of California Press. **1.3.5**

(1983) The demise of the demarcation problem. In *Physics, philosophy, and psychoanalysis*, ed. R. S. Cohen & L. Laudan, pp. 111–27. Boston: Reidel. **2.2.2**

Laudan, Larry. See Laudan, R.

Laudan, Rachel, Laudan, Larry, & Donovan, Arthur (1988) Testing theories of scientific change. In *Scrutinizing science*, ed. A. Donovan et al., pp. 3–44. Boston: Kluwer. **2.2.2**

Laul, J. C. See Ganapathy, R.; Keays, R. R.; Morgan, J. W.

Lebedinskii, A. I. See Gurevich, L. E.

Ledenev, B. N. See Al'tshuler, L. V.

Lee, Typhoon (1978) A local proton irradiation model for isotopic anomalies in the Solar System. *Astrophys. J.*, **224**, 217–26. **2.5.1**

(1979) New isotopic clues to Solar System formation. *Reviews of Geophysics and Space Physics*, **17**, 1591–1611. **2.5.1**

(1986) From big bang to bing bang (from the origin of the universe to the origin of the Solar System). In *The solar system*, ed. M. G. Kivelson, pp. 373–95. Englewood Cliffs, NJ: Prentice-Hall. **2.5.3**

Lee, Typhoon, & Papanastassiou, D. A. (1974) Mg isotopic anomalies in the Allende meteorite and correlations with O and Sr effects. *Georphys. Res. Lett.*, **1**, 225–8. **2.5.1**

Lee, Typhoon, Papanastassiou, D. A., & Wasserburg, G. J. (1975) Discovery of a large ^{26}Mg isotope anomaly and evidence for extinct ^{26}Al. *Bulletin of the American Physical Society*, **20**, 1486. **2.5.1**

(1976) Demonstration of ^{26}Mg excess in Allende and evidence of ^{26}Al. *Geophys. Res. Lett.*, **3**, 109–12. **1.1.5, 2.5.1**

(1977) Aluminum-26 in the early Solar System: Fossil or fuel? *Astrophys. J.*, **211**, L107– L110. **2.5.1**

(1978) Calcium isotope anomalies in the Allende meteorite. *Astrophys. J.*, **220**, L21– L25. **2.5.1**

Lee-Hu, C. See Gopalan, K.

Leitch, C. A., & Smith, J. V. (1981) Mechanical aggregation of enstatite chondrites from an inhomogeneous debris cloud. *Nature*, **290**, 228–30. **3.3.5**

Leith, C. K., Alden, William C., Penrose, R. A. F., Jr., Schuchert, Charles, MacMillan, William D., Willis, Bailey, & Moulton, F. R. (1929) (Issue on T. C. Chamberlin) *J. Geol.*, **37**, no. 4 (May). **1.1.1**

LeLevier, Robert. See Henyey, L. G.

Lerner, E. J. (1991) *The big bang never happened*. New York: Random House. **3.4.8**

Levee, R. D. See Henyey, L. G.

Levin, Aleksey E., & Brush, Stephen G., Editors (1995) *The origin of the Solar System: Soviet research, 1925–1991*. New York: American Institute of Physics. **1.1.3, 1.1.4, 3.1.1, 4.4.1**

Levin, Boris Yu. (1948) *Proiskhozhdenie zemli i planet*. Moscow: Pravda. **3.1.1**

(1953) Kosmogoniya planetnoi sistemy i evolutsiya solntsa (Cosmogony of the planetary system and the evolution of the sun) *Dokl. A. N*, **91**, 471–4. English translation in *The origin of the Solar System: Soviet research, 1925–1991*, ed. A. E. Levin & S. G. Brush. New York: American Institute of Physics (1995). **3.1.1**

(1956) *Origin of the Earth and planets.* Moscow: Foreign Languages Publishing House. Trans. from 2nd Russian edn. **3.1.1**

(1962) The origin of the Solar System. *New Scientist*, **13**, 323–5. **3.1.3, 3.4.3**

(1965/1966) The structure of the Moon. In *Proceedings of the Caltech–JPL Lunar and Planetary Conference (September 1965)*, ed. H. Brown et al., pp. 61–76. Pasadena: Jet Propulsion Laboratory Technical Memorandum 33-266 (publ. 1966).
4.4.1, 4.4.2

(1966) The structure of the Moon. *Sov. Astron. AJ*, **10**, 479–91. Trans. from *Astron. Zh.*, **43**, 606–21. **4.3.3**

(1972) Origin of the Earth. *Tectonophysics*, **13**, 7–29. **2.1.1.**

(1978a) Some problems concerning the accumulation of planets. *Sov. Astron. Letters*, **4**, 54–7. Trans. from *Pis'ma Astron. Zh.*, **4**, 102–7. **3.1.2**

(1978b) Relative velocities of planetesimals and the early accumulation of planets. *Moon and Planets*, **19**, 289–96. **3.1.2**

Levy E. H. (1985) Protostars and planets: Overview from a planetary perspective. In *Protostars and planets*. II, ed. D. C. Black & M. S. Matthews, pp. 3–16. Tucson: University of Arizona Press. **2.1.1, 3.4.8**

Levy, E. H., & Lunine, J. L. (1993) *Protostars and planets*. III. Tucson: University of Arizona Press. **1.1**

Levy, E. H. *See also* Black, D. C.

Lewin, Roger (1982) Biology is not postage stamp collecting. *Science*, **216**, 718–20.
2.2.2

Lewis, Isabel Martin (1919) *Splendors of the sky.* New York: Duffield. **1.2.6**

Lewis, John S. (1968) An estimate of the surface conditions on Venus. *Icarus*, **8**, 434–56. **3.3.5**

(1969) Geochemistry of the volatile elements on Venus. *Icarus*, **11**, 367–85.
3.3.5

(1972a) Low temperature condensations from the solar nebula. *Icarus* **16**, 241–52.
1.1.5

(1972b) Metal/silicate fractionation in the Solar System. *Earth Plan. Sci. Lett.*, **15**, 289–90. **3.3.3, 3.3.5**

(1973a) Chemistry of the outer Solar System. *Space Science Reviews*, **14**, 401–11.
3.3.4

(1973b) Chemistry of the planets. *Annual Review of Physical Chemistry*, **24**, 339–51.
3.3.4

(1973c) The origin of the planets and satellites. *Technology Review*, **76**, no. 1, 21–35.
3.3.5

(1974a) The chemistry of the Solar System. *Sci. Am.*, **230**, no. 3, 50–60, 65.
3.3.4

(1974b) Volatile element influx on Venus from cometary impacts. *Earth Plan. Sci. Lett.*, **22**, 239–44. **3.3.5**

(1981) Putting it all together. In *The new Solar System*, ed. J. K. Beatty et al., pp. 205–11, 218. Cambridge University Press/Cambridge, MA: Sky Publ. **3.3.4**

Lewis, John S., Barshay, Stephen S., & Noyes, Barbara (1979) Primordial retention of carbon by the terrestrial planets. *Icarus*, **37**, 190–206. **3.3.4, 3.3.5**

Lewis, John S., & Kreimendahl, Frank A. (1980) Oxidation state of the atmosphere and crust of Venus from Pioneer Venus results. *Icarus*, **42**, 330–7. **3.3.5**

Lewis, John S., & Prinn, R. G. (1984) *Planets and their atmospheres, origins and evolution.* New York: Academic Press. **3.3.5**

Lewis, John S. *See* Barshay, S. S.

Lewis, Richard (1974) The new Moon, in the light of Apollo. *Saturday Review/World,* 13 July, pp. 52–3. **4.3.4**

Lewis, Roy S., Anders, E., Shimamura, T., & Lugmair, G. W. (1983) Barium isotopes in Allende meteorite: Evidence against an extinct superheavy element. *Science,* **222,** 1013–15. **2.5.1**

Lewis, Roy S., Srinivasan, B., & Anders, Edward (1975) Host phase of a strange xenon component in Allende. *Science,* **190,** 1251–62. **2.5.2**

Liapunoff, A. *See* Lyapunov, A.

Lightner, B. D., & Marti, K. (1974) Lunar trapped xenon. *Proc. 5th Lunar Sci. Conf.,* 2023–31. **4.4.4**

Ligon, D. T., Jr. *See* Cuttita, F.

Lin, C. C. (1971) Theory of spiral structure. In *Highlights of astronomy* (XIV General Assembly, International Astronomical Union, 1970), vol. **2,** pp. 88–121. Dordrecht: Reidel. **2.5**

Lin, C. C., & Shu, F. (1964) On the spiral structure of disk galaxies. *Astrophys. J.,* **140,** 646–55. **2.5**

Lin, D. N. C., & Papaloizu, J. (1980) On the structure and evolution of the primordial solar nebula. *M. N. Roy. Astron. Soc.,* **191,** 37–48. **2.4.2**

   (1985) On the dynamical origin of the Solar System. In *Protostars and planets.* II, ed. D. C. Black & M. S. Matthews, pp. 980–1072. Tucson: University of Arizona Press. **2.4.2**

Lin, D. N. C. *See also* Stewart, G. R.

Lindblad, B. (1934) On the evolution of a rotating system of material particles, with applications to Saturn's rings, the planetary system and the galaxy. *M. N. Roy. Astron. Soc.,* **94,** 231–40. **1.1.3**

   (1935) A condensation theory of meteoric matter and its cosmological significance. *Nature,* **135,** 133–5. **1.1.3**

Lindqvist, P.-A. *See* Alfvén, H.

Lindsay, Robert Bruce (1973) *Julius Robert Mayer: Prophet of energy.* New York: Pergamon Press. **4.2.2**

Ling, J. C. *See* Mahoney, W. A.

Lingenfelter, Richard E., & Ramaty, Reuven (1978) Gamma-ray lines: A new window to the universe. *Physics Today,* **31,** no. 3 (March), 40–7. **2.5.3**

Lingenfelter, Richard E. *See also* Mahoney, W. A., Ramaty, R.

Lippolt, H. J. *See* Burnett, D. S.

Lissauer, Jack J., & Cuzzi, Jeffrey N. (1985) Rings and moons: clues to understanding the solar nebula. In *Protostars and planets.* II, ed. D. C. Black & M. S. Matthews, pp. 920–56. Tucson: University of Arizona Press. **3.4.4**

Lissauer, Jack J., & Safronov, Victor S. (1991) The random component of planetary rotation. *Icarus,* **93,** 288–97. **1.2.3**

Lissauer, Jack J. *See also* Cuzzi, J. N.

Little, S. *See* Aller, L. H.

Liu, M. K. *See* Urey, H. C.

Lodochnikov, V. N. (1939) Nekotorye obshie voprosy, sryazannye s magmoi, dayushei bazal'tovye porody (Some general questions concerning the magma basalt rocks). *Zapiski Vserosshiskogo Mineralogicheskogo Obshestva,* [2] **68,** 2, 428–42.
                                                              3.1.1, 4.1.3

Loewenstein, R. F. *See* Harper, D. A.

Logsdon, John M. (1970) *The decision to go to the Moon. Project Apollo and the national interest.* Cambridge, MA: MIT Press. **4.3.1**

Long, Alton L. *See* Simanton, J. R.

Longwell, Chester R. (1941) Geology. In *The development of the sciences* (2nd series), ed. L. L. Wodruff, pp. 147–96. New Haven: Yale University Press.   1.3.5

Lord, Harry C., III (1965) Molecular equilibria and condensation in a solar nebula and cool stellar atmospheres. *Icarus*, **4**, 279–88.   3.3.3

Love, A. E. H. (1889) On the oscillations of a rotating liquid and the genesis of the Moon. *Phil. Mag.*, [5] **27**, 254–64.   4.2.3

Loveridge, W. D. *See* Wanless, R. K.

Low, F. J. *See* Aumann, H. H.

Lowell, A. Lawrence (1935) *Biography of Percival Lowell.* New York: Macmillan.
                                                                                            1.2.4

Lowell, Percival (1895) *Mars.* Boston: Houghton Mifflin.   1.2.4

(1903) *The Solar System.* Boston: Houghton Mifflin.   1.2.4

(1909a) Mars as the abode of life. *Science*, **30**, 338–40.   1.2.4

(1909b) Planets and their satellite systems. *Astron. Nachr.*, **182**, 97–100.   1.2.4

(1909c) *The evolution of worlds.* New York: Macmillan.   1.2.4

(1909d) The revelation of evolution: A thought and its thinkers. *Atlantic Monthly*, **104**, 174–83.   1.2.4

(1916) The genesis of planets. *J. Roy. Astron. Soc. Canada*, 281–93.   1.2.4

Lowman, Paul D., Jr. (1966) The scientific value of manned lunar exploration. *Annals of the New York Academy of Sciences*, **140**, 628–35.   4.3.2

(1985) Lunar bases: A post-Apollo evaluation. In *Lunar bases and space activities of the 21st century*, ed. W. W. Mendell, pp. 35–46. Houston: Lunar and Planetary Institute.   4.4.4

Lüst, R. (1952) Die Entwicklung einer um einen Zentralkörper rotierenden Gasmasse. I. Lösungen der hydrodynamischen Gleichungen mit turbulenter Reibung. *Zeitschrift für Naturforschung*, **7a**, 87–98.   2.4.2

Lüst, R., & Schlüter, A. (1955) Drehimpulstransport durch Magnetfelder und die Abbremsung rotierender Sterne. *Z. f. Astrophysik*, **38**, 190–211.   3.4.2

Lugmair, G. W., Marti, K., & Scheinin, N. B. (1978) Incomplete mixing of products from r; p; and s-process nucleosynthesis: Sm-Nd systematics in Allende inclusion EK 1-04-1. *Lunar & Plan. Sci.*, **9**, 672–4.   2.5.3

Lugmair, G. W. *See also* Lewis, R. S.

Lunar Sample Preliminary Examination Team (1969) Preliminary examination of lunar samples from Apollo 11. *Science*, **165**, 1211–27.   4.4.3

(1970) Preliminary examination of lunar samples from Apollo 12. *Science*, **167**, 1325–39.   4.4.3

Lunine, Jonathan I., & Stevenson, David J. (1982) Formation of the Galilean satelllites in a gaseous nebula. *Icarus*, **52**, 14–39.   3.2.2

Lunine, Jonathan I. *See also* Stevenson, D. J.

Luyten, W. J., & Hill, E. L. (1937) On the origin of the Solar System. *Astrophys. J.*, **86**, 470–82.   1.3.4

Lyapunov, A. (1905) Sur un probleme de Tchebychef. *Memoires de l' Academie Imperiale des Sciences de St. Petersbourg* (8), **17**, no. 3, pp. 1–32.   4.2.4

(1908) Problème de minimum dans une question de stabilité des figures d'equilibre d'une masse fluide en rotation. *Memoires de l' Academie Imperiale des Sciences de St. Petersbourg* [8], **22**, no. 5.   4.2.4

Lynden-Bell, D., & Pringle, J. E. (1974) The evolution of viscous disks and the origin of the nebular variables. *M. N. Roy. Astron. Soc.*, **168**, 603–37.
                                                                1.1.5, 2.4.2, 4.1.5, 4.4.5

Lynn, W. T. (1904) The ninth satellite of Saturn. *J. Br. Astron. Assoc.*, **15**, 35–6.
                                                                                            1.2.4

Lyons, Richard D. (1969) Tiny glass spheres found in Moon dust: Scientists surprised. *New York Times*, **118** (29 July), 1, 16.   4.3.4

Lyttleton, R. A. (1936) The origin of the Solar System. *M. N. Roy. Astron. Soc.*, **96,** 559–68.   **1.3.4**
(1960) Dynamical calculations relating to the origin of the Solar System. *M. N. Roy. Astron. Soc.*, **121,** 551–69.   **1.3.4**
(1961) The birth of worlds. *Saturday Evening Post,* **234,** no. 50 (16 December), 54–7.   **2.2.3**
(1968) *Mysteries of the Solar System.* Oxford: Clarendon Press.
                                                                        2.2.1, 2.2.2, 3.1.1
Lyubimova, E. A. (1955) The heating up of the Earth's core during the formation of the Earth. (in Russian). *Izvestiya Akademii Nauk SSSR, Seriya Geofizika,* no. 5, 416–24.   **3.1.1**
MacDonald, Gordon J. F. (1964a) Tidal friction. *Reviews of Geophysics,* **2,** 467–541.
                                                                             4.3.3, 4.4.2
(1964b) Earth and Moon: Past and future. *Science,* **145,** 881–90.   **4.3.3, 4.4.2**
(1965) Origin of the Moon: Dynamical considerations. *Annals of the New York Academy of Sciences,* **118,** 739–82.   **4.4.2**
(1966) Origin of the Moon: Dynamical considerations. In *The Earth–Moon System,* ed. B. G. Marsden & A. G. W. Cameron, pp. 165–209. New York: Plenum Press.
MacDonald, Gordon J. F. *See also* Munk, W. H.; Urey, H. C.
MacGregor, I. *See* Arnold, J.
Mackey, Sampson Arnold (1825) *A new theory of the Earth, and of planetary motion; in which it is demonstrated that the sun is viceregent of his own system.* Norwich: Printed for the author by R. Walker.   **4.2.4**
MacMillan, William D. (1920) [Review of] Problems of cosmogony and stellar dynamics, by J. H. Jeans. *Astrophys. J.,* **51,** 309–33.   **1.3.3**
MacMillan, William D. *See also* Leith, C. K.
MacPherson, Glenn J., & Grossman, Lawrence (1981) A once-molten, coarse-grained, Ca-rich inclusion in Allende. *Earth Plan. Sci. Lett.,* **52,** 16–24.
                                                                                      3.3.5
MacPherson, Glenn J., Grossman, Lawrence, Hashimoto, Akihito, Bar-Matthews, Maryam, & Tanaka, Tsuyoshi (1984a) Petrographic studies of refractory inclusions from the Murchison meteorite. *Proc. 15th Lunar & Plan. Sci. Conf. (J. Geophys. Res.,* **89,** Supplement), C299–C312.   **3.3.5**
MacPherson, Glenn J., Paque, Julie M., Stolper, Edward, & Grossman, Lawrence (1984b) The origin and significance of reverse zoning in melilite from Allende type B inclusions. *J. Geol.,* **92,** 289–305.   **3.3.5**
MacPherson, Hector (1921) The place of the nebulae in stellar evolution. *Observatory,* **44,** 219–25.   **1.2.4**
MacPherson, Hector Copland (1906) *A century's progress in astronomy.* Edinburgh: Blackwood.   **1.2.2, 1.2.4**
MacPherson, Hector, Jr. (1909) Theories of celestial evolution. *Pop. Astron.,* **17,** 418–23.   **1.2.6**
Mahoney, W. A., Ling, J. C., Jacobson, A. S., & Lingenfelter, R. E. (1982) Diffuse galactic gamma-ray line emission from nucleosynthetic ^{60}Fe, ^{26}Al, and ^{22}Na: Preliminary results from HEAO 3. *Astrophys. J.,* **262,** 742–8.   **2.5.3**
Mahoney, W. A., Ling, J. C., Wheaton, Wm. A., & Jacobson, A. S. (1984) HEAO 3 discovery of ^{26}Al in the interstellar medium. *Astrophys. J.,* **286,** 578–85.
                                                                                      2.5.3
Malcuit, R. J., & Winters, R. R. (1987) Computer simulation model for early post-capture lunar orbital evolution: Implications for thermal history of Moon and Earth. *Lunar & Plan. Sci.,* **18,** 594–5.   **4.4.7**
Malin, Michael C. *See* Murray, B.; Phillips, R. J.
Mandelbaum, Leonard (1969) Apollo: How the United States decided to go to the Moon. *Science,* **163,** 649–54.   **4.3.1**

Manian, Samuel H., Urey, Harold C., & Bleakney, Walker (1934) An investigation of the relative abundances of oxygen isotopes, O^{16}:O^{18} in stone meteorites. *Journal of the American Chemical Society*, **56**, 2601–9.   1.1.5

Manuel, O. K. (1981) Heterogeneity in meteorite isotopic and elemental compositions: Proof of local element synthesis. *Geochemistry International*, **18**, no. 6, 101–25. Trans. from *Geokhimiya*, no. 12, 1776–800.   2.5.2

Manuel, O. K., Hennecke, E. W., & Sabu, D. D. (1972) Xenon in carbonaceous chondrites. *Nature Physical Science*, **240**, 99–101.   2.5.1

Manuel, O. K., & Hwaung, Golden (1983) Solar abundances of the elements. *Meteoritics*, **18**, 209–22.   2.5.2

Manuel, O. K., & Sabu, D. D. (1975/1976) Elemental and isotopic inhomogeneities in noble gases: The case for local synthesis of the chemical elements. *Transactions of the Missouri Academy of Science for 1975*, **9** (pub. 1976), 104–22.   2.5.2

Manuel, O. K. *See also* Sabu, D. D.; Srinivasan, B.

Marcy, Geoffrey. *See* Bodenheimer, P.

Margolis, Steven H. (1979) Grain motions in the solar nebula. *Moon and Planets*, **20**, 49–59.   2.5.2

Marouf, E. A. *See* Cuzzi, J. N.

Marsden, B. G. (1973) Lowell, Percival. *DSB*, **8**, 520–3.   1.2.4

Marsden, B. G., & Cameron, A. G. W., Editors (1964/1966) *The Earth–Moon system* (Proceedings of conference, January 1964). New York: Plenum Press (pub. 1966).   4.3.3, 4.4.2

Marten, A. *See* Courtin, R.; Gautier, D.

Marti, K. *See* Lightner, B. D.; Lugmair, G. W.; Urey, H. C.

Martin, R. *See* Taylor, S. R.

Marvin, Ursula B. (1973) The Moon after Apollo. *Technology Review*, **75**, no. 6, 12–23.   4.3.4

Marvin, Ursula B., Wood, J. A., & Dickey, J. S., Jr. (1970) Ca-Al rich phases in the Allende meteorite. *Earth Plan. Sci. Lett.*, **7**, 346–50.   3.3.3

Mascart, Jean (1902) Contribution à l'origine des petites planètes. *Bulletin de la Societe Astronomique de France*, **16**, 123–6.   1.2.2

Masursky, Harold. *See* Smith, B. A.

Mather, Kirtley F. (1928) *Old Mother Earth*. Cambridge, MA: Harvard University Press.   1.2.6

(1939) Earth structure and Earth origin. *Science*, **84**, 65–70.   1.3.3

Mathews, G. J., & Käppeler, F. (1984) Neutron-capture nucleosynthesis of neodymium isotopes and the s-process from A = 130 to 150. *Astrophys. J.*, **286**, 810–21.   2.5.3

Matsui, T., & Abe, Y. (1986) Origin of the Moon and its early thermal evolution. In *Origin of the Moon*, ed. W. K. Hartmann, R. J. Phillips, & G. J. Taylor, pp. 453–86. Copyright by Lunar and Planetary Institute.   4.4.5

Matthews, Mildred S. *See* Black, D. C.

Maull, Nancy. *See* Darden, L.

Mayeda, Toshiko K. *See* Clayton, R. N.; Grossman, L.; Onuma, N.

Mayer, J. R. (1848) *Beiträge zur Dynamik des Himmels in populärer Darstellung*. Heilbronn: Landherr.   4.2.2

(1851/1893) De l'influence des marees sur la rotation de la terre. Reported in *C. R. Acad. Sci. Paris*, **32** (1851), 652; published in *Kleinere Schriften und Briefe von Robert Mayer, nebst Mittheilungen aus seinem Leben*, ed. J. J. Weyrauch, pp. 282–5. Stuttgart: Verlag der J. G. Cotta'schen Buchhandlung (1893).   4.2.2

Mayr, Ernst (1965) Cause and effect in biology. In *Cause and effect*, ed. D. Lerner, pp. 33–50. Cambridge, MA: Harvard University Press.   2.2.2

(1985) How biology differs from the physical sciences. In *Evolution at a cross-roads*, ed. D. J. Depew & B. C. Weber, pp. 43–63. Cambridge, MA: MIT Press.

2.2.2

McCarthy, D. W., Jr. *See* Black, D. C.

McCarthy, J. (1911) The new astronomy. *Knowledge* (n.s.), **8**, 426.   **1.2.1**

McCauley, John F. *See* Smith, B. A.

McCord, Thomas B. *See* Chapman, C. R.

McCrea, W. H. (1960) The origin of the Solar System. *Proc. Roy. Soc.*, **A256**, 245–66.

2.2.2, 3.2.1

(1963) The origin of the Solar System. *Contemporary Physics*, **4**, 278–90.

2.2.1, 2.3.4, 3.2.1

(1974) Origin of the Solar System: Review of concepts and theories. In *On the origin of the Solar System*, ed. H. Reeves, pp. 2–20. Paris: CNRS.   **2.1.1, 2.2.1**

McCrea, W. *See also* De Groot, M.

McCulloch, M., & Wasserburg, G. J. (1978a) Barium and neodymium isotopic anomalies in the Allende meteorite. *Astropys. J.*, **220**, L15–L19.   **2.5.3**

(1978b) More anomalies from the Allende meteorite: Samarium. *Geophys. Res. Lett.*, **5**, 599–602.   **2.5.3**

McElroy, M. B. *See* Hoffman, J. H.

McFarland, J. *See* De Groot, M.

McGetchin, T. R., & Smyth, J. R. (1978) The mantle of Mars: Some possible geological implications of its high density. *Icarus*, **34**, 512–36.   **3.3.5**

McKinnon, William B., & Mueller, S. W. (1984) A reappraisal of Darwin's fission hypothesis and a possible limit to the primordial angular momentum of the Earth. *Origin of the Moon (Conf. abst.)*, p. 34.   **4.2.4**

McLaughlin, William I. (1980) Prediscovery evidence of planetary rings. *Journal of the British Interplanetary Society*, **33**, 287–94.   **3.4.5**

McMahon, Allen J. (1965) *Astrophysics and space science: An integration of the sciences.* Englewood Cliffs, NJ: Prentice-Hall.   **2.3.5, 2.4**

McMillan, R. S. *See* Black, D. C.

McNairn, William Harvey (1919) The story of cosmological theory. *Science*, **46**, 599–607.   **1.2.6**

McSween, Harry Y., Jr. (1984) Meteorites: Little rocks with lots of history. *Planetary Report*, **4**, no. 2 (March/April), 7–8.   **2.5.3**

(1987) *Meteorites and their parent planets.* Cambridge University Press.   **3.3.5**

(1989) Chondritic meteorites and the formation of planets. *American Scientist*, **77**, 146–53.   **3.3.5**

Melchior, P. (1974) Earth tides. *Geophysical Surveys*, **1**, 275–303.   **4.4.2**

Mellema, J. P. *See* Banerjee, S. K.

Melosh, H. J., & Sonet, C. P. (1986) When worlds collide: Jetted vapor plumes and the Moon's origin. In *Origin of the Moon*, ed. W. K. Hartmann et al., pp. 621–42. Houston: Lunar and Planetary Institute.   **4.4.5**

Menard, Henry W. (1971) *Science: Growth and change.* Cambridge, MA: Harvard University Press.   **4.1.5**

Mendis, D. A. *See* Houpis, H. L. F.; Ip, W.-H.

Menzel, Donald H. (1931) *Stars and planets.* New York: The University Society.

1.3.3

Mercer-Smith, J. A., Cameron, A. G. W., & Epstein, R. I. (1984) On the formation of stars from disk accretion. *Astrophys. J.*, **279**, 363–6.   **2.4.2**

Merleau-Ponty, Jacques (1983) *La science de l'univers à l'âge du positivisme. Etude sur les origines de la cosmologie contemporaine.* Paris: Vrin.   **2.1**

Merrill, G. P. (1909) The composition of stony meteorites compared with that of terrestrial igneous rocks and considered with reference to their efficacy in world-making. *Am. J. Sci.* [4], **27**, 469–74.   **1.2.5**

Mestel, L. (1968) Magnetic braking by a stellar wind. I. *M. N. Roy. Astron. Soc.*, **137**, 359–91.   **3.4.2**

(1974) Magnetohydrodynamics, hydrodynamics, dynamics of the solar system in the different models. In *On the origin of the Solar System*, ed. H. Reeves, pp. 21–7. Paris: CNRS.   **2.2.1**

Mestel, L. (D1962) Privately circulated manuscript and Institute for Advanced Study (Princeton) lecture notes, in possession of A. G. W. Cameron, Harvard University.   **2.4.1**

Meyer, Max Wilhelm (1906) *The making of the world*. Chicago: Kerr.   **1.2.4**

(1910) *Die Welt der Planeten*. Stuttgart: Kosmos, Gesellschaft der Naturfreunde & Franck'sche Verlagshandlung.   **1.2.4**

Michelson, A. A. (1914) Preliminary results of measurements of the rigidity of the Earth. Installation and observations by H. G. Gale assisted by Harold Alden. Calculations by W. L. Hart under the direction of F. R. Moulton. *J. Geol.*, **22**, 97–130. *Astrophys. J.*, **39**, 105–38.   **4.2.4**

Michelson, A. A., & Gale, Henry G. (1919) The rigidity of the Earth. *J. Geol.*, **27**, 585–601. *Astrophys. J.*, **50**, 330–45.   **4.2.4**

Milankovich, M. (1920) *Theorie mathematique des phénomènes thermiques produits par la radiation solaire*. Paris: Gauthier-Villars.   **4.2.2**

Miller, Samuel (1803) *Brief retrospect of the eighteenth century*. New York: Swords.   **1.1**

Millikan, R. A. (D1903) Letter to T. C. Chamberlin, 23 December 1903. box 5, C–UC.   **1.2.5**

Millis, R. L., Wasserman, L. H., & Birch, P. V. (1977) Detection of rings around Uranus. *Nature*, **267**, 330–1.   **3.4.4**

Milne, E. A. (1952) *Sir James Jeans: A biography*. Cambridge University Press.   **1.3**

Miner, Ellis D. (1990) *Uranus: the planet, rings and satellites*. New York: Horwood.   **3.4.5**

Miner, Ellis D. *See also* Stone, E. C.

Mink, D. *See* Elliot, J. L.

Mitler, H. E. (1975) Formation of an iron-poor Moon by partial capture, or: Yet another exotic theory of lunar origin. *Icarus*, **24**, 256–68.   **4.4.4, 4.4.5**

Mitler, H. E. *See also* Wood, J. A.

Mitroff, Ian I. (1974a) *The subjective side of science*. New York: American Elsevier.   **4.2.1, 4.3.4, 4.3.5, 4.4.4, 4.4.7**

(1974b) Science's Apollonic moon: A study in the psychodynamics of modern science. *Spring, An Annual of Jungian Psychology*, 102–12.   **4.2.1**

Mizuno, Hiroshi (1980) Formation of the giant planets. *Prog. Th. Phys.*, **64**, 544–57.   **1.1.5, 2.4.2, 3.2.2**

Mizuno, Hiroshi, & Boss, A. P. (1985) Tidal disruption of dissipative planetesimals. *Icarus*, **63**, 109–33.   **4.4.5**

Mizuno, Hiroshi, Nakazawa, Kiyoshi, & Hayashi, Chushiro (1978) Instability of a gaseous envelope surrounding a planetary core and formation of giant planets. *Prog. Th. Phys.*, **60**, 699–710.   **3.2.2**

Mizuno, Hiroshi. *See* Boss, A. P.

Mogro-Campero, A. (1975) Angular momentun transfer to the inner Jovian satellites. *Nature*, **258**, 692–3.   **3.4.2**

Molini-Velsko, Carol A. *See* Clayton, R. N.

Monaghan, J. J. *See* Gingold, R. A.

Monck, W. H. S. (1885) New stars and shooting-stars. *Observatory*, **8**, 335–6.   **1.2.4**

(1911) The collisions of stars. *Observatory*, **34**, 202–3.   **1.2.1**

Moore, R. *See* Pollack, J. B.

Mooser, J. (1904) *Theorie der Entstehung des Sonnensystems. Eine mathematische Behandlung der Kant-Laplace'schen Nebularhypothese*. Neue Bearbeitung. St Gallen: Fehr.   **1.2.2**

Morehouse, G. (1898) *The wilderness of worlds: A popular sketch of the evolution of matter from nebula to man and return. The life-orbit of a star.* New York: Ecker.   **1.2.2**

Moreux, Th. (1926) *Astronomy to-day.* London: Methuen.   **1.2.4**

Morfill, G. E., Tscharnuter, W. M., Völk, H. J. (1985) Dynamical and chemical evolution of the protoplanetary nebula. In *Protostars and planets,* II, ed. D. C. Black & M. S. Matthews, pp. 493–533. Tucson: University of Arizona Press.   **1.1.5**

Morfill, G. E. *See* Burns, J. A.; Wood, J. A.

Morgan, John W., Ganapathy, R., Higuchi, Hideo, Krähenbühl, U., & Anders, Edward (1974) Lunar basins: Tentative characterisation of projectiles, from meteoritic elements in *Apollo 17* boulders. *Proc. 5th Lunar Sci. Conf.,* 1703–36.

          4.4.3

Morgan, John W., Gros, Jacques, Takahashi, H., & Hertogen, Jan (1976) Lunar breccia 73215: Siderophile and volatile trace elements. *Proc. 7th Lunar Sci. Conf.,* 2189–99.   **4.4.3**

Morgan, John W., Hertogen, Jan, & Anders, Edward (1978) The Moon: Composition determined by nebular processes. *Moon and Planets,* **18,** 465–78.   **4.4.4**

Morgan, John W., Laul, J. C., Krähenbühl, Urs, Ganapathy, R., & Anders, Edward (1972) Major impacts on the Moon: Characterisation from trace elements in Apollo 12 and 14 samples. *Proc. 3rd Lunar Sci. Conf.,* 1377–95.   **4.4.3**

Morgan, John W. *See* Anders, E.; Ganapathy, R.; Gros, J.

Morrison, David. *See* Abell, G. O., Smith, B. A.

Moss, D. L. (1968) The influence of inhibition of convection on pre–main sequence stellar evolution. *M. N. Roy. Astron. Soc.,* **141,** 165–84.   **2.3.3**

Motylewski, Karen. *See* Wood, J. A.

Moulton, F. R. (1896) Some points which need to be emphasized in teaching general astronomy. *Pop. Astron.,* **4,** 400–407.   **1.2.2**

(1899/1910) Laplace's ring nebular hypothesis. (Paper presented at meeting, September 1899). *Publications of the Astronomical and Astrophysical Society of America,* I (pub. 1910), 98.   **1.2.3**

(1900) An attempt to test the nebular hypothesis by an appeal to the laws of dynamics. *Astrophys. J.,* **11,** 103–30.   **1.2.3, 1.2.5**

(1905) On the evolution of the Solar System. *Astrophys. J.,* **22,** 165–81.

          1.1.1, 1.2.5

(1906a) *Introduction to astronomy.* New York: Macmillan.   **1.1.1**

(1906b) Report. *Carn. Inst. Yb.,* **4,** 186–90.   **1.2.5**

(1907) On the probability of a near approach of two suns and the orbits of material ejected from them under the stimulus of their mutual tidal disturbances. *Carn. Inst. Yb.,* **5,** 168–9.   **1.2.5**

(1909a) Notes on the possibility of fission of a contracting rotating fluid mass. In *The tidal and other problems,* ed. T. C. Chamberlin et al., pp. 135–60. Carnegie Institution of Washington, Publication 107.   **4.2.4**

(1909b) Notes on the possibility of fission of a contracting rotating fluid mass. *Astrophys. J.,* **29,** 1–13. (abstract of Moulton, 1909a)   **4.2.4**

(1910) Inquiry into the fundamental problems of geology. *Carn. Inst. Yb.,* **8,** 225–6.

          1.2.5

(1928) The planetesimal hypothesis. *Science,* **68,** 549–59.   **1.3.3**

(1929) The planetesimal hypothesis. *Science,* **69,** 246–8.   **1.3.3**

(1935) *Consider the heavens.* Garden City, NY: Doubleday, Doran.   **1.2.3**

(1939) Composition of gaseous nebulae. *Scientific Monthly,* **48,** 485–6.   **1.2.6**

(D1897) Letter to T. C. Chamberlin, 25 November 1897. Box 5, folder "Correspondence with F. R. Moulton," C-UC.   **1.2.3**

(D1899a) Letter to T. C. Chamberlin, 21 June 1899. Box 5, C-UC.   **1.2.3**

(D1899b) Letter to G. E. Hale, 12 June 1899. Box 23 of letters to Hale, Yerkes Observatory.   **1.2.3**

(D1900a) Letter to G. E. Hale, 3 January 1900. Box 23 of letters to Hale, Yerkes
    Observatory.   **1.2.3**
(D1900b) Letter to G. E. Hale, 15 January 1900. Box 23 of letters to Hale, Yerkes
    Observatory.   **1.2.3**
(D1901) Letter to G. E. Hale, 26 March 1901. Box 27, Yerkes Observatory.
                                                                         1.2.3
(D1903) Letter to T. C. Chamberlin, 20 December 1903. Box 5, folder "Corre-
    spondence with F. R. Moulton," C-UC.   **1.2.5**
(D1904a) Letter to T. C. Chamberlin, 8 February 1904. Box 5, folder "Corre-
    spondence with F. R. Moulton," C-UC.   **1.2.5**
(D1904b) Letter to T. C. Chamberlin, 19 May 1904. Box 5, C-UC.   **1.2.5**
(D1904c) Letter to T. C. Chamberlin, 24 May 1904. Box 5, C-UC.   **1.2.5**
(D1905) Letter to R. S. Woodward, 3 August 1905. Chamberlin file, CIW.
                                                                         1.2.5
(D1914a) Letter to W. W. Cambell, 10 March 1914. LOA.   **4.2.4**
(D1914b) Letter to W. W. Campbell, 23 December 1914. LOA.   **4.2.4**
(D1915) Letter to W. W. Campbell, 30 April 1915. LOA.   **4.2.4**
(D1926a) Letter to H. N. Russell, 4 January 1926. Box 31, R-PU.   **1.3.4**
(D1926b) Letter to H. N. Russell, 10 April 1926. Box 31, R-PU.   **1.3.4**
(D1929a) Letter to H. N. Russell, 6 June 1929. Box 33, R-PU.   **1.2.5**
(D1929b) Letter to H. N. Russell, 12 June 1929. Box 33, R-PU.   **1.3.4**
Moulton, F. R. *See also* Chamberlin, T. C.; Leith, C. K.
Müller, Edith A. *See* Goldberg, Leo
Mueller, George E. (1964) Manned space flight. In *NASA authorization for fiscal year
    1965,* Hearings before the Committee on Aero-Space Sciences, U.S. Senate, 88th
    Congress, 2nd Session, 4 March 1964, Part 1, pp. 5–99.   **4.3.2**
Mueller, S. W. *See* McKinnon, W. B.
Münch, G. *See* Ingersoll, A. P.
Muir, Patricia. *See* Taylor, S. R.
Mullan, D. J. *See* De Groot, M.
Mumford, N. W. (1912) A question on the capture theory. *Pop. Astron.,* **20,** 228–32.
                                                                         1.2.4
Munk, Walter H. (1964/1966) Variation of the Earth's rotation in historical time. In
    The Earth–Moon system (proceedings of 1964 conference), ed. B. G. Marsden &
    A. G. W. Cameron, pp. 52–69. New York: Plenum Press (1966).   **4.4.2**
Munk, Walter H., & MacDonald, G. J. F. (1960) *The rotation of the Earth: A geophysical
    discussion.* Cambridge University Press.   **4.4.2**
Munson, R. (1971) *Man and nature: Philosophical issues in biology.* New York: Dell.
                                                                         2.2.2
Murray, Bruce, Malin, Michael C., & Greeley, Ronald (1981) *Earthlike planets: Surfaces
    of Mercury, Venus, Earth, Moon, Mars.* San Francisco: Freeman.
                                                         3.1.2, 3.3.4, 4.4.4
Murray, Bruce C., Strom, R. G., Trask, N. J., & Gault, D. E. (1975) Surface history of
    Mercury: Implications for terrestrial planets. *J. Geophys. Res.,* **80,** 2508–14.
                                                                         4.4.5
Murthy V. R. (1960) Isotopic composition of silver in an iron meteorite. *Phys. Rev.
    Letters,* **5,** 539–41.   **2.3.2**
Murthy, V. R., & Sandoval, P. (1965) Chromium isotopes in meteorites. *J Geophys.
    Res.,* **70,** 4379–82.   **2.3.4**
Murthy, V. R., Schmitt, R. A., & Rey, P. (1970) Rubidium-strontium age and ele-
    mental and isotopic abundances of some trace elements in lunar samples. *Science,*
    167, 476–9.   **4.4.4**
Murthy, V. R., & Urey, H. C. (1962) The time of formation of the solar system
    relative to nucleosynthesis. *Astrophys. J.,* **135,** 626–31.   **2.3.2, 2.3.4, 2.5.1**

Musson, W. Balfour (1909) Development in the stellar universe. *J. Roy. Astron. Soc. Canada,* **3**, 5–27.   **1.2.4**

Myers, P. C. *See* Cameron, A. G. W.

Nagata, Takesi (1979) Meteorite magnetism and the early Solar System magnetic field. *Phys. Earth Plan. Int.,* **20**, 324–41.   **2.3.5**

Nagel, Brigitte (1991) *Die Welteislehre. Ihre Geschichte und ihre Rolle in "Dritten Reich."* Stuttgart: Verlag für Geschichte der Naturwissenschaften und der Technik.
                                                                                                                                                4.2.4

Nakagawa, Y. *See* Hayashi, C.

Nakamura, Y., Latham, G., Lammlein, D., Ewing, M., Duennebier, F., & Dorman, J. (1974) Deep lunar interior inferred from recent seismic data. *Geophys. Res. Lett.,* **1**, 137–40.   **4.3.4**

Nakamura, Yosio, Latham, Gary, & Dorman, H. J. (1982) Apollo lunar seismic experiment: Final summary, April 1982. *Proc. 13th Lunar & Plan. Sci. Conf. (J. Geophys. Res.,* **87**, Supplement), A117–A123.   **4.3.4**

Nakazawa, K., Komuro, T., & Hayashi, C. (1983) Origin of the Moon: Capture by gas drag of the Earth's primordial atmosphere. *Moon and Planets,* **28**, 311–27.
                                                                                                                                                4.4.4

Nakazawa, Kiyoshi. *See* Hayashi, C.; Mizuno, H.

Nance, W. *See* Taylor, S. R.

Napier, W. McD., & Dodd, R. J. (1974) On the origin of the asteroids. *M. N. Roy. Astron. Soc.,* **166**, 466–89.   **3.4.7**

Napier. W. McD. *See* Dodd, R. J.

Ness, Norman F. *See* Acuna, M. H.

Neugebauer, G. *See* Aumann, H. H.; Ingersoll, A. P.

Newcomb, Simon (1903) *Astronomy for everybody.* London: Pitman.   **1.2.2**

Newell, Homer (1961) Statement of Homer E. Newell, Deputy Director, Space Flight Programs, NASA (1 March 1961). In *NASA scientific and technical programs,* Hearings before the Committee on Aeronautical and Space Sciences, U.S. Senate, 87th Congress, 1st Session, 28 February and 1 March 1961, pp. 207–27. Washington, DC: Government Printing Office.   **4.3.2**

(1973) Harold Urey and the Moon. *Moon,* **7**, 1–5.   **4.3.1**

(1981) *Beyond the atmosphere: Early years of space science.* Washington, DC: Government Printing Office. National Aeronautics and Space Administration report NASA SP-4211.   **4.3.1**

(D1959) Notes on meeting with H. Urey, R. Jastrow, J. O'Keefe, 16 January 1959. Notebook in Newell files 28, NASA-A.   **4.3.1**

Newell, Homer. *See also* Jastrow, R.

Newell, Patrick T. (1985) Review of the critical ionization velocity effect in space. *Reviews of Geophysics,* **23**, 93–104.   **3.4.5**

Newhall, X. X. *See* Yoder, C. F.

Newsom, Horton E. (1984a) The abundance of molybdenum in lunar samples. *Lunar & Plan. Sci.,* **15**, 605–6.   **4.3.4**

(1984b) The lunar core and the origin of the Moon. *Eos,* **65**, 369–70, 461.
                                                                                                                                                4.3.4

Newsom, Horton E., & Drake, Michael J. (1983) Experimental investigation of the partitioning of phosphorus between metal and silicate phases: Implications for the Earth, Moon, and eucrite parent body. *Geoch. Cosmoch. Acta,* **47**, 93–100.
                                                                                                                                                4.4.4

(1987) Formation of the Moon and terrestrial planets: Constraints from V, Cr, and Mn abundances in planetary mantles and from new partitioning experiments. *Lunar & Plan. Sci.,* **18**, 716–17.   **4.4.4**

Newsom, Horton E., & Taylor, S. R. (1989) Geochemical implications of the formation of the Moon by a single giant impact. *Nature,* **338**, 29–34, 360.   **4.1.6**

Nicholson, P. D. *See* Elliot, J. L.

Nölke, Friedrich (1908) *Das Problem der Entwicklung unseres Planetensystems. Aufstellung einer neuen Theorie nach vorgehender Kritik der Theorien von Kant, Laplace, Poincaré, Moulton, Arrhenius, u.a.* Berlin: Springer.   **1.2.6**

   (1922) Uber die Entstehung der Oberflächenformationen des Mondes. *Astron. Nachr.*, **215**, 217–28.   **4.2.4**

   (1924) Muss die Darwinsche Erklärung der Entwicklung des Erdmondes aufgegeben werden? *Astron. Nachr.*, **220**, 269–72.   **4.2.4**

   (1930) *Der Entwicklungsgang unseres Planetensystems; eine kritische Studie.* Berlin: Dümmler.   **1.3.4, 4.2.4**

   (1932) Die vorgeologische Entwicklung der Erde als Schlüssel zum Verständnis der geologischen Entwicklung. *Gerlands Beitr.*, **37**, 252–70.   **4.2.4**

   (1934) Der Ursprungsort des Mondes. *Gerlands Beitr.*, **41**, 86–91.   **4.2.4**

Nolan, James (1885) *Darwin's theory of the genesis of the Moon.* Melbourne: Robertson.   **4.2.4**

   (1886) Tidal friction and the evolution of a satellite. *Nature*, **34**, 286–7.   **4.2.4**

   (1887) Tidal friction and the evolution of a satellite. *Nature*, **35**, 75.   **4.2.4**

   (1895) *Satellite evolution: The evident scope of tidal friction. The meaning of Saturn's rings.* Melbourne: Robertson.   **4.2.4**

Nordmann, J. C. *See* Turcotte, D. L.

Nørgaard, Henry (1980) ^{26}Al from red giants. *Astrophys. J.*, **236**, 895–8.   **2.5.3**

Nørgaard, Henry. *See also* Arnould, M.

Noyes, Barbara. *See* Lewis, J. S.

Numbers, Ronald L. (1977) *Creation by natural law: Laplace's nebular hypothesis in American thought.* Seattle: University of Washington Press.   **1.2.2, 2.1**

Nyquist, L. E. *See* Bogard, D. D.

Öpik, E. J. (1953) Stellar associations and supernovae. *Irish Astron. J.*, **2**, 219–33.   **2.3.2**

   (1955a) Dust and star formation in supernova explosions. In *Les particules solides dans les astres* (Communications prés. 6e Coll. Int. Astrophys., Liège, July 1954) = *Memoires de la Société Royale Scientifique, Liège* [4], **15**, 634–7.   **2.3.2**

   (1955b) The origin of the Moon. *Irish Astron. J.*, **3**, 245–8.   **4.3.2**

   (1961) Tidal deformations and the origin of the Moon. *Astron, J.*, **66**, 60–7.   **4.4.4, 4.4.6**

   (1962a) Jupiter: Chemical composition, structure and origin of a giant planet. *Icarus*, **1**, 200–57.   **3.2.1**

   (1962b) Surface properties of the Moon. In *Progress in the astronautical sciences*, ed. S. F. Singer, vol. 1, pp. 219–60. Amsterdam: North-Holland.   **4.3.2, 4.4.6**

   (1962/1963) Dissipation of the solar nebula. In *Origin of the Solar System*, ed. R. Jastrow & A. G. W. Cameron, pp. 73–5. New York: Academic Press.   **2.3.3**

   (1967) Evolution of the Moon's surface. I. *Irish Astron. J.*, **8**, 38–52.   **4.4.2**

   (1972) Comments on lunar origin. *Irish Astron. J.*, **10**, 190–238.   **4.4.2, 4.4.4, 4.4.6**

   (1977) The rings of Uranus. *Irish Astron, J.*, **13**, 45–8.   **3.4.5**

O'Halloran, Sylvester N. E. (1912) The second law of thermodynamics and Professor Bickerton's theory. *Knowledge* (n.s.), **9**, 361.   **1.2.1**

O'Hara, J. G. (1975) George Johnstone Stony, F.R.S. and the concept of the electron. *Notes Rec. Roy. Soc.*, **29**, 265–76.   **1.2.3**

Okamoto, Isao (1969) On the loss of angular momentum from the protosun and the formation of the Solar System. *Pub. Astron. Soc. Japan*, **21**, 25–53.   **2.3.5**

O'Keefe, John A. (1963) Two avenues from astronomy to geology. In *The Earth sciences: Problems and progress in current research*, ed. T. W. Donnelly, pp. 43–58. Chicago: University of Chicago Press.   **4.1.3, 4.3.3**

   (1964/1966) The origin of the Moon and the core of the Earth. In *The Earth–*

Moon system (Proceedings of conference held in January 1964), ed. B. G. Marsden & A. G. W. Cameron, pp. 224–33. New York: Plenum Press (1966).   **4.1.3**

(1968) Fission hypothesis for the origin of the Moon. *Astron. J.*, **73**, 195–6.
                                                                                 4.3.3

(1969a) Manned landings and theories of lunar formation. *Bulletin of the Atomic Scientists*, **25**, no. 7 (September), 56–60.   **4.1.3, 4.1.4, 4.3.3**

(1969b) Origin of the Moon. *J. Geophys. Res.*, **74**, 2758–67.
                                                      4.1.3, 4.1.4, 4.3.3, 4.4.3

(1969c) Manned landings and theories of lunar formation. In *Man on the Moon*, ed. E. Rabinowitch & R. S. Lewis, pp. 136–46. New York: Basic.   **4.3.3**

(1970) The origin of the Moon. *J. Geophys. Res.*, **75**, 6565–74.   **4.3.4**

(1972a) Geochemical evidence for the origin of the Moon. *Naturwissenschaften*, **59**, 45–52.   **4.3.4**

(1972b) The origin of the Moon: Theories involving joint formation with the Earth. *Astrophys. Space Sci.*, **16**, 201–11.   **4.3.4**

(1972c) Inclination of the Moon's orbit: The early history. *Irish Astron. J.*, **10**, 241–50.   **4.3.4, 4.4.2**

(1973) After Apollo: Fission origin of the Moon. *Bulletin of the Atomic Scientists*, **29**, no. 9 (November), 26–9.   **4.3.4**

(D1978) Letter to H. Newell, 22 June 1978. Newell Papers, box 40, NASA-A.
                                                                                 4.3.1

(D1984) History of the Earth–Moon system. Colloquium lecture at Goddard Space Flight Center, Greenbelt, MD, 28 September 1984.   **4.4.4**

O'Keefe, John A., & Sullivan, Edward C. (1978) Fission origin of the Moon: Cause and timing. *Icarus*, **35**, 272–83.   **4.3.5, 4.4.4**

O'Keefe, John A., & Urey, Harold C. (1977) The deficiency of siderophile elements in the Moon. *Phil. Trans.*, **A285**, 569–75.   **4.3.4, 4.3.5**

O'Keefe, John D., & Ahrens, T. J. (1977) Impact-induced energy partitioning, melting, and vaporization on terrestrial planets. *Proc. 8th Lunar Sci. Conf.*, 3357–74.   **4.4.5**

Oldham, R. D. (1918) The constitution of the interior of the Earth as revealed by earthquakes. *Nature*, **102**, 235–6.   **1.3.3**

O'Leary, B. (1970) *The making of an ex-astronaut*. Boston: Houghton Mifflin.
                                                                                 4.1.3

Olness, Robert J. *See* Graboske, Harold C., Jr.

Olsen, Edward J., & Grossman, Lawrence (1974) A scanning electron microscope study of olivine crystal surfaces. *Meteoritics*, **9**, 243–54.   **3.3.3**

Olsen, Edward J. *See also* Grossman, L.

O'Mara, B. J. *See* Aller, L. H.

O'Neill, James R. *See* Friedman, I.

Onuma, Naoki, Clayton, Robert N., & Mayeda, Toshiko K. (1970) Oxygen isotope fractionation between minerals and an estimate of the temperature of formation. *Science*, **167**, 536–8.   **4.3.4, 4.4.3**

(1972) Oxygen isotope cosmothermometer. *Geoch. Cosmoch. Acta*, **36**, 169–88.
                                                                                 4.3.4

Onuma, N. *See also* Clayton, R. N.

Orowan, E. (1969) Density of the Moon and nucleation of planets. *Nature*, **222**, 867.
                                                      4.1.3, 4.4.3, 4.4.6

Orton, G. S. *See* Ingersoll, A. P.

Ostic, R. G., Russell, R. D., & Reynolds, P. H. (1963) A new calculation for the age of the Earth from abundances of lead isotopes. *Nature*, **199**, 1150–2.   **4.3.3**

O'Toole, Thomas (1979) Evidence points to solar system born of stellar explosion. *Washington Post* (January 1), A3.   **2.5.2**

Owen, Richard (1857) *Key to the geology of the globe*. New York: Barnes/Boston: Gould & Lincoln.   **4.2.1**

Owen, Tobias. *See* Gautier, D.; Goldsmith, D.; Smith, B. A.

Page, Thornton (1948) The origin of the Earth. *Physics Today*, **1**, no. 6, 12–24.   **1.3.4**
    (1980) Review of *Cosmogony of the Solar System* by Hoyle. *Sky & Tel.*, **59**, 324–5.
                                                                    2.3.5

Pagel, B. E. J. (1973) Stellar and solar abundances. In *Cosmochemistry*, ed. A. G. W.
    Cameron, pp. 1–21.   **3.3.3**

Painter, George Stephen (1940) *Science and evolutionary theory*. Washington, DC: Wash-
    ington College Press.   **1.3.5**

Palme, Christl. *See* Wänke, H.

Palme, Herbert. *See* Hertogen, J., Wänke, H.

Pannella, Giorgio (1972) Paleontological evidence on the Earth's rotational history
    since early Precambrian. *Astrophys. Space Sci.*, **16**, 212–37.   **4.3.3**

Paolicchi, P. *See* Farinella, P.

Papaloizu, J. *See* Lin, D. N. C.

Papanastassiou, D. A., Huneke, J. C., Esat, T. M., & Wasserburg, G. J. (1978) Pan-
    dora's box of the nuclides. *Lunar & Plan. Sci.*, **9**, 859–61.   **2.5.3**

Papanastassiou, D. A. *See also* Albee, A. L., Lee, T.

Paque, Julie M. *See* MacPherson, G. J.

Pariiski, N. N. (1960) The influence of Earth tides on the secular retardation of the
    Earth's rotation. *Sov. Astron. AJ*, **4**, 515–22. Trans. from *Astron. Zh.*, **37** (1960),
    543–49.   **4.4.2**

Parker, Peter D. *See* Champagne, A. E.

Parson, Alfred Lauck (1944) Vapour pressure of solids at low temperature (and the
    origin of the planets). *Nature*, **154**, 707–8.   **1.1.3**
    (1945) The vapour pressures of planetary constituents at low temperatures, and their
    bearing on the question of the origin of the planets. *M. N. Roy. Astron. Soc.*, **105**,
    244–5.   **1.1.3**

Patterson, Andrew H. (1909) The origin of the Moon. *Science*, **29**, 936–7.   **4.2.4**

Payne, Cecilia H. (1925a) *Stellar atmospheres: A contribution to the observational study of
    high temperature in the reversing layers of stars*. Cambridge, Mass.: Harvard University
    Press (Harvard Observatory Monographs, No. 1).   **1.1.2**
    (1925b) Astrophysical data bearing on the relative abundance of elements. *Proc. N.
    A. S.*, **11**, 192–8.   **1.1.2**

Peale, S. J. *See* Boss, A. P.

Pearce, G. W. *See* Lanoix, M.

Pearl, J. C. *See* Hanel, R. A.

Pechernikova, G. V. *See* Safronov, V. S.

Peebles, P. J. E. (1964) The structure and composition of Jupiter and Saturn. *Astrophys.
    J.*, **140**, 328–7.   **3.2.1**

Peirce, Benjamin (1871) The contraction of the Earth. *Nature*, **3**, 315.   **4.2.1**
    (1873) [On the formation of the shell of the Earth by shrinkage.] *Proc. Am. Acad.*,
    8, 106–8.   **4.2.1**

Pellas, Paul (1977) Citation on the award of the Leonard Medal of the Meteoritical
    Society to Dr. Hans E. Suess, Dept. of Chemistry, Univ. of California, San Di-
    ego. *Meteoritics*, **12**, 161–4.   **3.3.3**

Penrose, R. A. F., Jr. *See* Leith, C. K.

Pepin, R. O. (1964) Isotopic analyses of xenon. In *The origin and evolution of atmo-
    spheres and oceans* (proceedings of a conference, April 1963), ed. P. J. Brancazio &
    A. G. W. Cameron, pp. 191–233. New York: Wiley.   **2.4.1**

Pepin, R. *See* Arnold, J.

Perri, Fausto, & Cameron, A. G. W. (1974) Hydrodynamic instability of the solar
    nebula in the presence of a planetary core. *Icarus*, **22**, 416–25.   **3.2.1, 3.2.2**

Persson, H. (1963) Electric field along a magnetic line of force in a low-density
    plasma. *Physics of Fluids*, **6**, 1756–9.   **3.4.4**

(1966) Electric field parallel to the magnetic field in a low-density plasma. *Physics of Fluids*, **9**, 1090–8.   **3.4.4**

Petelski, E. F., Fahr, H. J., Ripken, H. W., Brenning, N., & Axnäs, I. (1980) Enhanced interaction of the solar wind and the interstellar neutral gas by virtue of a critical velocity effect. *Astronomy and Astrophysics*, **87**, 20–30.   **3.4.5**

Petford, A. D. *See* Blackwell, D. E.

Phillips, Roger J., & Malin, M. C. (1983) The interior of Venus and tectonic implications. In *Venus*, ed. D. M. Hunten et al., pp. 159–214. Tucson: University of Arizona Press.   **3.3.5**

Phillips, Roger J. *See* Hartmann, W. K.

Picciotto, E. *See* Hanappe, F.

Pickering, W. H. (1893a) The rotation of Jupiter's outer satellites. *Astronomy and Astrophysics*, **12**, 481–94.   **1.2.4**

(1893b) Polar inversion of the planets and satellites. *Astronomy and Astrophysics*, **12**, 692–3.   **1.2.4**

(1899) Entdeckung eines neuen Saturnmondes. *Astron. Nachr.*, **149**, 31–2.   **1.2.4**

(1901) Explanation of the inclination of the planetary axes. *Astron. J.*, **22**, no. 511 (14 November), 56–7.   **1.2.4**

(1903) *The Moon: A summary of the existing knowledge of our satellite, with a complete photographic atlas*. New York: Doubleday, Page.   **4.2.4**

(1904a) Direct and retrograde rotation of the planets. *Astron. Nachr.*, **164**, 201–4.
                                                                                        1.2.4

(1904b) The ninth satellite of Saturn. *Harvard College Observatory Annals*, **53**, 45–73.
                                                                                        1.2.4

(1905) Note on the evolution of the Solar System. *Astrophys. J.*, **22**, 354–5.
                                                                                 1.2.4, 1.2.5

(1905/1910) Planetary inversion. *Publications of the Astronomy and Astrophysics Society of America*, **1**, 250–1 (abstract of paper at meeting, December 1905, published 1910).   **1.2.4**

(1907) The place of origin of the Moon. *Pop. Astron.*, **15**, 274–87.   **4.2.4**

(1917) Why the axes of the planets are inclined. *Pop. Astron.*, **25**, 487–9.   **1.3.5**

(1920) Various suggestions relating to stellar evolution, planetary genesis, and hyperbolic comets. *Pop. Astron.*, **28**, 77–80.   **1.3.5**

Pierce, Michael J., & Tully, R. Brent (1988) Distances to the Virgo and Ursa Major clusters and a determination of H_0. *Astrophys. J.*, **330**, 579–95.   **2.1.2**

Pillinger, C. T. (1984) Light element stable isotopes in meteorites — from grams to picograms. *Geoch. Cosmoch. Acta*, **48**, 2739–66.   **2.5**

Pine, M. R. *See* Cameron, A. G. W.

Pirraglia, J. A. *See* Hanel, R. A.

Podolak, Morris, & Cameron, A. G. W. (1974a) Models of the giant planets. *Icarus*, **22**, 123–48.   **3.2.1**

(1974b) Possible formation of meteoritic chondrules and inclusions in the precollapse protoplanetary atmosphere. *Icarus*, **23**, 326–33.   **3.2.1**

Podolak, Morris, & Reynolds, Ray. T. (1984) Consistency tests of cosmogonic theories from models of Uranus and Neptune. *Icarus*, **57**, 102–11.   **3.2.3**

(1985) What have we learned from modeling giant planet interiors? In *Protostars and planets*. II, ed. D. C. Black & M. S. Matthews, pp. 847–72. Tucson: University of Arizona Press.   **2.4.2**

Podosek, F. A. *See* Albee, A. L.

Pogodin, M. A. *See* Gnedin, Yu. N.

Poincaré, Henri (1911) Leçons sur les hypothèses cosmogoniques. Paris: Hermann et fils.   **1.2.6**

(1913) Leçons sur les hypothèses cosmogoniques. 2nd edn. Paris: Hermann.
                                                                               1.1.1, 1.2.4, 1.2.6

Pollack, James B. (1975) The rings of Saturn. *Space Science Reviews*, **18**, 3–93.   **3.4.4**
   (1978) The rings of Saturn. *American Scientist*, **66**, 30–7.   **3.4.4**
   (1979) Climatic change on the terrestrial planets. *Icarus*, **37**, 479–553.   **3.3.5**
   (1984) Origin and history of the outer planets: Theoretical models and observational constraints. *Ann. Rev. Astron. Astrophys.*, **22**, 389–424.   **3.2.2, 3.2.3**
   (1985) Formation of the giant planets and their satellite-ring systems: An overview.
   In *Protostars and planets*, ed. D. C. Black & M. S. Matthews, pp. 791–831. Tucson:
   University of Arizona Press.   **1.1.5, 3.2.3, 3.4.8**
Pollack, James B., & Black, David C. (1979) Implications of the gas compositional
   measurements of Pioneer Venus for the origin of planetary atmospheres. *Science*,
   205, 56–9.   **3.3.5**
Pollack, James B., Burns, J. A., & Tauber, M. E. (1979) Gas drag in primordial cir-
   cumplanetary envelopes: A mechanism for satellite capture. *Icarus*, **37**, 587–611.
       3.2.1
Pollack, James B., & Cuzzi, Jeffrey N. (1981) Rings in the Solar System. *Sci. Am.*,
   245, no. 5 (November), 104–29.   **3.4.4**
Pollack, James B., Grossman, Allen S., Moore, Ronald, & Grasboske, Harold C., Jr.
   (1976) The formation of Saturn's satellites and rings as influenced by Saturn's
   contraction history. *Icarus*, **29**, 35–48.   **3.3.5**
   (1977) A calculation of Saturn's gravitational contraction history. *Icarus*, **30**, 111–28.
       3.2.1
Pollack, James B. *See also* Bodenheimer, P.; Burns, J. A.; Cabot, W.; Cameron,
   A. G. W.; Donahue, T. M.; Grasboske, H. C., Jr.; Grossman, A. S.
Pomeroy, J. H., & Hubbard, N. J., Editors (1974/1977) *The Soviet–American Conference
   on the Cosmochemistry of the Moon and planets* (Moscow, June 1974). Washington,
   DC: National Aeronautics & Space Administration, report SP-370 (pub. 1977).
       2.2.5
Poor, Charles Lane (1908) *The Solar System: A study of recent observations.* New York:
   Putnam's.   **1.2.6**
   (1910) Review of Lowell's *The evolution of worlds. Science*, **31**, 506–7.   **1.2.4**
Popper, Karl (1934/1959) *The logic of scientific discovery.* London: Hutchinson. Trans.
   from *Logik der Forschung* (pub. 1934).   **1.3.5, 2.2.2**
   (1962) *Conjectures and refutations.* New York: Basic Books.   **1.3.5, 4.4.7**
   (1974) Darwinism as a metaphysical research programme. In *The Philosophy of Karl
   Popper*, ed. P. A. Schilpp, pp. 133–43. LaSalle, IL: Open Court.
       1.3.5, 2.2.2, 4.4.7
   (1978) Natural selection and the emergence of mind. *Dialectica*, **32**, 339–55.
       1.3.5, 4.4.7
   (1980) Evolution. *New Scientist*, **87**, 611.   **1.3.5, 4.4.7**
Pottasch, Stuart R. (1963) The lower solar corona: The abundance of iron. *M. N.
   Roy. Astron. Soc.*, **125**, 543–56.   **3.3.3**
Pratap, R. (1977) The asteroidal belt and Kirkwood gaps. II. Kinematical theory.
   Pramana, **8**, 447–56.   **3.4.7**
Prentice, A. J. (1978) Origin of the Solar System. *Moon and Planets*, **19**, 341–98.
       3.4.2
Pringle, J. E. *See* Lynden-Bell, D.
Prinn, Ronald G. *See* Lewis, J. S.
Pyne, Stephen J. (1980) *Grove Karl Gilbert: A great engine of research.* Austin: University
   of Texas Press.   **4.2.4**
Quijano-Rico, M. *See* Wänke, H.
Radl, E. (1930) *The history of biological theories.* London: Oxford University Press.
       1.2.2

Radzievsky, V. V. (1952) The origin of the Moon in the light of the cosmogonic theory of O. Schmidt. (in Russian). *Byuleten, Vsesoyuznoe Astronomo-Geodezicheskoe Obshchestvo*, no. 11 (18), 3–8.   **4.3.2**

Radzievsky, V. V. *See* Artemev, A. V.

Raffety, Charles W. (1912) The spectroscopic aspect of the impact theory. *Knowledge* (n.s.), **9,** 106.   **1.2.1**

Rainger, Ronald (1993) Biology, geology, or neither, or both: Vertebrate paleontology at the University of Chicago, 1892–1950. *Perspectives on Science*, **1,** 478–519.
                                                                         1.1.1

Rajan, R. Sundar. *See* Herbst, W.

Ramaty, Reuven (1978) Mechanisms and sites for astrophysical gamma ray line production. In *Gamma ray spectroscopy in astrophysics*, ed. T. L. Cline & R. Ramaty, pp. 6–41. Greenbelt, MD: Goddard Space Flight Center, GSFC NASA Technical Memorandum 79619.   **2.5.3**

Ramaty, Reuven, & Lingenfelter, R. E. (1977) ^{26}Al: A galactic source of gamma-ray line emission. *Astrophys. J. Lett.*, **213,** L5–L7.   **2.5.3**

Ramaty, Reuven. *See also* Lingenfelter, R. E.

Rammensee, W., & Wänke, H. (1977) On the partition coefficient of tungsten between metal and silicate and its bearing on the origin of the Moon. *Proc. 8th Lunar Sci. Conf.*, 399–409.   **4.3.4, 4.4.4**

Rammensee, W. *See* Wänke, H.

Ramsey, W. H. (1948) On the constitution of the terrestrial planets. *M. N. Roy. Astron. Soc.*, **108,** 406–13.   **3.1.1, 4.1.3, 4.3.2**

(1949) On the nature of the Earth's core. *Mon. Not. Roy. Astron. Soc. G. S.*, **5,** 409–26.   **3.1.1, 4.1.3, 4.3.2**

(1951) On the constitution of the major planets. *M. N. Roy. Astron. Soc.*, **111,** 427–47.   **3.2.1**

Randic, L. (1950) Schmidt's theory of the origin of visual binary stars and of the Solar System. *Observatory*, **70,** 217–22.   **1.1.3**

Razbitnaya, E. P. (1954) *On the origin of the Moon* (in Russian). Doctoral dissertation, Leningrad State Paedagogical Institute.   **4.3.2**

Red'kina, N. P. *See* Gnedin, Yu. N.

Redman, Leander A. (1919) *Professor Montgomery's discoveries in celestial mechanics.* San Francisco: Pernau-Walsh.   **1.2.4**

Reed, G. W., Kigoshi, K., & Turkevich, A. (1960) Determinations of concentrations of heavy elements in meteorites by activation analysis. *Geoch. Cosmoch. Acta,* **20,** 122–40.   **3.3.3**

Reeves, H. (1968) Astrophysical implications of the stellar abundances of Li, Be and B. In *Origin and distribution of the elements,* ed. L. H. Ahrens, pp. 117–23. New York: Pergamon Press.   **2.3.4**

Reeves, H., Editor (1972/1974) *Origin of the Solar System/L'Origine du système solaire* (proceedings of symposium at Nice, 1972). Paris, CNRS (pub. 1974).
                                                                   2.2.5, 2.3.4

Reeves, H. (1978a) The origin of the Solar System. In *The origin of the Solar System,* ed. S. F. Dermott, pp. 1–17. New York: Wiley.   **1.1.4**

(1978b) The "bing bang" theory of the origin of the Solar System. In *Protostars and planets,* ed. T. Gehrels, pp. 399–426. Tucson: University of Arizona Press.
                                                                   2.5.1, 2.5.2

Reeves, H., & Audouze, J. (1968) Early heat generation in meteorites. *Earth Plan. Sci. Lett.*, **4,** 135–41.   **2.5.1**

Reeves, H., Fowler, W. A., & Hoyle, F. (1970) Galactic cosmic ray origin of Li, Be and B in stars. *Nature,* **226,** 727–9.   **2.3.4, 2.5**

Reeves, H. *See also* Bernas, R.

Reid, Harry Fielding (1924) The planetesimal hypothesis and the solar system. *Am. J. Sci.* [5], **7**, 37–64.   1.3.4

Reid, M. J. *See* Ward, W. R.

Reingold, Nathan (1979) National science policy in a private foundation: The Carnegie Institution of Washington. In *The organization of knowledge in modern America, 1860–1920,* ed. A. Oleson & J. Voss, pp. 313–41. Baltimore: Johns Hopkins University Press.   1.2.5

Reuyl, Dirk, & Holmberg, Erik (1943) On the existence of a third component in the system 70 Ophiuchi. *Astrophys J.,* **97,** 41–5.   1.1.3, 2.1.2

Rey, P. *See* Murphy, V. R.

Reynolds, J. H. (1960) Isotopic composition of primordial xenon. *Phys. Rev. Lett.,* **4,** 351–4.   2.3.1, 2.3.2

Reynolds, J. H., & Turner, G. (1964) Rare gases in the chondrite Renazzo. *J. Geophys. Res.,* **69,** 3263–81.   2.5.1

Reynolds, P. H. *See* Ostic, R. G.

Reynolds, Ray T. *See* Grossman, A. S.; Podolak, M.

Richards, J. R. *See* Cooper, J. A.

Richer, J. *See* Garz, T.

Rieder, R. *See* Wänke, H.

Rigden, S. M., & Ahrens, T. J. (1981) Impact vaporization and lunar origin. *Lunar & Plan. Sci.,* **12,** 885–7.   4.4.5

Rightmire, Robert A. *See* Simanton, J. R.

Ringwood, A. E. (1960) Some aspects of the thermal evolution of the Earth. *Geoch. Cosmoch. Acta,* **20,** 241–59.   3.3.2, 4.1.3, 4.1.4, 4.3.3, 4.4.3

(1962) Present status of the chondritic Earth model. In *Researches on Meteorites,* ed. C. B. Moore, pp. 198–216. New York: Wiley.   3.3.2

(1963) The origin of high-temperature minerals in carbonaceous chondrites. *J. Geophys. Res.,* **68,** 1141–3.   3.3.2

(1964/1966) The chemical composition and origin of the Earth. In *Advances in Earth Sciences* (proceedings of a conference held in 1964), ed. P. M. Hurley, pp. 287–356. Cambridge, MA: MIT Press.   3.3.2

(1966a) Chemical evolution of the terrestrial planets. *Geoch. Cosmoch. Acta,* **30,** 41–104.   3.3.1, 3.3.2, 4.1.3, 4.1.4, 4.3.3, 4.4.3, 4.4.4

(1966b) Genesis of chondritic meteorites. *Reviews of Geophysics,* **4,** 113–74.   3.3.3

(1970) Origin of the Moon: The precipitation hypothesis. *Earth Plan. Sci. Lett.,* **8,** 131–40.   4.1.5, 4.4.4

(1971) Reply [to S. F. Singer]. *J. Geophys. Res.,* **76,** 8075–6.   4.4.4

(1974) The early chemical evolution of the planets. In *In the Beginning . . . ,* ed. J. P. Wild, pp. 48–84. Canberra: Australian Academy of Science.   3.3.3

(1977) Mare basalt petrogenesis and the composition of the lunar interior. *Phil. Trans.,* **A285,** 577–86.   4.4.4

(1978) Origin of the Moon. *Lunar & Plan. Sci.,* **9,** 961–3.   4.3.3, 4.4.4, 4.4.5

(1979) *Origin of the Earth and Moon.* New York: Springer.   3.3.1, 3.3.4, 4.3.1, 4.3.3, 4.3.4, 4.4.4, 4.4.5

(1984) The Earth's core: Its composition, formation and bearing upon the origin of the Earth. *Proc. Roy. Soc.,* **A395,** 1–46.   4.3.3, 4.4.4

(1986a) Composition and origin of the Moon. In *Origin of the Moon,* ed. W. K. Hartmann et al., pp. 673–98. Houston: Lunar and Planetary Institute.   4.4.4, 4.4.5

(1986b) Terrestrial origin of the Moon. *Nature,* **322,** 323–8.   4.4.4, 4.4.5

(1986c) Non-constraints on the origin of the Moon. (Comments on "Origin of the Earth's Moon: Constraints from alkali volatile trace elements" by M. L. Kreutzberger, M. J. Drake, and J. H. Jones.) *Geoch. Cosmoch. Acta,* **50,** 1825.

                                                   4.4.4, 4.4.5

(1987) Gordian knots and lunar origin. *Lunar & Plan. Sci.,* **18,** 838–9.   **4.4.5**

(D1985) Composition and origin of the Moon. Preprint. Research School of Earth Sciences, Australian National University, Canberra.   **4.4.5**

Ringwood, A. E., & Anderson, D. L. (1977) Earth and Venus: A comparative study. *Icarus,* **30,** 243–53.   **3.3.3, 3.3.5**

Ringwood, A. E., Delano, J. W., Kesson, S. E., & Hibberson, W. (1980) More on lunar siderophiles: The strange case of rhenium. *Lunar & Plan. Sci.,* **11,** 929–31.
                                                   4.4.4

Ringwood, A. E., & Essene, E. (1970a) Petrogenesis of Apollo 11 Basalts, internal constitution and origin of the Moon. *Proceedings of the Apollo 11 Lunar Science Conference,* ed. A. A. Levinson, vol. 1, 769–99. New York: Pergamon Press.
                                    4.1.4, 4.3.3, 4.4.3

(1970b) Petrogenesis of lunar basalts and the internal constitution and origin of the Moon. *Science,* **167,** 607–10.   **4.1.4, 4.3.3, 4.4.4**

Ringwood, A. E., & Kesson, S. E. (1976) A dynamic model for mare basalt petrogenesis. *Proc. 7th Lunar Sci. Conf.,* 1697–722.   **4.4.4**

(1977) Basaltic magmatism and the bulk composition of the Moon. II. Siderophile and volatile elements in Moon, Earth and chondrites: Implications for lunar origin. *Moon,* **16,** 425–64.   **4.4.3, 4.4.4**

(1981) Siderophile elements in mare basalts revisited. *Lunar & Plan. Sci.,* **12,** 888–90.   **4.4.4**

Ringwood, A. E. *See also* Delano, J. W.; Kato, K.

Ripken, H. W. *See* Petelski, E. F.

Risteen, Allan Douglas (1895) Molecules and the molecular theory of matter. Boston: Ginn.   **1.2.3**

Ritter, Gustav A. (1906) *Das Weltall und die Entwicklungsgeschichte der Erde.* Berlin: Merkur.   **1.2.2**

(1909) *Die Wunder der Urwelt.* Berlin: Merkur.   **1.2.2**

Robertson, H. P. (1937) Dynamical effects of radiation in the Solar System. *M. N. Roy. Astron. Soc.,* **97,** 423–38.   **1.1.2**

Robinson, Karen L. *See* Wasson, J. T.

Roche, Edouard (1873) Essai sur la constitution et l'origine du système solaire. *Mem. Acad. Montpellier,* **8,** 235–324.   **4.1.1, 4.2.1, 4.2.4**

Rochester, M. G. *See* Urey, H. C.

Rolston, William E. (1910) Planetology. *Nature,* **84,** 99–10.   **1.2.4**

Rose, H. J., Jr. *See* Cuttita, F.

Rosholt, J. N. *See* Tatsumoto, M.

Ross, John E., & Aller, Lawrence H. (1976) The chemical composition of the sun. *Science,* **191,** 1223–9.   **3.3.3**

Rouse, C. A. (1983) Calculation of stellar structure. III. Solar models that satisfy the necessary conditions for a unique solution to the stellar structure equation. *Astronomy and Astrophysics,* **126,** 102–10.   **2.5.2**

(1985) Evidence for a small, high-Z, iron-like solar core. *Astronomy and Astrophysics,* **149,** 65–72.   **2.5.2**

Rubin, Allen E. (1984) Whence came the Moon? *Sky & Tel.,* **68,** 389–93.   **4.4.5**

Rubicam, David Parry (1975) Tidal friction and the early history of the Moon's orbit. *J. Geophys. Res.,* **80,** 1537–48.   **4.4.2**

Rudowski, R. *See* Taylor, S. R.

Runcorn, S. K. (1978) The ancient lunar core dynamo. *Science,* **199,** 771–3.
                                                   4.3.4

(1979) An iron core in the Moon generating an early magnetic field? *Proc. 10th Lunar & Plan. Sci. Conf.,* 2325–33.   **4.3.4**

(1982a) Primeval displacements of the lunar pole. *Phys. Earth Plan. Int.*, **29,** 135– 47. **4.4.6**

(1982b) The Moon's deceptive tranquility. *New Scientist,* **96** (October 21), 174–80. **4.4.6**

(1984) Implications of lunar palaeomagnetism for the origin of the Moon. *Origin of the Moon Conf. Abst.*, p. 10. **4.3.4**

Runcorn, S. K., Collinson, D. W., & Stephenson, A. (1985) Lunar palaeomagnetism: Core dynamo and the primeval satellite system. *Eos,* **66,** 253. **4.3.4**

Ruskol, E. L. (1958/1964) The formation of a protoplanet. *Prob. Cosmog.,* **7,** 5–13. Presented at IAU meeting, August 1958. Trans. from *Vopr. Kosmog.,* 7 (1964), 8– 14. **3.1, 4.4.1**

(1960) The origin of the Moon. I. Formation of a swarm of bodies around the Earth. *Sov. Astron. AJ,* **4,** 657–68. Trans. from *Astron. Zh.,* **37,** 690–702. **4.3.2**

(1960/1962) The origin of the Moon. In *The Moon* (IAU Symposium No. 14, Pulkovo Observatory, December 1960), ed. Z. Kopal & A. K. Mikhailov, pp. 149–55. New York: Academic Press (pub. 1962). **4.1.3, 4.4.1**

(1963a) On the origin of the Moon. II. The growth of the Moon in the circumter- restrial swarm of satellites. *Sov. Astron. AJ,* **7,** 221–7. Trans. from *Astron. Zhur.,* **40,** 288–96. **4.1.3, 4.4.1**

(1963b) The tidal evolution of the Earth–Moon system. *Bulletin of the Academy of Science of the USSR, Geophysics Series,* **2,** 129–33. Trans. from *Izvestiya Akademii Nauk SSSR, Seriya Geofizika,* 1963, no. 2, 216–22. **4.1.3, 4.4.1**

(1966a) On the past history of the Earth–Moon system. *Icarus,* **5,** 221–7. **4.1.3, 4.4.1, 4.4.2**

(1966b) The tidal history and origin of the Earth–Moon system. *Sov. Astron. AJ,* **10,** 659–5. Trans. from *Astron. Zh.,* **43,** 829–36. **4.4.1, 4.4.2**

(1971/1972a) The origin of the Moon. III. Some aspects of the dynamics of the circumterrestrial swarm. *Sov. Astr. AJ,* **15** (1972), 646–54. Trans. from *Astron. Zh.,* **48** (1971), 819–29. **4.1.4, 4.4.6**

(1971/1972b) Possible differences in the chemical composition of the Earth and Moon, for a Moon formed in the circumterrestrial swarm. *Sov. Astr. AJ,* **15** (1972), 1061–3. Trans. from *Astron. Zh.,* **48** (1971), 1336–8. **4.1.4, 4.4.6**

(1972) Formation of the Moon from a cluster of particles encircling the Earth. *Izvestiya Physics of the Solid Earth,* no. 7, 483–88. Transl. from *Izv. A. N. Fiz. Zemli,* no. 7, 99–108. **4.1.4, 4.4.4, 4.4.6**

(1973) On the model of the accumulation of the Moon compatible with the data on the composition and the age of lunar rocks. *Moon,* **6,** 190–201. **4.4.4**

(1974/1977) The origin of the Moon. In *Soviet–American Conference on Cos- mochemistry of the Moon and Planets, Moscow, June 1974,* ed. J. H. Pomeroy & N. J. Hubbard, pp. 815–22. Washington, DC: National Aeronautics and Space Ad- ministration, report NASA Sp-370 (pub. 1977). **4.4.4**

(1975) *Origin of the Moon.* Washington, DC: National Aeronautics and Space Ad- ministration, report NASA TT 16,623. Trans. from *Proiskhozhdeniye Luny* (Mos- cow: Nauka). **4.1.3, 4.4.1, 4.4.4**

(1982) Origin of planetary satellites. *Izvestiya Academy of Science of the USSR, Physics of the Solid Earth,* **18,** 425–33. Trans. from *Izv. A. N. Fiz. Zemli,* no. 6, 40–51. **4.4.4**

Ruskol, E. L. *See also* Safronov, V. S.

Russ, G. Price, II. *See* Albee, A. L.

Russell, C. T. (1984) On the Apollo subsatellite evidence for a lunar core. In *Origin of the Moon conf. abst.*, p. 7. **4.3.4**

Russell, C. T., Coleman, P. J., Jr., & Goldstein, B. E. (1981) Measurements of the lunar induced magnetic moment in the geomagnetic tail: Evidence for a lunar core? *Proc. 12th Lunar & Plan. Sci. Conf.*, 831–6. **4.3.4**

Russell, H. N. (1924) The applications of modern physics to astronomy. *J. Roy. Astron. Soc. Canada*, **18**, 137–64, 201–23, 233–63.   **1.2.6**

(1929) Astronomical books. (Review of *The two solar families* by T. C. Chamberlin) *Saturday Review of Literature*, **6**, no. 1 (July 27), 7.   **1.3.4, 1.3.5**

(1931) Worlds from a catastrophe. *Sci. Am.*, **145**, no. 2 (August), 92–3.   **1.3.4**

(1935) *The Solar System and its origin.* New York: Macmillan.   **1.1.2, 1.3.4, 1.3.5**

(1943) Anthropocentrism's demise. New discoveries lead to the probability that there are thousands of inhabited planets in our galaxy. *Sci. Am.*, **169** (July), 260–1.   **1.1.3**

(D1925) Letter to F. R. Moulton, 29 December 1925. Box 35, R-PU.   **1.3.4**

(D1926a) Letter to F. R. Moulton, 15 March 1926. Box 35, R-PU.   **1.3.4**

(D1926b) Letter to F. R. Moulton, 3 May 1926. Box 35, R-PU.   **1.3.4**

(D1927) Letter to L. H. Adams, 25 February 1927. Box 34, R-PU.   **1.3.4**

(D1929a) Letter to F. R. Moulton, 4 June 1929. Box 35, R-PU.   **1.3.4**

(D1929b) Letter to F. R. Moulton, 8 June 1929. Box 35, R-PU.   **1.3.4**

(D1932a) Letter to W. J. Luyten, 26 November 1932. Box 35, R-PU.   **1.3.2**

(D1932b) Letter to W. J. Luyten, 10 December 1932. Box 35, R-PU.   **1.3.4**

(D1933?) Letter to S. A. Mitchell, dated 24 February 1933 (probably written in 1934). Box 23, R-PU.   **1.3.4**

(D1934a) Letter to H. Jeffreys, 19 March 1934. Box 34, R-PU.   **1.3.4**

(D1934b) Letter to H. Jeffreys, 18 April 1934. Box 34, R-PU.   **1.3.4**

(D1939) Letter to L. Spitzer, 18 May 1939. Box 35, R-PU.   **1.3.4**

Russell, H. N., Dugan, R. S., & Stewart, J. Q. (1926) *Astronomy: A revision of Young's Manual of Astronomy.* I. The solar system. Boston: Ginn.   **1.2.6, 1.3.4**

(1930) *Astronomy: A revision of Young's Manual of Astronomy.* 2nd edn. Boston: Ginn.   **1.2.6**

(1935) *Astronomy: A revision of Young's Manual of Astronomy.* 3rd edn. Boston: Ginn.   **1.2.6**

Russell, Jane. *See* Black, D. C.

Ruzmaikina, T. V. *See* Safronov, V. S.

Sabu, D. D., & Manuel, O. K. (1976) Xenon record of the early Solar System. *Nature*, **262**, 28–32.   **2.5.2**

Sabu, D. D. *See also* Manuel, O. K.

Safronov, V. S. (1958/1964) On the growth of the Earth-type planets. *Prob. Cosmog.*, **7**, 63–70. Trans. from *Vopr. Kosmog.*, **6**, 63–77.   **4.4.1, 4.4.5**

(1959) On the primeval temperature of the Earth. *Bulletin (Izvestiya), Academy of Science of the USSR, Geophysics Series*, no. 1 (January), 85–9. Trans. from *Izvestiya Akademi Nauk SSSR, Seriya Geofizika*, no. 1, 139–43.   **3.1.1**

(1960/1964) The formation of protoplanetary dust clouds. *Prob. Cosmog.*, **7** (1964), 120–45. Trans. from *Vopr. Kosmog.*, **7** (1960).   **4.4.5**

(1962a) The temperature of the dust component of the protoplanetary cloud. *Sov. Astron. AJ*, **6**, 217–25. Trans. from *Astron. Zh.*, **39**, 278–89.   **3.1.1**

(1965/1966) Sizes of the largest bodies falling onto the planets during their formation. *Sov. Astron. AJ*, **9** (1966), 987–91. Trans. from *Astron. Zh.*, **42** (1965), 1270–6.   **3.1.1, 4.4.5**

(1966/1967) The protoplanetary cloud and its evolution. *Sov. Astron. AJ*, **10** (1967), 650–8. Trans. from *Astron. Zh.*, **43** (1966), 817–8.   **3.1.1**

(1969/1972) *Evolution of the protoplanetary cloud and formation of the Earth and planets.* Jerusalem: Israel Program for Scientific Translations, document NASA TT-F-677 (1972). Trans. from *Evolyutsiya doplanetnogo oblaka i obrazovanie zemli i planet.* Moscow: Izdatel'stvo "Nauka" (1969).   **1.2.3, 3.1.1, 3.1.2, 4.4.1, 4.4.5**

(1970/1972). Ejection of bodies from the Solar System in the course of the accumulation of the giant planets and the formation of the cometary cloud. In *The*

Motion, evolution of orbits, and origin of comets (IAU symposium no. 45), ed. G. A. Chebotarev et al., pp. 329–34. Dordrecht: Reidel.   **3.1.1**

(1974) Accumulation of the planets. In *On the origin of the Solar System,* ed. H. Reeves, pp. 89–113. Paris: CNRS.   **3.2.1**

(1978) The heating of the Earth during its formation. *Icarus,* **33,** 3–12.   **3.1.2**

(1979) On the origin of asteroids. In *Asteroids,* ed. T. Gehrels & M. S. Matthews, pp. 975–91. Tucson: University of Arizona Press.   **3.4.7**

(1981) Initial state of the Earth and its early evolution. In *Evolution of the Earth,* ed. R. J. O'Connell & W. S. Fyfe, pp. 249–55. Boulder: Geological Society of America.   **3.1.2**

(1983/1995) The development of Soviet planetary cosmogony. In *The origin of the Solar System: Soviet research, 1925–1991,* ed. A. E. Levin & S. G. Brush. New York: American Institute of Physics (1995). Trans. from *O. Iu. Shmidt i sovetskayia geofizika 80-kh godov,* pp. 41–57. Moscow: Izdatelstvo "Nauka" (1983).   **3.1.1**

Safronov, V. S., & Kozlovskaya, S. V. (1977) Heating of the Earth by the impact of accreted bodies. *Izvestiya Physics of the Solid Earth,* **13,** 677–84. Trans. from *Izv. A. N. Fiz. Zemli,* no. 10.   **3.1.2**

Safronov, V. S., Pechernikova, G. V., Ruskol, E. L., & Vityazev, A. V. (1986) Protosatellite swarms. In *Satellites,* ed. J. A. Burns & M. S. Matthews, pp. 89–116. Tucson: University of Arizona Press.   **4.4.6**

Safronov, V. S., & Ruskol, E. L. (1977) The accumulation of satellites. *Sov. Astron. AJ,* **21,** 211–17. Trans. from *Astron. Zh.,* **54,** 378–87.   **4.4.4**

(1982) On the origin and initial temperature of Jupiter and Saturn. *Icarus,* **49,** 284–96.   **3.2.2**

Safronov, V. S., & Ruzmaikina, T. V. (1985) Formation of the solar nebula and the planets. In *Protostars and Planets.* II, ed. D. C. Black & M. S. Matthews, pp. 959–80. Tucson: University of Arizona Press.   **3.4.8**

Safronov, V. S., & Vityazev, A. V. (1983/1985) Origin of the Solar System. *Soviet Scientific Reviews,* **4** (1985), 1–98. Trans. from *Itogi Nauki i Tekhniki,* Seriya Astronomiya, **24** (1983), 5–93.   **3.1.1**

Safronov, V. S. *See also* Lissauer, J. J.

Sagan, Carl (1975) The Solar System. *Sci. Am.,* **233,** no. 3 (September), 23–31.
                                                                3.4.5

(1981) Obituary. Harold Clayton Urey: 1893–1981. *Icarus,* **48,** 348–52.   **3.3.1**

(1988) Some mysteries of planetary science. *Planetary Report,* **8,** no. 3 (May/June), 12–16.   **3.4.5**

Sagan, Carl. *See also* Shklovskii, I. S.; Smith, B. A.

Sagaret, Jules (1931) *Le système du monde de Pythagore à Eddington.* Paris: Payot.
                                                                1.2.6

Salisbury, R. C. *See* Chamberlin, T. C.

Salpeter, E. E. (1974) Dying stars and reborn dust. *Rev. Mod. Phys.,* **46,** 433–6.
                                                                2.5.1

Sandage, Allan, & Tammann, G. A. (1986) The dynamical parameters of the expanding universe as they constrain homogeneous world models with and without Λ. In *Inner space/outer space,* ed. E. W. Kolb et al., pp. 41–64. Chicago: University of Chicago Press.   **2.1.2**

Sandoval, P. *See* Murthy, V. R.

Sanz, H. G. *See* Albee, A. L.

Sargood, D. G. *See* Skelton, R. T.

Savchenko, K. N. (1953) Some questions of non-classical celestial mechanics and cosmogony. (In Russian). *Trudy Odesskogo Gos.,* **5,** 59–147. (Cited by Ruskol, 1960; see also *Astronomisches Jahresbericht,* **55** [1955], 120.)   **4.3.2**

Schaeberle, J. M. (1906) The probable volcanic origin of nebulous matter. *Nature* **73,** 296–7.   **1.2.6**

Schatzman, Evry. (1962) A theory of the role of magnetic activity during star formation. *Annales d'astrophysique,* **25,** 18–29.   **3.4.2**

Schatzman, Evry. *See* Bernas, R.

Scheiner, J. (1899) Ueber das Spectrum des Andromedanebula. *Astron. Nachr.,* **148,**
325–8.   **1.2.5**

Scheinin, N. B. *See* Lugmair, G. W.

Schempp, William V. *See* Baines, K. H.

Schlüter, A. *See* Lüst, R.

Schmidt, O. J. (1944) A meteoric theory of the origin of the Earth and planets.
Comptes Rendus (Doklady) Academie des Sciences de l'URSS, **45,** 229–33.

                                                                     1.1.3, 3.1.1

  (1949) *Four lectures on the theory of the origin of the Earth* (in Russian). Moscow:
  Izdatelstvo Akad. Nauk SSSR.   **3.1.1**

  (1958) *A theory of Earth's origin: Four lectures.* Trans. from 3rd Russian edn. by G. H.
  Hanna. Moscow: Foreign Languages Publishing House.   **1.1.3, 3.1.1**

Schmitt, Harrison (1975) Evolution of the Moon: The 1974 model. *Space Science
Reviews,* **18,** 259–79.   **4.4.4**

Schmitt, Roman A. *See* Gooding, J. L.; Murthy, V. R.

Schneeberger, T. J. *See* Worden, S. P.

Schofield, N., & Woolfson, M. M. (1982a) The early evolution of Jupiter in the
absence of solar tidal forces. *M. N. Roy. Astron. Soc.,* **198,** 947–59.   **3.2.1**

  (1982b) The early evolution of Jupiter in the tidal field of the sun. *M. N. Roy.
  Astron. Soc.,* **198,** 961–73.   **3.2.1**

Schramm, D. N. (1971) Nucleosynthesis of ^{26}Al in the early Solar System and in
cosmic rays. *Astrophys. Space Sci.,* **13,** 249–66.   **2.5.1**

  (1977) Isotopic evidence for a supernova origin of the Solar System. *Bulletin of the
  American Astronomical Society,* **9,** 544.   **2.5.1**

  (1978) Supernovae and the formation of the Solar System. In *Protostars and planets,*
  ed. T. Gehrels, pp. 384–98. Tuscon: University of Arizona Press.   **2.5.1**

Schramm, D. N., Tera, F., & Wasserburg, G. J. (1970) The isotopic abundance of ^{26}Mg
and limits on ^{26}Al in the early Solar System. *Earth Plan. Sci. Lett.,* **10,** 44–59.

                                                                                  2.5.1

Schramm, D. N. *See* Falk, S. W.

Schuchert, C. *See* Leith, C. K.

Schultz, Susan F. (1976). *Thomas C. Chamberlin, An intellectual biography of a geologist
and educator.* Ph.D. dissertation, University of Wisconsin-Madison.   **1.1.1**

Schwarcz, H. P. *See* Clarke, W. B.

Schwarzschild, K. (1898) Die Poincaresche Theorie des Gleichgewichtes einer homo-
genen rotierenden Flüssigkeitsmasse. *Neue Annalen Kgl. Sternwarte in München,* **3,**
231–99.   **4.2.4**

Schwarzschild, Martin (1958) *Structure and evolution of the stars.* Princeton: Princeton
University Press.   **2.3.2**

  (D1984) Characters behind the theory of stellar evolution. Colloquium at Goddard
  Space Flight Center, 8 June 1984.   **3.4.8**

Scott, E. H. *See* Durisen, R. H.

Scott, N. *See* Gautier, D.

Scrutton, Colin T., & Hipkin, Roger G. (1973) Long-term changes in the rotation
rate of the Earth. *Earth-Science Reviews,* **9,** 259–74.   **4.3.3**

Secco, L. *See* Dallaporta, N.

Sedgwick, W. F. (1898) On the oscillations of a heterogeneous compressible liquid
sphere and the genesis of the Moon; and on the figure of the Moon. *Messenger of
Mathematics,* **27,** 159–73.   **1.3.3**

See, T. J. J. (1896) *Researches on the evolution of the stellar systems.* Lynn, MA: Nichols.

                                                                     1.2.2, 1.2.4

(1897) Recent discoveries respecting the origin of the universe. *Atlantic Monthly*, **80**, 484–92.   **1.2.4**

(1906a) Miss Clerke's Modern Cosmogonies. *Pop. Astron.*, **14**, 313–314.   **1.2.4**

(1906b) Significance of the spiral nebulae. *Pop. Astron.*, **14**, 614–616.
                                                                                1.2.4, 1.2.6

(1909a) The planar arrangement of the planetary system. *Nature*, **81**, 275.
                                                                                         1.2.4

(1909b) On the cause of the remarkable circularity of the orbits of the planets and satellites and on the origin of the planetary system. *Astron. Nachr.*, **180**, 185–94; *Pop. Astron.*, **17**, 263–72.   **1.2.4**

(1909c) The laws of cosmical evolution and the extension of the Solar System beyond Neptune. *Pub. Astron. Soc. Pac.*, **21**, 60–71.   **1.2.4**

(1909d) Origin of the lunar terrestrial system by capture with further considerations on the theory of satellites and on the physical cause which has determined the directions of the rotations of the planets about their axes. *Astron. Nachr.*, **181**, 365–86; *Pop. Astron.*, **17**, 634–40, **18**: 24–31, 106–110, 155–61.
                                                                             1.2.4, 4.1.1, 4.2.4

(1909e) The past history of the Earth as inferred from the mode of formation of the Solar System. *Proc. Am. Phil. Soc.*, **48**, 119–25.   **1.2.4**

(1909f) Dynamical theory of the capture of satellites and of the division of nebulae under the secular action of a resisting medium. *Astron. Nachr.*, **181**, 333–50; *Pop. Astron.*, **17**, 481–94, 534–44.   **1.2.6, 4.2.4**

(1909g) The capture theory of satellites. *Pub. Astron. Soc. Pac.*, **21**, 167–73.
                                                                                         1.2.6

(1909h) Further considerations on the theory of the rotation of the principal planets and on the growth of the minor globes which have finally become satellites. *Astron. Nachr.*, **182**, 213–18.   **1.2.6**

(1909i) The terrestrial origin of the Moon – A protest. *Sci. Am.*, **101**, no. 6 (August 7), 91.   **4.2.4**

(1910a) *Researches on the evolution of the stellar systems. II. The capture theory of cosmical evolution.* Lynn, MA: Nichols.   **1.2.4, 4.2.4**

(1910b) Some recent discoveries in cosmical evolution. *Pop. Astron.*, **18**, 532–7.
                                                                                1.2.6, 4.2.4

(1910c) Results of some recent researches in cosmical evolution. *Proc. Am. Phil. Soc.*, **49**, 207–21.   **1.2.6**

(1910d) The origin of the so-called craters on the Moon by the impact of satellites, and the relation of these satellite indentations to the obliquities of the planets. *Pub. Astr. Soc. Pac.*, **22**, 13–20.   **4.2.4**

(1911a) Professor E. W. Brown's verification of the capture of satellites. *Pop. Astron.*, **19**, 422–5; *Pub. Astron. Soc. Pac.*, **23**, 164–7.   **1.2.6**

(1911b) Remarks on the problems of cosmogony. *Astron. Nachr.*, **189**, 317–20.
                                                                                         1.2.6

(1912) The new science of cosmogony. *Scientia*, **11**, 18–35.   **1.2.6**

(1914) The law of nature in celestial evolution. *Scientia*, **15**, 169–86.   **1.2.6**

(1915) The origin of the Moon. *J. Br. Astron. Ass.*, **25**, 282–4.   **4.2.4**

Seeley, D. *See* Berendzen, R.

Seemann, Friedrich (1900) Zur Kant-Laplaceschen Theorie. *Prometheus*, **11**, 753–6, 772–6.   **1.2.2**

Shabad, T. (1979) Soviet claims joint credit for finding Jupiter's ring. *New York Times* (April 18), B20.   **3.4.6**

Shaler, N. S. (1898) *Outlines of the Earth's history.* New York: Appleton.   **1.2.2, 1.2.4**

Shane, K. C. *See* Clarke, W. B.

Shapley, Harlow (1926) *Starlight.* New York: Doran.   **1.2.6**

Sharma, L. K. (1977) Uranus rings: Indian scientist's claim. *Times of India* (April 18).
    3.4.5

Shields, Janet A. *See* Goettel, Kenneth A.

Shimamura, T. *See* Lewis, R. S.

Shklovskii, I. (1988) Nevydumanniye rasskazy (Uninvented stories). *Energiya*, no. 7,
    52–3.   **1.1.4**

Shklovskii, I., & Sagan, Carl (1966) *Intelligent life in the universe.* San Francisco:
    Holden-Day.   **3.3.1**

Shoemaker, Eugene M. (1977) Why study impact craters? In *Impact and explosion
    cratering: Planetary and terrestrial implications,* ed. D. J. Roddy et al., pp. 1–10. New
    York: Pergamon Press.   **4.3.4**

Shoemaker, Eugene M. *See also* Smith, B. A.

Showalter, M. R. *See* Burns, J. A.

Shu, F. *See* Lin, C. C.

Shukhman, I. G. (1984) Collisional dynamics of particles in Saturn's rings. *Soviet As-
    tron. AJ,* **28,** 574–85. Trans. from *Astron. Zh.,* **61,** 985–1004.   **3.4.7**

Sill, G. T. (1972) Sulfuric acid in the Venus clouds. *Communications of the Lunar and
    Planetary Laboratory (University of Arizona),* no. 171, 191–8.   **3.3.5**

Silver, Leon T. (1970) Uranium-thorium-lead isotopes in some Tranquillity Base sam-
    ples and their implications for lunar history. *Proceedings of the Apollo 11 Lunar
    Science Conference,* ed. A. A. Levinson, vol. 2, 1533–74. New York: Pergamon
    Press.   **4.4.4**

Simanton, James R., Rightmire, Robert A., Long, Alton L., & Kohman, Truman P.
    (1954) Long-lived radioactive aluminum 26. *Phys. Rev* [2], **96,** 1711–12.
    2.5.1

Singer, S. Fred (1967) The origin and dynamical evolution of the Moon. In *Physics of
    the Moon,* ed. S. F. Singer, pp. 207–37. Tarzana, CA: American Astronautical
    Society.   **4.3.3**

    (1968) The origin of the Moon and geophysical consequences. *Geophys. J. Roy.
    Astron. Soc.,* **15,** 205–26.   **4.4.2**

    (1970) Origin of the Moon by capture and its consequences. *Eos,* **51,** 637–41.
    4.3.3, 4.4.4

    (1971) Discussion of paper by A. E. Ringwood, "Petrogenesis of Apollo basalts and
    implications for lunar origin." *J. Geophys. Res.,* **76,** 8071–4.   **4.4.4**

    (1977) The early history of the Earth–Moon system. *Earth-Science Reviews,* **13,** 171–
    89.   **3.1.2, 4.4.4**

    (1986) Origin of the Moon by capture. In *Origin of the Moon,* ed. W. K. Hartmann
    et al., pp. 471–85. Houston: Lunar and Planetary Institute.   **4.4.2**

Singer, S. Fred. *See also* De Groot, M.

Skelton, R. T., Kavanagh, R. W., & Sargood, D. G. (1983) $^{26}Mg(p,n)^{26}Al$ cross sec-
    tion measurements. *Astrophys. J.,* **271,** 404–7.   **2.5.3**

Slattery, Wayne L. (1977) The structure of the planets Jupiter and Saturn. *Icarus,* **32,**
    58–72.   **3.2.2**

Slattery, Wayne L. *See also* Benz, W.

Slichter, Louis B. (1941) Cooling of the Earth. *Bull. Geol. Soc. Am.,* **52,** 561–600.
    3.3.1

    (1963) Secular effects of tidal friction upon the Earth's rotation. *J. Geophys. Res.,* **68,**
    4281–8.   **4.4.2**

Slipher, V. M. (D1914) Letter to M. Wolf, 17 July 1914. Roll 5, frames 1450–1, Early
    Correspondence of the Lowell Observatory, Microfilm Edition (Flagstaff, AZ,
    1973).   **1.2.6**

Smart, W. M. (1928) *The sun, the stars, and the universe.* London: Longmans, Green.
    1.3.3

Smith, Bradford A. (1983) Voyager to the giant planets. In *Astronomy from space,* ed. J. Cornell & P. Gorenstein, pp. 81–98. Cambridge, MA: MIT Press.   **3.4.6**

Smith, Bradford A., Soderblum, Laurence A., Johnson, Torrence V., Ingersoll, Andrew P., Collins, Stewart A., Shoemaker, Eugene M., Hunt, G. E., Masursky, Harold, Carr, Michael H., Davies, Merton E., Cook, Allan F., II, Boyce, Joseph, Danielson, G. Edward, Owen, Tobias, Sagan, Carl, Beebe, Reta F., Veverka, Joseph, Strom, Robert G., McCauley, John F., Morrison, David, Briggs, Geofrey A., Suomi, Verner E. (1979) The Jupiter system through the eyes of Voyager 1. *Science,* **204,** 951–72.   **3.4.6**

Smith, Bradford A., & Terrile, Richard J. (1984) A circumstellar disk around β Pictoris. *Science* **226,** 1421–4.   **2.1.2**

Smith, J. V. (1974) Origin of Moon by disintegrative capture with chemical differentiation followed by sequential accretion. *Lunar Sci.,* **5,** 718–20.   **4.4.6**

(1977a) Possible controls on the bulk composition of the Earth: Implications for the origin of the Earth and Moon. *Proc. 8th Lunar Sci. Conf.,* 333–69.
                    4.4.3, 4.4.4

(1977b) Chemical evidence on origin of Earth and Moon. *Lunar Sci.,* **8,** 881–3.
                    4.4.4

(1979) Mineralogy of the planets: A voyage in space and time. *Mineralogical Magazine,* **43,** 1–89.   **3.3.5**

(1982) Heterogeneous growth of meteorites and planets, especially the Earth and Moon. *J. Geol.,* **90,** 1–48.   **3.3.4, 4.4.4**

Smith, J. V. *See also* Leitch, C. A.

Smith, Kirk R., Fesharaki, Fereidun, & Holdren, John P., Editors (1986) *Earth and the human future: Essays in honor of Harrison Brown.* Boulder, CO: Westview Press.
                    3.3.1

Smith, Robert W. (1982) *The expanding universe: Astronomy's "great debate," 1900–1931.* Cambridge University Press.   **1.2.6**

(1990) Edwin P. Hubble and the transformation of cosmology. *Physics Today,* **43,** no. 4 (April), 52–8.   **1.2.6**

Smith, William Hayden. *See* Baines, K. H.

Smoluchowski, Roman (1983a) *The solar system: The sun, planets, and life.* New York: Freeman.   **4.4.4**

(1983b) The interiors of the giant planets – 1983. *Moon and Planets,* **28,** 137–54.
                    3.2.2

Smoluchowski, Roman. *See* Torbett, M.

Smyth, J. R. *See* McGetchin, T. R.

Snelder, H. A. M. (1970) Arrhenius, Svante August. *DSB,* **1,** 296–302.   **1.2.4**

Snyder, Carl (1907) *The world machine, the first phase: The cosmic mechanism.* London: Longmans, Green.   **1.2.4**

(D1906) Letter to T. J. J. See, 19 March 1906. Container 15. See Papers, Library of Congress.   **1.2.6**

Soderblum, Laurence A. *See* Smith, B. A.

Sollas, W. J. (1900) Presidential address to Geology Section. *Nature,* **62,** 481–9.
                    1.2.2

Solomon, Sean C. (1976) Some aspects of core formation in Mercury. *Icarus,* **28,** 509–21.   **3.3.5**

(1986) On the early thermal state of the Moon. In *Origin of the Moon,* ed. W. K. Hartmann et al., pp. 435–52. Houston: Lunar and Planetary Institute.   **4.4.5**

Sonett, C. P. (1978) Evidence for a primordial magnetic field during the meteorite parent body era. *Geophys. Res. Lett.,* **5,** 151–4.   **2.3.5**

Sonett, C. P., Colburn, D. S., & Schwartz, K. (1975) Formation of the lunar crust: An electrical source of heating. *Icarus,* **24,** 231–55.   **2.3.5**

Sonett, C. P. *See also* Melosh, H. J.; Wiskerchen, M. J.

Sorokin, N. A. (1965) The relative orientation of the Earth's equator and the Moon's orbit in the remote past. *Sov. Astron. AJ*, **9**, 826–9. Trans. from *Astron. Zh.*, **42**, 1070–4.   4.4.2

Space Science Board (D1959) Interim report of the Ad Hoc Committee on Chemistry of Space and Exploration of the Moon and Planets, 8 January 1959. National Academy of Sciences of the U.S.A.   4.3.2

(D1960) Minutes of Fourth Meeting, 10–11 March 1960. National Academy of Sciences of the U.S.A.   4.3.2

Spettel, B. *See* Dreibus, G.; Wänke, H.

Spitzer, Lyman, Jr. (1939) The dissipation of planetary filaments. *Astrophys. J.*, **90**, 675–88.   1.3.4

(1948) The formation of cosmic clouds. In *Centennial symposia, December 1946*, pp. 87–108. Cambridge, MA: Harvard College Observatory, Monograph no. 7.   1.1.3

Spruch, Grace Marmor (1977) New evidence supports supernova origin of Solar System. *Physics Today*, **30**, no. 5 (May), 17–19.   2.5.2

Srinivasan, B., Alexander, F. C., Jr., Manuel, O. K., & Troutner, D. E. (1969) Xenon and krypton from the spontaneous fission of californium-252. *Phys. Rev.* [2], **179**, 1166–9.   2.5.1

Srinivasan, B., & Anders, Edward (1978) Noble gases in the Murchison meteorite: Possible relics of s-process nucleosynthesis. *Science*, **201**, 51–5.   2.5.3

Srinivasan, B. *See* Lewis, R. S.

Stacey, F. D. *See* Cooper, J. A.

Steele, Ian M. *See* Grossman, L.

Stein, J. W. *See* Black, D. C.

Stephenson, A. *See* Runcorn, S. K.

Stevens, R. D. *See* Wanless, R. K.

Stevenson, David J. (1978) The outer planets and their satellites. In *The origin of the Solar System*, ed. S. F. Dermott, pp. 395–431. New York: Wiley.   3.4.5

(1982a) Interiors of the giant planets. *Annual Review of Earth and Planetary Science*, **10**, 257–95.   3.2.2

(1982b) Formation of the giant planets. *Plan. Space Sci.*, **30**, 755–64.   3.2.2

(1987) Origin of the Moon – The collision hypothesis. *Annual Review of Earth and Planetary Science*, **15**, 271–315.   4.1.5, 4.1.6, 4.4.5

Stevenson, David J., Harris, A. W., & Lunine, J. I. (1986) Origins of satellites. In *Satellites*, ed. J. A. Burns & M. S. Matthews, pp. 39–88. Tuscon: University of Arizona Press.   4.4.4

Stevenson, David J. *See* Lunine, J. I.; Thompson, A. C.

Stewart, G. R., Lin, D. N. C., & Bodenheimer, P. (1984) Collision-induced transport processes in planetary rings. In *Planetary rings*, ed. R. Greenberg & A. Brahic, pp. 47–512. Tuscon: University of Arizona Press.   3.4.7

Stewart, John Quincy. *See* Russell, H. N.

Stocking, G. W., Jr. (1968) *Race, culture, and evolution: Essays in the history of anthropology*. New York: Free Press.   1.2.2

Stolper, Edward. *See* MacPherson, G. J.

Stone, E. C., & Miner, E. D. (1981) Voyager 1 encounter with the Saturnian system. *Science*, **212**, 159–62.   3.4.4

(1982) Voyager 2 encounter with the Saturnian system. *Science*, **215**, 499–504.
                                                                                3.4.4

Stone, E. C. *See also* Breneman, H. H.

Stoney, G. J. (1897) Of atmospheres upon planets and satellites. *Transactions of the Scientific Society of Dublin*, **6**, 305–38.   1.2.3

(1898) Of atmospheres upon planets and satellites. *Astrophys. J.*, **7**, 25–55.
                                                                                1.2.3

(1900) Escape of gases from planetary atmospheres. *Nature*, **61**, 515; **62**, 78.   1.2.3

Strand, K. Aa. (1943) 61 Cygni as a triple system. *Pub. Astron. Soc. Pac.*, **55**, 29–32.
     1.1.3, 2.1.2

Strangway, D. W. *See* Arnold, J.; Lanoix, M.; Sugiura, N.

Stratton, F. J. M. (1906) Planetary inversion. *M. N. Roy. Astron. Soc.*, **66**, 374–402.
     1.2.4, 1.3.5

   (1909) Tidal problems. *Nature*, **81**, 102–3.   **1.2.6**

   (1910) Cosmogony. *M. N. Roy. Astron. Soc.*, **70**, 366–8.   **1.2.6**

Strauss, David (1993) "Fireflies flashing in unison": Percival Lowell, Edward Morse and the birth of planetology. *J. Hist. Astron.*, **24**, 157–69.   **1.2.4**

Strom, R. G. *See* Murray, B. C.; Smith, B. A.

Struve, Otto (1945) The cosmogonical significance of stellar rotation. *Pop. Astron.*, **53**, 201–18, 259–76.   **1.1.3**

   (1950) *Stellar evolution*. Princeton: Princeton University Press.   **1.1.3, 3.4.2**

   (1958) Henri Poincaré and his cosmogonical studies. *Sky & Tel.*, **7**, 226–8.
     1.2.6

Stubbs, Peter (1970) The ashes of the sun. *New Scientist*, **47**, 465–7.   **2.3.5**

Sündermann, J. (1982) The resonance behaviour of the world ocean. In *Tidal friction and the Earth's rotation*. II, ed. P. Brosche & J. Sündermann, pp. 165–74. New York: Springer-Verlag.   **4.4.2**

Suess, Hans E. (1962/1963) Properties of chondrules. I. In *Origin of the Solar System* (proceedings of 1962 conference), ed. R. Jastrow & A. G. W. Cameron, pp. 143–6. New York: Academic Press (pub. 1963).   **3.3.3**

   (1964) On element synthesis and the interpretation of the abundance of heavy elements. *Proc. N. A. S.*, **52**, 387–97.   **3.3.3**

   (1965) Chemical evidence bearing on the origin of the Solar System. *Ann. Rev. Astron. Astrophys.*, **3**, 217–34.   **3.3.3, 3.3.5**

   (1987) *Chemistry of the Solar System: An elementary introduction to cosmochemistry*. New York: Wiley.   **3.3.3**

Sugiura, N., Lanoix, M., & Strangway, D. W. (1979) Magnetic fields of the solar nebula as recorded in chondrules from the Allende meteorite. *Phys. Earth Plan. Int.*, **20**, 342–9.   **2.3.5**

Sullivan, Edward C. *See* O'Keefe, J. A.

Sullivan, Walter (1961) Urey holds Moon predated Earth. *New York Times*, 27 April, p. 23.   **3.3.1, 4.3.2**

   (1977a) Physicists say bang began Solar System. *New York Times*, 10 February, p. 17.
     2.5.2

   (1977b) Death of giant star could create others. *New York Times*, 14 March, p. 13.
     2.5.2

   (1983) New confidence in Moon rocks safety. *New York Times*, 13 December, p. C15.   **4.4.4**

Suomi, Verner E. *See* Smith, B. A.

Summer, Audrey L. *See* Grossman, A. S.

Sundberg, L. L. *See* Baedecker, P. A.

Swenson, Loyd S., *See* Brooks, C. G.

Swindle, T. D., Caffee, M. W., Hohenberg, C. M., & Taylor, S. R. (1986) I-Pu-Xe dating and the relative ages of the Earth and Moon. In *Origin of the Moon*, ed. W. K. Hartmann et al., pp. 331–57. Houston: Lunar and Planetary Institute.
     4.4.4

Sztrokay, K. I., Tolnay, V., & Foldvari-Vogl, M. (1961) Mineralogical and chemical properties of the carbonaceous meteorite from Kaba, Hungary. *Acta Geologica Academia Scientia Hungarica*, **7**, 57–103. (*Chemical Abstracts*, **57**, 4386h).   **3.3.2**

Tai, Wen-Sai, & Chen, Dao-Han (1977) Critical review of theories on the origin of the Solar System. *Chinese Astronomy*, **1**, 165–82. Trans. from *Acta Astronomica Sinica*, **17**, 93–105.   **3.4.3, 3.4.7**

Takahashi, Hiroshi. *See* Gros, J.; Hertogen, J.

Tammann, G. A. *See* Sandage, A.

Tanaka, Tsuyoshi. *See* MacPherson, G. J.

Tarling, D. H. (1978) Plate tectonics: Present and past. In *Evolution of the Earth's crust,* ed. D. H. Tarling, pp. 361–408. New York: Academic Press.   **4.3.3**

Tatarewicz, J. N. (1986) Federal funding and planetary astronomy, 1950–75: A case study. *Social Studies of Science,* **16,** 79–103.   **3.3.1**

    (1990) *Space technology and planetary astronomy.* Bloomington: Indiana University Press.   **3.3.1**

Tatsumoto, Mitsunobu (1979) Age of the Moon: An isotopic study of U-Th-Pb systematics of Apollo 11 lunar samples. II. *Proceedings of the Apollo 11 Lunar Science Conference,* ed. A. A. Levinson, Vol. 2, pp. 1595–1612. New York: Pergamon Press.   **4.4.4**

Tatsumoto, Mitsunobu, & Rosholt, John N. (1970) Age of the Moon: An isotopic study of uranium–thorium–lead systematics of lunar samples. *Science,* **167,** 461–3.   **4.4.4**

Tauber, M. E. *See* Pollack J. B.

Tayler, R. J. (1968) Stellar evolution. *Reports on Progress in Physics,* **31,** 167–223.   **2.3.3**

Taylor, Frank Bursley (1898) *An endogenous planetary system.* Fort Wayne, IN: Archer.   **1.2.4, 4.2.4**

Taylor, G. I. (1919) Tidal friction in the Irish Sea. *Phil. Trans.,* **A220,** 1–33.   **4.2.4**

Taylor, G. J. *See* Hartmann, W. K.

Taylor, Hugh P., Jr. *See* Epstein, S.

Taylor, Stuart Ross (1965) Enrichment of iron during accretion in the solar nebula. *Nature,* **208,** 886–7.   **3.3.3**

    (1969) *Lunar Science Institute Newsletter,* **1,** no. 4 (August).   **4.4.3**

    (1975) *Lunar science: A post-Apollo view.* Elmsford, NY: Pergamon Press.   **4.4.4**

    (1984) Tests of the lunar fission hypothesis. In *Origin of the Moon Conf. Abst.,* p. 25.   **4.4.7**

    (1986a) The origin of the Moon: Geochemical considerations. In *Origin of the Moon,* ed. W. K. Hartmann et al., pp. 125–43. Houston: Lunar and Planetary Institute.   **4.4.5**

    (1986b) Cutting the Gordian knot: Lunar compositions and Mars-sized impactors. *Lunar & Plan. Sci.,* **17,** 881–2.   **4.4.5**

    (1987a) The origin of the Moon. *American Scientist,* **75,** 468–77.   **4.4.5**

    (1987b) Loss of volatile elements during impact events in relation to lunar composition and origin. *Lunar & Plan. Sci.,* **18,** 1002–3.   **4.4.5**

    (1987c) The unique lunar composition and its bearing on the origin of the Moon. *Geoch. Cosmoch. Acta,* **51,** 1297–1309.   **4.1.6, 4.4.5, 5.4.4**

    (D1980) Chance and design in science, invention, technology. In *Pieces of another world.* Canberra: Research School of Earth Sciences, Australian National University.   **3.3.1, 4.4.3**

Taylor, Stuart Ross, & Jakeŝ, P. (1974) The geochemical evolution of the Moon. *Proc. 5th Lunar Sci. Conf.,* 1287–1305.   **4.4.3**

Taylor, Stuart Ross, Johnson, P. H., Martin, R., Bennett, D., Allen, J., & Nance, W. (1970) Preliminary chemical analysis of Apollo 11 lunar samples. *Proceedings of the Apollo 11 Lunar Science Conference,* ed. A. A. Levinson, Vol. 2, pp. 1627–35. New York: Pergamon Press.   **4.4.3**

Taylor, Stuart Ross, Rudowski, R., Muir, Patricia, Graham, A., & Kaye, Maureen (1971) Trace element chemistry of lunar samples from the Ocean of Storms. *Proc. 2nd Lunar Sci. Conf.,* 1083–1100.   **4.4.4**

Taylor, Stuart Ross. *See* Newsom, H. E.; Swindle, T. D.

Tera, F. *See also* Albee, A. L.; Schramm, D. N.

ter Haar, D. *See* Haar, D. ter

Terrile, Richard J. *See* Smith, B. A.

Teschke, F. *See* Wänke, H.

Thacker, R. *See* Wänke, H.

Theys, John C. *See* Chevalier, R. A.

Thielemann, F.-K. *See* Arnould, M.; Hillebrandt, W.

Thompson, A. C., & Stevenson, D. J. (1983) Two-phase gravitational instabilities in thin disks with application to the origin of the Moon. *Lunar & Plan. Sci.*, **14**, 787–8.   4.4.5

Thompson, W. B. *See* Baxter, D. C.

Todd, David P. (1989) *Stars and telescopes.* Boston: Little, Brown.   1.2.2

(1922) *Astronomy.* New York: Harper.   1.2.6

Tolnay, V. *See* Sztrokay, K. I.

Torbett, M., Greenberg, R., & Smoluchowski, R. (1982) Orbital resonances and planetary formation sites. *Icarus*, **49**, 313–26.   3.2.3

Tornow, Eugen (1902) Die Entstehung des Sonnensystems. *Weltall*, **3**, 69.   1.2.4

Townley, S. D. (1905) Evolution of the Solar System. *Pub. Astron. Soc. Pac.*, **17**, 199–200.   1.2.6

Tozer, D. C. (1968) Fractionation of iron in the Solar System. *Nature*, **217**, 437.   3.3.3

(1976/1978) Terrestrial planet evolution and the observational consequences of their formation. In *The origin of the Solar System* (NATO Advanced Study Institute, 1976), ed. S. F. Dermott, pp. 433–62. New York: Wiley (pub. 1978).   3.3.4

Tozer, D. C. *See also* Harris, P. G.

Trask, N. J. *See* Murray, B. C.

Treash, Robert (1972) Magnetic remanence in lunar rocks. *Pensee*, **2**, no. 2 (May), 21–3.   4.3.4

Treder, H.-J. (1987) Die Entwicklung des Sonnensystems – Beziehungen zur Physik und zur Kosmologie. *Gerlands Beitr.*, Sonderheft, 4–12.   2.1

Tropp, Henry S. (1974) Moulton, Forest Ray. *DSB*, **9**, 552–3.   1.1.1, 1.2.3

Troutner, D. E. *See* Srinivasan, B.

Trulsen, Jan (1971) Towards a theory of jet streams. *Astrophys. Space Sci.*, **12**, 329–48.   3.4.7

(1971/1972) Theory of jet streams. In *From plasma to planet* (21st Nobel Symposium, 1971), ed. A. Elvius, pp. 179–94. Stockholm: Almqvist & Wiksell/New York: Wiley (pub. 1972).   3.4.7

(1972a) Numerical simulations of jetstreams. I. The three-dimensional case. *Astrophys. Space Sci.*, **17**, 241–62.   3.4.7

(1972b) Numerical simulations of jetstreams. II. The two-dimensional case. *Astrophys. Space Sci.*, **18**, 3–20.   3.4.7

Truran, J. W., & Cameron, A. G. W. (1978) [26]Al production in explosive carbon burning. *Astrophys. J.*, **219**, 226–9.   2.5.3

Truran, J. W. *See also* Cameron, A. G. W.

Tscharnuter, W. M. (1984) The formation of the planetary system. *Celestial Mechanics*, **34**, 289–96.   3.4.2

Tscharnuter, W. M. *See also* Morfill, G. E.

Tully, R. Brent (1988) Origin of the Hubble constant controversy. *Nature*, **334**, 209–12.   2.1.2

Tully, R. Brent. *See also* Pierce, M. J.

Turcotte, D. L., Nordmann, J. C., & Cisne, J. L. (1974) Evolution of the Moon's orbit and the origin of life. *Nature*, **251**, 124–5.   4.3.3

Turcotte, D. L., & Kellogg, L. H. (1986) Implications of isotope data for the origin of the Moon. In *Origin of the Moon*, ed. W. K. Hartmann et al., pp. 311–29. Houston: Lunar and Planetary Institute.   4.4.5

Turekian, K. K., & Clark, S. P., Jr. (1969) Inhomogeneous accumulation of the Earth from the primitive solar nebula. *Earth Plan. Sci. Lett.*, **6**, 346–8.   **3.3.4, 4.4.3**
(1975) The non-homogeneous accumulation model for terrestrial planet formation and the consequences for the atmosphere of Venus. *Journal of the Atmospheric Sciences*, **32**, 1257–61.   **3.3.4**

Turekian, K. K. *See also* Clark, S. P., Jr.

Turkevich, Anthony. *See* Reed, G. W.

Turner, G. *See* Reynolds, J. H.

Turner, H. H. (1903/1970) The new star in Gemini (lecture at the Royal Institution, 5 June 1903). In *The Royal Institution Library of Science – Astronomy*, vol. **2**, pp. 36–46. New York: American Elsevier (pub. 1970).   **1.2.4**
(1909) A vital charge in our theory of the solar system. *Current Literature*, **47**, 445–7.   **1.2.5**

Tyler, G. L. *See* Cuzzi, J. N.

Uchida, Yutaka (1986) Magnetodynamic phenomena in the solar and stellar outer atmospheres. *Astrophys. Space Sci.*, **118**, 127–148.   **2.3.5**

United States Office of Education (1936) Have you heard? (radio script). Copy at U. S. Naval Observatory Library, Washington, DC.   **4.2.4**

Unsöld, A. *See* Garz, T.

Upham, Warren (1905) The nebular and planetesimal theories of the Earth's origin. *American Geologist*, **35**, 202–20.   **1.2.6**

Urey, Harold C. (1951) The origin and development of the Earth and other terrestrial planets. *Geoch. Cosmoch. Acta*, **1**, 209–77; **2**, 263–8.   **3.1.1, 3.3.1**
(1951/1952) Condensation processes and the origin of the major and terrestrial planets. In *L. Farkas memorial volume, 1951*, ed. A. Farkas & E. P. Wigner, pp. 3–12. Jerusalem: Research Council of Israel (pub. 1952).   **3.3.1, 3.3.3**
(1952) *The planets: Their origin and development*. New Haven: Yale University Press.   **1.1.4, 3.3.1, 4.3.1, 4.3.2**
(1953) On the concentration of certain elements at the Earth's surface. *Proc. Roy. Soc.*, **A219**, 281–92.   **3.3.1**
(1954a) On the dissipation of gas and volatilized elements from protoplanets. *Astrophys. J. Supplement*, **1**, 147–73.   **3.3.1**
(1955a) The cosmic abundances of potassium, uranium, and thorium and the heat balances of the Earth, the Moon, and Mars. *Proc. N. A. S.*, **41**, 127–44.   **2.5.1**
(1955b) Some criticisms of "On the origin of the lunar surface features" by G. P. Kuiper. *Proc. N. A. S.*, **41**, 423–8.   **4.3.2**
(1956a) Diamonds, meteorites, and the origin of the Solar System. *Astrophys. J.*, **124**, 623–7.   **3.1, 3.3.1, 4.3.2**
(1956b) The origin and significance of the Moon's surface. *Vistas in Astronomy*, **2**, 1667–80.   **3.3.1, 4.3.2**
(1956c) The origin of the Moon's surface features. *Sky & Tel.*, **15**, 108–111, 161–3.   **4.3.2**
(1958) Composition of the Moon's surface. *Zeitschrift für Physikalische Chemie* (N. F.), **16**, 346–57.   **4.3.2**
(1959) Primary and secondary objects. *J. Geophys. Res.*, **64**, 1721–37.   **3.3.1, 4.3.2**
(1960a) The duration of intense bombardment processes on the Moon. *Astrophys. J.*, **132**, 502–3.   **4.1.2, 4.3.2**
(1960b) The origin and nature of the Moon. *Endeavour*, **19**, 87–99.   **4.1.2, 4.3.2**
(1960/1962) The origin of the Moon and its relationship to the origin of the Solar System. In *The Moon* (IAU Symposium No. 14 at Pulkowo Observatory, 1960), ed.

Z. Kopal & Z. K. Mikhailov, pp. 133–48. New York: Academic Press (pub. 1962).
4.1.2, 4.3.2

(1961a) The Moon. In *Science in space*, ed. L. V. Berkner & H. Odishaw, pp. 185–98. New York: McGraw-Hill. **4.3.2**

(1961b) The planets. In *Science in space*, ed. L. V. Berkner & H. Odishaw, pp. 199–217. New York: McGraw-Hill. **4.3.2**

(1962) Origin and history of the Moon. In *Physics and astronomy of the Moon*, ed. Z. Kopal, pp. 481–523. New York: Academic Press. **4.1.2, 4.3.2**

(1963a) The origin and evolution of the Solar System. In *Space science*, ed. D. P. LeGalley, pp. 123–68. New York: Wiley. **2.3.5, 3.4.3**

(1963b) Statement of Dr. Harold C. Urey, Professor of Chemistry at Large, University of California, San Diego, La Jolla, California. In *Scientists' testimony on space goals,* Hearings of the Senate Committee on Aeronautical and Space Sciences, 88th Congress, 1st Session, 10 and 11 June 1963, pp. 50–62. Washington, DC: U.S. Government Printing Office. **4.3.2**

(1964/1966) The capture hypothesis of the origin of the Moon. In *The Earth–Moon system* (conference held in January 1964), ed. B. G. Marsden & A. G. W. Cameron, pp. 210–12. New York: Plenum Press (pub. 1966). **3.3.3**

(1966a) Biological material in meteorites: A review. *Science,* **151,** 157–66.
4.3.2

(1966b) "Dust" on the Moon. *Science,* **153,** 1419–20. **4.2.4**

(1966c) Observations on the Ranger VIII and IX pictures. *Proceedings of the Caltech–JPL Lunar and Planetary Conference,* ed. H. Brown et al., pp. 1–23. Pasadena, CA: Jet Propulsion Laboratory, Technical Memorandum 33-266. **4.3.2**

(1967) The origin of the Moon. III. In *Mantles of the Earth and terrestrial planets,* ed. S. K. Runcorn, pp. 251–60. New York: Interscience.
3.3.3, 4.1.2, 4.3.2, 4.4.2

(1969a) The space program and problems of the origin of the Moon. *Bulletin of the Atomic Scientists,* **25,** no. (April), 24–6. **4.3.1, 4.4.7**

(1969b) As I see it. (Interview). *Forbes,* **104,** no. 2 (5 July), 44–8. **4.3**

(1969c) Water on the Moon. *Science,* **164,** 1088. **4.3.2**

(1972) Evidence for objects of lunar mass in the early solar system and for capture as a general process for the origin of satellites. *Astrophys. Space Sci.,* **16,** 311–23.
4.3.4

(1973) The Moon and its origin. *Science and Public Affairs [Bulletin of the Atomic Scientists],* **29,** no. 9 (November), 5–10. **4.3.4**

(1974) Evidence for lunar-type objects in the early solar system. In *Highlights of Astronomy as presented at the XVth General Assembly and the Extraordinary General Assembly of the International Astronomical Union, 1973,* ed. G. Contopoulos, pp. 475–81. Boston: Reidel. **4.3.4**

(1976) Acceptance speech. (On receiving Goldschmidt Medal.) *Geoch. Cosmoch. Acta,* **40,** 570. **4.3.4, 4.3.5**

(D1949) Letter to E. C. Bullard, 28 December 1949. Box 11, folder "Bullard, E. C.," U-UCSD. **4.4.1**

(D1950a) Letter to E. C. Bullard, 4 January 1950. Box 11, folder "Bullard, E. C.," U-UCSD. **4.4.1**

(D1950b) Letter to A. E. Benfield, 11 January 1950. Box 11, folder "Benfield," U-UCSD. **4.4.1**

(D1950c) Letter to E. C. Bullard, 3 August 1950. Box 11, folder "Bullard, E. C.," U-UCSD. **4.4.1**

(D1952a) Letter to K. E. Bullen, 14 March 1952. Box 11, folder "Bullen, K. E.," U-UCSD. **4.4.1**

(D1952b) Letter to K. E. Bullen, 31 July 1952. Box 11, folder "Bullen, K. E.," U-UCSD. **4.4.1**

(D1953) Letter to G. Gamow, 8 May 1953. Box 13, folder "Gamow, G.," U-UCSD. 4.3.2

(D1954) Letter to D. Alter, 9 June 1954. Box 10, folder "Alter, D.," U-UCSD. 4.3.2

(D1955a) Letter to L. B. Slichter, 6 May 1955. Box 23, U-UCSD. 4.3.2

(D1955b) Letter to D. H. Menzel, 26 October 1955. Box 17, folder "Menzel, D. H.," U-UCSD. 4.3.2

(D1957) Letter to W. DeMarcus, 29 October 1957. Box 12, folder "DeMarcus, W. C.," U-UCSD. 4.4.1

(D1959) The face of the Moon: Two-thirds of a century later. Text of lecture at Philosophical Society, Washington, DC, December 1959. U-UCSD. 4.2.4

(D1961a) Letter to Homer Newell, 19 June 1961. In *Documents in the history of NASA*, pp. 439–41. Washington, DC: NASA History Office (1975). 4.3.1, 4.3.2

(D1961b) Letter to S. Chandrasekhar, 20 December 1961. Box 11, folder "Chandrasekhar, S.," U-UCSD. 4.3.3

(D1962a) Letter to S. Chandrasekhar, 13 April 1962. Box 11, folder "Chandrasekhar, S.," U-UCSD. 4.3.3

(D1962b) Letter to S. Chandrasekhar, 22 May 1962. Box 11, folder "Chandrasekhar, S.," U-UCSD. 4.3.3

(D1963) Letter to S. Chandrasekhar, 17 September 1963. Box 11, folder "Chandrasekhar, S.," U-UCSD. 4.3.2

(D1965) Statement before Committee of the House of Representatives on the objectives of the exploration of the Moon and planets, 11 March 1965. Unpublished manuscript. NASA-A. 4.3.2

(D1969a) Letter to G. E. Mueller, 26 September 1969. Urey file, NASA-A. 4.3.2

(D1969b) Letter to J. Findlay, 27 October 1969. Urey file, NASA-A. 4.3.2

(D1969c) Letter to C. H. Townes, 27 October 1969. Urey file, NASA-A. 4.3.2

(D1977) Letter to H. Newell, 16 March 1977. Newell Papers, container 41, folder "Planetary science," NASA-A. 4.4.4

(D19XX) Observations on Ranger VII pictures. Undated manuscript, NASA-A. 4.3.1

Urey, Harold C., Elsasser, W. M., & Rochester, M. G. (1959) Note on the internal structure of the Moon. *Astrophys. J.*, **129**, 842–48. 4.3.2, 4.4.2

Urey, Harold C., & MacDonald, Gordon J. F. (1971) Origin and history of the Moon. In *Physics and astronomy of the Moon*, ed. Z. Kopal, pp. 213–89. New York: Academic Press. 4.3.4, 4.4.4

Urey, Harold C., Marti, K., Hawkins, J. W., & Liu, M. K. (1971) Model history of the lunar surface *Proc. 2nd Lunar Sci. Conf.*, **2**, 987–98. 4.3.4

Urey, Harold C. *See also* Manian, S. H.; Murthy, V. R.; O'Keefe, J. A.

Vail, S. M. *See* Hartmann, W. K.

Valley, George E., & Anderson, Herbert H. (1947) A comparison of the abundance ratios of the isotopes of terrestrial and meteoritic iron. *Journal of the American Chemical Society*, **69**, 1871–5. 1.1.5

Van de Kamp, Peter (1956) Planetary companions of stars. *Vistas in Astronomy*, **2**, 1040–8. 1.1.3, 2.1.2

(1986) Dark companions of stars. *Space Science Reviews*, **43**, 211–327. 2.1.2

Van den Bergh, Sidney (1981) Size and age of the universe. *Science*, 825–30. 2.1.2

Van Dyke, V. (1964) *Pride and power: The rationale of the space program.* Urbana: University of Illinois Press. 4.3.1

Vangioni-Flam, E., Audouze, J., & Chièze, J.-P. (1980) ^{22}Ne and ^{26}Al nucleosynthesis in novae and supernovae. *Astronomy and Astrophysics*, **82**, 234–7. 2.5.3

Van Maanen, A. (1916a) Preliminary evidence of internal motion in the spiral nebula Messier 101. *Proc. N. A. S.*, **2**, 386–90.   **1.2.6**

(1916b) Preliminary evidence of internal motion in the spiral nebula Messier 101. *Astophys. J.*, **44**, 210–28.   **1.2.6**

(1923) Investigations on proper motion, tenth paper: Internal motions in the spiral nebula Messier 33, NGC 598. *Astrophys. J.*, **57**, 264–78.   **1.3.3**

Vasyliunas, Vytenis (1987) Non-existence of gravitationally controlled partial corotation in cosmogonic plasmas. *Geophys. Res. Lett.*, **14**, 171–3.   **3.4.4**

Vernon, M. J. *See* Compston, W.

Veronnet, Alexandre (1914) *Les hypothèses cosmogoniques modernes.* Paris: Hermann et fils.   **1.2.6**

Very, Frank W. (1903) An inquiry into the cause of the nebulosity around Nova Persei. *Am. J. Sci.* [4], **16**, 49–60.   **1.2.4**

Veverka, Joseph. *See* Smith, B. A.

Vickery, Ann M., & Melosh, H. Jay (1987) Orbital evolution of the vapor jet from a giant impact. *Lunar & Plan. Sci.*, **18**, 1042–3.   **4.4.5**

Vilcsek, E. *See* Wänke, H.

Vityazev, A. V. *See* Safronov, V. S.

Vizgirda, Joanna, & Anders, Edward (1976) Composition of the eucrite parent body. *Lunar Sci.*, **7**, 898–900.   **4.4.4**

Völk, H. J. *See* Morfill, G. E.

Von Braumüller (1898) Geschichtliche Darstellung der hauptsächlichsten Theorien über die Entstehung des Sonnensystem. *Himmel und Erde*, **10**, 289–300, 357–74.   **1.2.2**

Vosters, M. *See* Hanappe, F.

Vsekhsvyatskii, S. K. (1962) Possible existence of a ring of comets and meteorites around Jupiter. *Sov. Astron. AJ*, **6**, 226–35. Trans from *Astron. Zh.*, **39**, 290–302.   **3.4.6**

Wacker, J. F. *See* Greenberg, R.

Wackerbarth, A. D. (1867) On an astronomical presentiment of Immanuel Kant relative to the constancy of the Earth's sidereal period of rotation on its axis. *M. N. Roy. Astron. Soc.*, **27**, 200.   **4.2.2**

Wänke, Heinrich (1974) Chemistry of the Moon. *Topics in Current Chemistry*, **44**, 115–54.   **4.3.4**

(1981a) Chemistry and accretion history of the Earth and the inner planets. In *The Solar System and its exploration*, ed. W. R. Burke, pp. 141–50. Noordwijk, Netherlands: ESA Scientific and Technical Publications Branch, ESTEC.   **4.3.4**

(1981b) Constitution of terrestrial planets. *Phil. Trans.*, **A303**, 287   **4.3.4**

Wänke, Heinrich, Baddenhausen, Hildegard, Blum, K., Cendales, M., Dreibus, Gerlind, Hofmeister, H., Kruse, H., Jagoutz, E., Palme, Christl, Spetel, B., Thacker, R., & Vilcsek, E. (1977) On the chemistry of lunar samples and achondrites – Primary matter in the lunar highlands: A re-evaluation. *Proc. 8th Lunar Sci. Conf.*, 2191–213.   **4.3.4**

Wänke, Heinrich, & Dreibus, Gerlind (1977/1979) The Earth–Moon system: Chemistry and origin. In *Origin and distribution of the elements* (proceedings of symposium, May 1977), ed. L. H. Ahrens, pp. 99–109. New York: Pergamon Press (pub. 1979).   **4.3.4**

(1982) Chemical and isotopic evidence for the early history of the Earth–Moon system. In *Tidal friction and the Earth's rotation.* II, ed. P. Brosche & J. Sündermann, pp. 322–44. New York: Springer Verlag.   **4.4.4, 4.4.5**

(1984) Chemistry and accretion of Earth and Mars. *Lunar & Plan. Sci.*, **15**, 884.   **4.4.5**

(1986) Geochemical evidence for the formation of the Moon by impact-induced fission of the proto-Earth. In *Origin of the Moon,* ed. W. K. Hartmann et al., pp. 649–72. Houston: Lunar and Planetary Institute. **4.4.5**

Wänke, H., Dreibus, G., Palme, H., Rammensee, W. & Weckwirth, G. (1983) Geochemical evidence for the formation of the Moon from material of the Earth's mantle. *Lunar & Plan. Sci.,* **14,** 818–9. **4.4.4**

Wänke, H., Palme, H., Baddenhausen, H., Dreibus, G., Jagoutz, E., Kruse, H., Palme, C., Spettel, B., Teschke, F., & Thacker, R. (1975) New data on the chemistry of lunar samples: Primary matter in the lunar highlands and the bulk composition of the Moon. *Proc. 6th Lunar Sci. Conf.,* 1313–40. **4.3.4**

Wänke, H., Wlotzka, F., Baddenhausen, H., Balacesa, A., Spettel, B., Teschke, F., Jagoutz, E., Kruse, H., Quijano-Rico, M., & Rieder, R. (1971) Apollo 12 samples: Chemical composition and its relation to sample locations and exposure ages, the two component origin of the various soil samples and studies on lunar metallic particles. *Proc. 2nd Lunar Sci. Conf.,* 1187–208. **4.4.4**

Wänke, Heinrich. *See also* Dreibus, G.; Rammensee, W.

Walcott, C. D. (D1902a) Letter to T. C. Chamberlin, 25 January 1902. CIW. **1.2.5**

(D1902b) Letter to T. C. Chamberlin, 28 January 1902. CIW. **1.2.5**

(D1902c) Letter to T. C. Chamberlin, 12 March 1902. CIW. **1.2.5**

Waldrop, M. M. (1981) Mauna Kea. (I): Halfway to space. *Science,* **214,** 1010–13. **1.1.4**

(1982) The origin of the Moon. *Science,* **216,** 606–7. **4.3.4, 4.4.4**

Walker, James (1929) Svante Arrhenius. *Annual Report of the Smithsonian Institution for 1928,* 715–35. Reprinted from the *Journal of the Chemical Society* (London). **1.2.4**

Walker, James C. G. (1977) *Evolution of the atmosphere.* New York: Macmillan. **3.3.4**

Walker, R. *See* Arnold, J.

Walker, R. G. *See* Aumann, H. H.

Wallace, A. R. (D1913) Letter to W. L. Webb, 28 October 1913. T. J. J. See Papers, Library of Congress **1.2.4**

Wanless, R. K., Loveridge, W. D., & Stevens, R. D. (1970) *Proceedings of the Apollo 11 Lunar Science Conference,* ed. A. A. Levinson, vol. 2, pp. 1729–39. New York: Pergamon Press. **5.4.4**

Ward, Richard A., & Fowler, W. A. (1980) Thermalization of long-lived nuclear isomeric states under stellar conditions. *Astrophys. J.,* **238,** 266–86. **2.5.3**

Ward, William R., & Cameron, A. G. W. (1978) Disc evolution within the Roche limit. *Lunar & Plan. Sci.,* **9,** 1205–7. **4.1.5, 4.4.5**

Ward, William, R., & Reid, M. J. (1973) Solar tidal friction and satellite loss. *M. N. Roy. Astron. Soc.* **164,** 21–32. **4.4.6**

Ward, William, R. *See also* Cameron, A. G. W.; Goldreich, P.

Wark, D. A. (1979) Birth of the presolar nebula: The sequence of condensation revealed in the Allende meteorite. *Astrophys. Space Sci.,* **65,** 275–95. **2.5.2**

Warren, Paul H. (1986) The bulk-Moon MgO/FeO ratio: A highlands perspective. In *Origin of the Moon,* ed. W. K. Hartmann et al., pp. 279–310. Houston: Lunar and Planetary Institute. **4.4.5**

Warren, P. H., & Wasson, J. T. (1978) Compositional-petrographic investigation of pristine nonmare rocks. *Proc. 9th Lunar & Plan. Sci. Conf.,* 185–217. **4.4.4**

(1979) Effects of pressure on the crystallization of a "chondritic" magma ocean and implications for the bulk composition of the Moon. *Proc. 10th Lunar & Plan. Sci. Conf.,* 2051–83. **4.4.4**

Warren, Paul H. *See also* Wasson, J. T.

Wasserburg, G. J. (1985) Short-lived nuclei in the early solar system. In *Protostars and planets.* II, ed. D. C. Black & M. S. Matthews, pp. 703–37. Tucson: University of Arizona Press. **3.4.8**

Wasserburg, G. J. *See also* Albee, A. L.; Arnold, J.; Burnett, D. S.; Kelly, W. R.; Lee, T.; McCulloch, M.; Papanastassiou, D. A.; Schramm, D. N.

Wasserman, L. H. *See* Millis, R. L.

Wasson, John T. (1971) Volatile elements on the Earth and the Moon. *Earth Plan. Sci. Lett.*, **11**, 219–25.   4.4.3

(1978) Maximum temperatures during the formation of the solar nebula. In *Protostars and planets*, ed. T. Gehrels, pp. 488–501. Tucson: University of Arizona Press.   3.3.5

(1985) *Meteorites: Their record of early Solar-System history*. New York: Freeman.
                                                                          2.2.3, 3.3.5, 4.4.4

Wasson, John T., Chou, Chen-Lin, Robinson, Karen L., & Baedecker, Philip A. (1975) Siderophiles and volatiles in the Apollo 16 rocks and soils. *Geoch. Cosmoch. Acta*, **39**, 1475–85.   4.4.3

Wasson, John T., & Warren, Paul H. (1979) Formation of the Moon from differentiated planetesimals of chondritic composition. *Lunar & Plan. Sci.*, **10**, 1310–12.
                                                                                                4.4.6

(1984) The origin of the Moon. *Origin of the Moon Conf. Abst.*, 57.   4.4.6

Wasson, John T. *See also* Baedecker, P. A.; Warren, P. H.

Waterfield, Reginald L. (1938) *A hundred years of astronomy*. New York: Macmillan.
                                                                                                1.2.4

Weaver, Thomas A. *See* Woosley, S. E.

Webb, William Larkin (1913) *Brief biography and popular account of the unparalleled discoveries of T. J. J. See*. Lynn, MA: Nichols.   1.2.4

Weckwerth, G. *See* Wänke, H.

Wefel, John P. *See* Arnett, W. E.

Wegener, Alfred (1912) Die Entstehung der Kontinente. *Petermanns Geographische Mitteilungen*, **58**, 185–95, 253–6, 305–9.   4.2.4

Weidenschilling, S. J. (1975) Close encounters of small bodies and planets. *Astron. J.*, **80**, 145–53.   3.1.2

(1978) Iron/silicate fractionation and the origin of Mercury. *Icarus*, **35**, 99–111.
                                                                                                3.3.3

(1983) Progress toward the origin of the Solar System. *Reviews of Geophysics and Space Physics*, **21**, 206–13.   3.2.2, 3.2.3

Weidenschilling, Stuart J., & Davis, D. R. (1985) Orbital resonances in the solar nebula: Implications for planetary accretion. *Icarus*, **62**, 16–29.   3.2.2

Weidenschilling, Stuart J., Greenberg, R., Chapman, C. R., Herbert, F., Davis, D. R., Drake, M. J., Jones, J. H., & Hartmann, W. K. (1986) Origin of the Moon from a circumterrestrial disk. In *Origin of the Moon*, ed. W. K. Hartmann et al., pp. 731–62. Houston: Lunar and Planetary Institute.   4.4.6

Weidenschilling, Stuart J. *See* Greenberg, R.; Herbert, F.

Weizsäcker, C. F. von (1944) Über die Entstehung des Planetensystems. *Z. f. Astrophysik.*, **22**, 319–55.   1.1.4

(1948) Das Spektrum der Turbulenz bei grossen Reynoldschen Zahlen. *Z. Physik*, **124**, 614–27.   1.1.4

(1951) Anwendungen der Hydrodynamik auf Probleme der Kosmogonie. In *Festschrift zur Feier des 200 jährigen Bestehens der Akademie der Wissenschaften in Göttingen, Mathematisch-physikalische Klasse*, pp. 86–122. Berlin: Springer.   1.1.4

(1988) Eine Erinnerung zur Planetentheorie. *Mitteilungen, Arbeitskries Geschichte der Geophysik*, **7**, no. 2 (March), 7–10.   1.1.4

Wesselius, P. R. *See* Aumann, H. H.

Wetherill, George W. (1974/1977) Pre-mare cratering and early Solar System history. In *The Soviet–American Conference on Cosmochemistry of the Moon and Planets (Mos-*

cow, 1974), ed. J. H. Pomeroy & N. J. Hubbard, pp. 553–67. Washington, DC: National Aeronautics and Space Administration, Report NASA SP-370 (published 1977).   **3.1.2**

(1975a) Late heavy bombardment of the Moon and terrestrial planets. *Proc. 6th Lunar Sci. Conf.*, 1539–61.   **3.1.2, 4.4.5**

(1975b) Radiometric chronology of the early Solar System. *Annual Review of Nuclear Science*, **25**, 285–328.   **2.5**

(1976) The role of large bodies in the formation of the Earth. *Proc. 7th Lunar Sci. Conf.*, 3245–57.   **3.1.2, 4.4.5**

(1980a) Numerical calculations relevant to the accumulation of the terrestrial planets. In *The continental crust and its mineral deposits*, ed. D. W. Strangway, pp. 3–24. Toronto: Geological Association of Canada, Special Papers No. 20.
                                                                           3.1.1

(1980b) Could the solar wind have been the source of the high concentration of ^{36}Ar in the atmosphere of Venus? *Lunar Sci.*, **11**, 1239–41.   **3.3.5**

(1981a) The formation of the Earth from planetesimals. *Sci. Am.*, **244**, no. 6 (June), 163–74.   **3.1.2**

(1981b) Solar wind origin of ^{36}Ar on Venus. *Icarus*, **46**, 70–80.   **3.3.5**

(1985) Occurrence of giant impacts during the growth of the terrestrial planets. *Science*, **228**, 877–9.   **3.1.2**

(1986) Accumulation of the terrestrial planets and implications concerning lunar origin. In *Origin of the Moon*, ed. W. K. Hartmann et al., pp. 519–50. Houston: Lunar and Planetary Institute.   **4.1.6, 4.4.5**

(1988) Remarks on the conference. In *Workshop on the origins of Solar Systems*, ed. J. A. Nuth & O. Sylvester, pp. 81–6. Houston: Lunar and Planetary Institute, Technical Report 88-04.   **2.2.4**

Wetherill, George W. *See* Gopalan, K.

Wheaton, William A. *See* Mahoney, W. A.

Whipple, F. L. (1942/1946) Concentrations of the interstellar medium. (presented at Inter-American Congress of Astrophysics, Mexico, 1942, under the title "Theory of the interstellar Medium.") *Astrophys. J.*, **104** (1946), 1–11.   **1.1.3**

(1948a) The dust cloud hypothesis. *Sci. Am.*, **178**, no. 5 (May), 34–44.   **1.1.3**

(1948b) Kinetics of cosmic clouds. In *Centennial Symposia, December 1946*, pp. 109–42. Cambridge, MA: Harvard College Observatory, Monograph no. 7.
                                                                           1.1.3

(1964) The history of the Solar System. *Proc. N. A. S.*, **52**, 565–94.   **2.3.5, 2.4**

(1974) [Discussion remark]. In *The origin of the Solar System*, ed. H. Reeves, pp. 86–7. Paris: CNRS.   **3.4.7**

(1976) Background of modern comet theory. *Nature*, **263**, 15–19.   **4.4.7**

(1979) Origin of the Solar System. *Nature*, **278**, 819.   **2.3.5**

(1981) Triumphs in space. *Harvard Magazine*, **84**, no. 1 (September–October), 27–34.   **4.4.5**

Whitaker, Ewen A. (1985) *The University of Arizona's Lunar and Planetary Laboratory: Its founding and early years.* Tucson: University of Arizona.   **3.3.1**

White, Marvin L. (1972) An asymmetrically rotating fluid disc with applications. *Astrophys. Space Sci.*, **16**, 295–310.   **3.4.7**

Whitney, Charles A. (1971) *The discovery of our galaxy.* New York: Knopf.
                                                                           1.3.1, 2.1

Wickramasinghe, N. C. *See* Hoyle, F.

Wild, J. P., Editor (1974) *In the beginning . . . Symposium on the origin of planets and life, held as part of the Copernicus 500th birthday celebration at Canberra on 27 April 1973.* Canberra: Australian Academy of Science.   **2.2.5**

Wildt, Rupert (1958) Inside the planets. *Pub. Astron. Soc. Pac.,* **70,** 237–50.   **3.2.1**
    (1961) Planetary interiors. In *Planets and satellites,* ed. G. P. Kuiper & B. M. Middle-
    hurst, pp. 159–212. Chicago: University of Chicago Press.   **3.2.1**
Wilhelms, Don E. *See* Baldwin, R. B.
Williams, I. P. (1975) *The origin of the planets.* New York: Crane, Russak.   **1.1.4**
    (1979) A survey of current problems in planetary cosmogony. *Moon and Planets,* **20,**
    3–13.   **2.2.1, 3.2.1, 3.4.3**
Williams, I. P., & Cremin, A. W. (1968) A survey of theories relating to the origin of
    the solar system. *Q. J. Roy. Astron. Soc.,* **9,** 40–62.   **2.2.1**
Williams, I. P. *See* Donnison, J. R.
Williams, James G. *See* Gradie, J. C.; Yoder, C. F.
Williams, M. B. (1981) Similarities and differences between evolutionary theory and
    theories of physics. In *PSA 1980* (proceedings of the biennial meeting of the
    Philosophy of Science Association), vol. **2,** pp. 385–96. East Lansing, MI: Philos-
    ophy of Science Association.   **2.2.2**
Willis, Bailey (1929) Memorial of Thomas Chrowder Chamberlin (1843–1928). *Bull.
    Geol. Soc. Am.,* **40,** 23–44.   **1.1.1**
Willis, Bailey. *See also* Leith, C. K.
Wilson, R. H., Jr. *See* De Groot, M.
Winnik, H. C. (1970) Science and morality in Thomas C. Chamberlin. *Journal of the
    History of Ideas,* **31,** 441–56.   **1.1, 1.2.3**
Winters, R. R. *See* Malcuit, R. J.
Wise, D. U. (1963) An origin of the Moon by rotational fission during formation of
    the Earth's core. *J. Geophys. Res.,* **68,** 1547–54.   **4.1.3, 4.3.2, 4.3.3**
    (1964/1966) Origin of the Moon by fission. In *The Earth–Moon system* (conference
    held in 1964), ed. B. G. Marsden & A. G. W. Cameron, pp. 213–23. New York:
    Plenum Press (pub. 1966).   **4.1.3**
    (1969) Origin of the Moon from the Earth: Some new mechanisms and compari-
    sons. *J. Geophys. Res.,* **74,** 6034–45.   **4.4.3, 4.4.4**
Wiskerchen. M. J., & Sonett, C. P. (1977) A lunar metal core? *Proc. 8th Lunar Sci.
    Cong.,* 515–35.   **4.3.4**
Witting, J. M. (1966) *Scientific objectives of deep space investigations: The origin and evolu-
    tion of the solar system.* Chicago: Illinois Institute of Technology, Report No.
    IITRI TR P-18; NASA Report CR 79724 = N67 10880.
        2.1.1, 2.2.1, 3.1.1, 3.4.8
    (1969) Relevance of future space missions to the origin of the Solar System. In
    Advanced space experiments, ed. O. L. Tiffany & E. Zaitzeff, pp. 195–236. Tarazana,
    CA: American Astronautical Society. (*Advances in the Astronautical Sciences,* vol.
    25.)   **2.1.1, 2.2.1**
Wlotzka, F. *See* Wänke, H.
Wolf, Rainer, & Anders, Edward (1980) Moon and Earth: Compositional differences
    inferred from siderophiles, volatiles, and alkalis in basalts. *Geoch. Cosmoch. Acta,*
    44, 2111–24.   **4.4.3, 4.4.4**
Wolf, Rainer, Woodrow, Alicia, & Anders, Edward (1979) Lunar basalts and pristine
    highland rocks: Comparison of siderophile and volatile elements. *Proc. 10th Lunar
    & Plan. Sci. Conf.,* 2107–30.   **4.4.3, 4.4.4**
Wolfe, C. W. (1969) Secondary relief features as clues to planetary formation. *Annals
    of the New York Academy of Science,* **163,** 81–9.   **4.3.3**
Wolff, Sidney C. *See* Abell, G. O.
Wood, Clement (1927) *The outline of man's knowledge.* New York: Grosset & Dunlap.
        1.2.6
Wood, John A. (1958) *Silicate meteorite structures and the origin of the meteorites.* Cam-
    bridge, MA: Smithsonian Institution Astrophysical Observatory, Technical Re-
    port no. 10.   **3.3.3**

(1962) Chondrules and the origin of the terrestrial planets. *Nature,* **194,** 127–30.
                                                            2.3.5, 2.4, 3.3.3

(1962/1963) Properties of chondrules. II. In *Origin of the Solar System,* ed. R. Jastrow & A. G. W. Cameron, pp. 147–54. New York: Academic Press.   **3.3.3**

(1963) On the origin of chondrules and chondrites. *Icarus,* **2,** 152–80.   **3.3.3**

(1977) Origin of Earth's Moon. In *Planetary satellites,* ed. J. A. Burns, pp. 513–29. Tucson: University of Arizona Press.   **4.4.1, 4.4.4**

(1979) *The Solar System.* Englewood Cliffs, NJ: Prentice-Hall.   **3.3.4, 3.3.5**

(1981) The interstellar dust as a precursor of Ca, Al-rich inclusions in carbonaceous chondrites. *Earth Plan. Sci. Lett.,* **56,** 32–44.   **3.3.5**

(1982a) As the world turns: Angular momentum and the solar nebula. *Meteoritics,* **17,** 298–9.   **3.3.5**

(1982b) Citation on the award of the Leonard Medal of the Meteoritical Society to Robert N. Clayton, Enrico Fermi Institute, University of Chicago, Chicago, Illinois. *Meteoritics,* **17,** 171–6.   **2.5.1**

(1983) Exploration of the Moon. In *Astronomy from space,* ed. J. Cornell & P. Gorenstein, pp. 61–80. Cambridge, MA: MIT Press.   **4.4.4**

(1985) Meteoritic constraints on processes in the solar nebula. In *Protostars and planets.* II, ed. D. C. Black & M. S. Matthews, pp. 687–702. Tucson: University of Arizona Press.   **2.2.3, 2.2.4, 3.4.8**

(1986) Moon over Mauna Loa: A review of hypotheses of formation of Earth's Moon. In *Origin of the Moon,* ed. W. K. Hartmann et al., pp. 17–55. Houston: Lunar and Planetary Institute.   **4.4.4, 4.4.5**

Wood, J. A. et al., (1981) Cosmochemical constraints. (other authors not named) In *Basaltic volcanism on the terrestrial planets,* by the Basaltic Volcanism Study Project, pp. 538–57. New York: Pergamon Press.   **3.3.4**

Wood, John A., & Mitler, H. E. (1974) Origin of the Moon by a modified capture mechanism, *or half a loaf is better than a whole one. Lunar Sci.,* **5,** 851–3.
                                                            4.4.4, 4.4.6

Wood, John A., & Morfill, Gregor E. (1988) A review of solar nebula models. In *Meteorites and the early Solar System,* ed. J. F. Kerridge & M. S. Matthews, pp. 329–47. Tucson: University of Arizona Press.   **3.3.5**

Wood, John A., & Motylewski, Karen (1979) Meteorite research. *Reviews of Geophysics and Space Physics,* **17,** 912–25.   **3.3.5**

Wood, John A. *See also* Marvin, U. B.

Woodrow, Alicia. *See* Wolf, R.

Woodward, R. S. (D1904) Letter to C. D. Walcott, 2 November 1904. Chamberlin file, CIW.   **1.2.5**

(D1906) Letter to T. C. Chamberlin, 16 November 1906. Chamberlin file, CIW.
                                                            1.2.5

(D1909a) Letter to T. C. Chamberlin, 8 September 1909. Chamberlin file, CIW.
                                                            1.2.5

(D1909b) Letter to T. C. Chamberlin, 21 October 1909. Chamberlin file, CIW.
                                                            1.2.5

(D1910a) Letter to T. C. Chamberlin, 8 March 1910. Chamberlin file, CIW.
                                                            1.2.5

(D1910b) Letter to T. C. Chamberlin, 21 January 1910. Chamberlin file, CIW.
                                                            1.2.5

(D1910c) Letter to T. C. Chamberlin, 3 January 1910. Chamberlin file, CIW.
                                                            1.2.5

Woolfson, M. M. (1969) The evolution of the Solar System. *Rep. Prog. Phys.,* **32,** 135–85.   **2.2.1, 2.3.5**

(1971) The origin of planetary systems. *Physics Bulletin,* **22,** 266–72.
                                                            2.3.5, 4.4.5

Woolfson, M. M. *See also* Dormand, J. R.; Schofield, N.

Woosley, S. E., & Weaver, Thomas E. (1980) Explosive neon burning and [26]Al gamma-ray astronomy. *Astrophys. J.*, **238**, 1017–25.   **2.5.3**

Woosley, S. E. *See also* Clayton, D. D.

Worden, Simon P., Schneeberger, Timothy J., Kuhn, Jeffrey R., & Africano, John L. (1981) Flare activity on T Tauri stars. *Astrophys. J.*, **244**, 520–7.   **2.5.3**

Wright, R. L. *See* Cuttita, F.

Yiou, F. *See* Bernas, R.

Yoder, C. F. (1984) The size of the lunar core. *Origin of the Moon Conf. Abst.*, p. 6.
                                                                               4.3.4

Yoder, C. F., Williams, J. G., Dickey, J. O., & Newhall, X. X. (1984) Tidal dissipation in the Earth and Moon from lunar laser ranging. *Origin of the Moon Conf. Abst.*, 31.   **4.4.2**

Young, A. T. (1973) Are the clouds of Venus sulfuric acid? *Icarus*, **18**, 564–82.
                                                                               3.3.5

Young, Anne Sewell (1901) Density of the solar nebula. *Astrophys. J.*, **13**, 338–43.
                                                                               1.2.3

Young, Charles A. (1895) *A text-book of general astronomy for colleges and scientific schools.* Boston: Ginn.   **1.2.3**

   (1900) *A text-book of general astronomy*, revised edition. Boston: Ginn.   **1.2.2**

Zellner, B. *See* Degewij, J.; Gradie, J. C.

Zharkov, V. N. (1983) Models of the internal structure of Venus. *Moon and Planets*, **29**, 139–75.   **3.3.5**

Zhuchikhin, V. I. *See* Al'tshuler, L. V.

Zill, P. *See* Arnold, J.

Index